Smart Sensors, Measurement and Instrumentation

Volume 8

Series editor

S. C. Mukhopadhyay, Palmerston North, New Zealand

For further volumes:
http://www.springer.com/series/10617

Alex Mason · Subhas Chandra Mukhopadhyay
Krishanthi Padmarani Jayasundera
Nabarun Bhattacharyya
Editors

Sensing Technology: Current Status and Future Trends II

Springer

Editors
Alex Mason
School of Built Environment
Built Environment and Sustainable Technologies Research Institute
Liverpool John Moores University
Liverpool
UK

Subhas Chandra Mukhopadhyay
School of Engineering and Advanced Technology
Massey University (Manawatu Campus)
Palmerston North
New Zealand

Krishanthi Padmarani Jayasundera
Institute of Fundamental Sciences
Massey University (Manawatu Campus)
Palmerston North
New Zealand

Nabarun Bhattacharyya
Centre for Development in Advanced Computing
Kolkata
India

The book editors are "Guest Editors".

ISSN 2194-8402 ISSN 2194-8410 (electronic)
ISBN 978-3-319-02314-4 ISBN 978-3-319-02315-1 (eBook)
DOI 10.1007/978-3-319-02315-1
Springer Cham Heidelberg New York Dordrecht London

Library of Congress Control Number: 2013953628

Printed on acid-free paper

Springer is part of Springer Science+Business Media (www.springer.com)

Preface

The applications of Sensing Technology include medical diagnostics, industrial manufacturing, defense, national security, and prevention of natural disaster. The correct detection of events by high performance sensors, and appropriate analysis of sensor signals can lead to early warning of phenomena, such as "Superstorm Sandy" which hit the eastern coast of the United States in 2012, and help to prevent deaths from these types of catastrophic incident. There is a need for interaction between researchers across technologically advanced and developing countries working on design, fabrication, and development of different sensors.

This book contains a collection of selected works stemming from the 2012 International Conference on Sensing Technology (ICST), which was held in Kolkata, India. This was the sixth time the conference had been held, and over the years it has become an incredibly successful event—in 2012 it attracted over 245 papers and provided a forum for interaction between researchers across technologically advanced and developing countries working on design, fabrication, and development of different sensors.

The conference was jointly organized by the Center for the Development of Advanced Computing (CDAC), India, and the School of Engineering and Advanced Technology, Massey University, New Zealand. We wholeheartedly thank the members of CDAC for extending their support to the conference, as well as the authors and the Technical Program Committee: without the support of these people the conference would not be possible.

Since ICST provides a platform for a wide range of sensing technologies, however of late there has been significant interest in sensors which mimic the human or other biological sensors, this book focuses specifically on work in this area. The first volume of this book, available separately, considers a broader range of sensors and their applications.

Chapter 1 discusses the implementation of improved cochlea implants. Such implants seeks provide aid to those who have suffered hearing loss, a problem which impacts over 36 million people. This work seeks to improve the process through which the implants are constructed, moving from wire-based systems to microfabricated electrode arrays, which provides highly localized stimulation and recording of the neural tissue. In addition to it, electronics and sensor integration enhances working performances with added functionality.

Chapters 2–4 focus on machine vision systems, considering applications of machine vision for sorting in the food industry (Chap. 2), methods for processing of data produced by vision systems (Chap. 3), and the implementation of a stereovision system for tracking of a moving object.

Chapter 5 considers the development and validation of an electronic tongue-based sensor, seeking to compare the performance of biochemical systems against the subjective evaluation of human testers.

Chapters 6–11 study the growingly popular electronic nose type sensors, used for a variety of applications; tea quality estimation (Chaps. 7 and 10); food lifetime (Chap. 8) and vapor toxicity (Chap. 11).

The final chapter (Chap. 12) discuss the implementation of a sensor testing system, specifically aimed at physiological sensing systems. This is an important consideration; over time sensor characteristics may alter, fail or become otherwise compromised. It is therefore important to have standard methods to ensure that integrity of sensors, particularly those which are relied upon to provide life critical information.

This book is written for academic and industry professionals working in the field of sensing, instrumentation, and related fields, and is positioned to give a snapshot of the current state of the art in sensing technology, particularly from the applied perspective. The book is intended to give broad overview of the latest developments, in addition to discussing the process through which researchers go through in order to develop sensors, or related systems, which will become more widespread in the future.

We would like to express our appreciation to our distinguished authors of the chapters whose expertise and professionalism has certainly contributed significantly to this book.

Alex Mason
Subhas Chandra Mukhopadhyay
Krishanthi Padmarani Jayasundera
Nabarun Bhattacharyya

Contents

1 **Cochlear Implant Electrode Improvement for Stimulation and Sensing** 1
N. S. Lawand, P. J. French, J. van Driel, J. J. Briaire and J. H. M. Frijns

2 **Machine Vision Based Techniques for Automatic Mango Fruit Sorting and Grading Based on Maturity Level and Size** . . . 27
C. S. Nandi, B. Tudu and C. Koley

3 **Region Adaptive, Unsharp Masking Based Lanczos-3 Interpolation for 2-D Up-Sampling: Crisp-Rule Versus Fuzzy-Rule Based Approach** 47
A. Acharya and S. Meher

4 **Gaze-Controlled Stereo Vision to Measure Position and Track a Moving Object: Machine Vision for Crane Control** 75
Yasuo Yoshida

5 **Integrated Determination of Tea Quality Based on Taster's Evaluation, Biochemical Characterization and Use of Electronics** 95
P. Biswas, S. Chatterjee, N. Kumar, M. Singh, A. Basu Majumder and B. Bera

6 **Electronic Nose and Its Application to Microbiological Food Spoilage Screening** 119
M. Falasconi, E. Comini, I. Concina, V. Sberveglieri and E. Gobbi

7 **Multiclass Kernel Classifiers for Quality Estimation of Black Tea Using Electronic Nose** 141
P. Saha, S. Ghorai, B. Tudu, R. Bandyopadhyay and N. Bhattacharyya

8 **Electronic Nose Setup for Estimation of Rancidity in Cookies** . . . 161
D. Chatterjee, P. Bhattacharjee, H. Lechat, F. Ayouni, V. Vabre and N. Bhattacharyya

9 **Optimization of Sensor Array in Electronic Nose by Combinational Feature Selection Method** 189
P. Saha, S. Ghorai, B. Tudu, R. Bandyopadhyay and N. Bhattacharyya

10 **Exploratory Study on Aroma Profile of Cardamom by GC-MS and Electronic Nose** 207
D. Ghosh, S. Mukherjee, S. Sarkar, N. K. Leela, V. K. Murthy, N. Bhattacharyya, P. Chopra and A. M. Muneeb

11 **High Frequency Surface Acoustic Wave (SAW) Device for Toxic Vapor Detection: Prospects and Challenges** 217
T. Islam, U. Mittal, A. T. Nimal and M. U. Sharma

12 **Electronic and Electromechanical Tester of Physiological Sensors** 243
E. Sazonov, T. Haskew, A. Price, B. Grace and A. Dollins

About the Editors 263

Chapter 1
Cochlear Implant Electrode Improvement for Stimulation and Sensing

N. S. Lawand, P. J. French, J. van Driel, J. J. Briaire and J. H. M. Frijns

Abstract Electrode array, an important component of the Cochlear Implant (CI) design holds a key position in restoring the hearing process to the deaf patients. It represents a direct interface between the auditory nerve (biological tissue) and the electronic system of the CI. Electrode arrays are available in different design, material, shape and size depending upon the requirement and the application of the device. The traditional fabrication method of the device restricts the electrode usability and its performance. In this chapter we investigate and explore capable materials for CI electrode array fabrication used for stimulation purposes. Here we discuss the CMOS compatible electrode material Titanium Nitride (TiN) as one of the possible candidate for electrical stimulation in electrode array. Electrical characterization in terms of current density, Electromigration, Impedance, and Temperature Coefficient of resistance (TCR) for different materials were performed to demonstrate electrical compatibility. Micro-fabrication process for electrode array is discussed which exhibits the future advantageous manufacturing technique in comparison with the traditional method used nowadays.

N. S. Lawand (✉) · P. J. French · J. van Driel
EI Lab, EWI Faculty, Delft University of Technology,
2628 CD Delft, The Netherlands
e-mail: n.s.lawand@tudelft.nl

P. J. French
e-mail: p.j.french@tudelft.nl

J. van Driel
e-mail: J.W.vanDriel@student.tudelft.nl

J. J. Briaire · J. H. M. Frijns
ENT Department, Leiden University Medical Centre,
P.O. Box 9600, 2300 RC Leiden, The Netherlands
e-mail: J.J.Briaire@lumc.nl

J. H. M. Frijns
e-mail: J.H.M.Frijns@lumc.nl

A. Mason et al. (eds.), *Sensing Technology: Current Status and Future Trends II*,
Smart Sensors, Measurement and Instrumentation 8, DOI: 10.1007/978-3-319-02315-1_1,

Keywords Cochlear implants (CI's) · Microelectrode material · Neural stimulation and sensing · Titanium Nitride (TiN)

1 Introduction

Hearing loss is observed in people of all ages. Approximately 36 million people are affected to some extent of hearing loss in the United States alone [1]. Hearing aid devices benefit many of these patients by sound amplification but those with severe profound loss of sensory hair cells (sensorineural hearing loss), are not benefited from these devices. By studies it has been found that 85 % of hearing loss is due to damage of sensory hair cells inside cochlea. This damage can be genetic or caused by diseases such as meningitis injury, measles disease, and ageing or with intake of improper drugs causing adverse effect on the functioning of the hair cells. Over the last two decades auditory neural prosthesis known as cochlear implants (CI's) have benefited these patients. CI's are implantable devices which by-pass the non-functional inner ear and directly stimulate the auditory nerve with electric currents thus enabling deaf people to experience speech and sound again. It actually overlooks the damaged or the missing hair cells within the cochlea which normally would do the decoding of the sound. The CI consists of a receiver-stimulator package, which receives power and decodes the instructions for controlling the electrical stimulation, and an electrode array, which has electrodes placed inside the cochlea near the hair cells in order to stimulate them which are in turn connected to the auditory nerve fibres which further connect to the auditory cortex of the brain as seen in Fig. 1 [2].

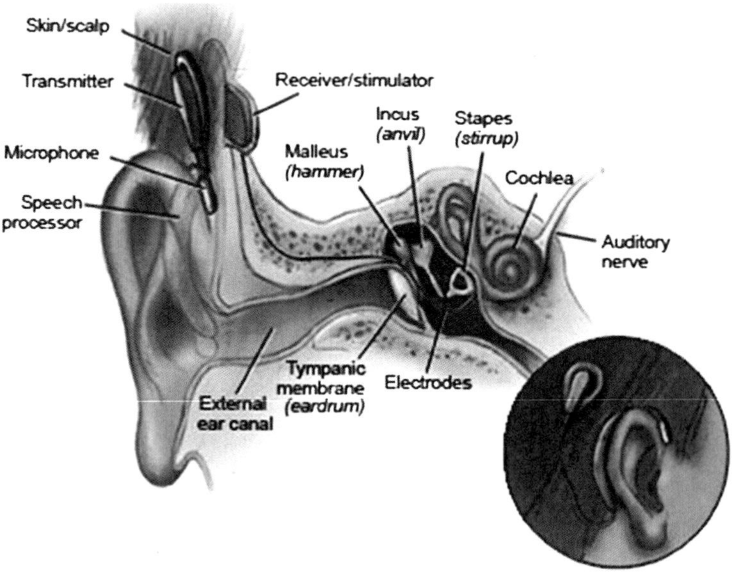

Fig. 1 The CI with microphone, transmitter, receiver/stimulator and the electrode array [2]

1.1 Cochlear Implant Electrode Array

CI today has not only provided a functional hearing to more than 120,000 hearing-impaired people, but also an important industry attracting researchers to contribute in their respective fields. CI electrode array is one of the important components of the implant which is in close proximity with the auditory neurons passing the external auditory information in terms of electrical signal through the auditory nerve to the auditory cortex. The electrode array is placed with the help of an insertion tool, which is straight. The electrode array is available in the pre-curled and flexible options. During the placement, the array is pushed off the insertion tool, curling itself into the scala tympani. Contact with the walls should be avoided to minimize trauma. However, 10–20 % of the patients receiving a CI loose their residual hearing during the surgery due to unwanted physical contact [3]. The placement of a CI is illustrated by a picture in Fig. 2.

The present electrodes arrays are simply a bundle of platinum/iridium (Pt/Ir) wires welded to platinum strips acting as stimulation sites. The amount of electrodes and material choices are different per manufacturer and model. Some of them have implants made of titanium wires, attached to approximately 20 titanium electrodes, which are coated with a soft bio-compatible polymer. There are three major CI manufacturers: Advanced Bionics, Cochlear and MED-EL. An electrode array from MED-EL is shown in Fig. 3.

The electrode arrays manufactured with the current technology are limited in electrode count, due to their large size relative to the size of the scala tympani (ST). Also

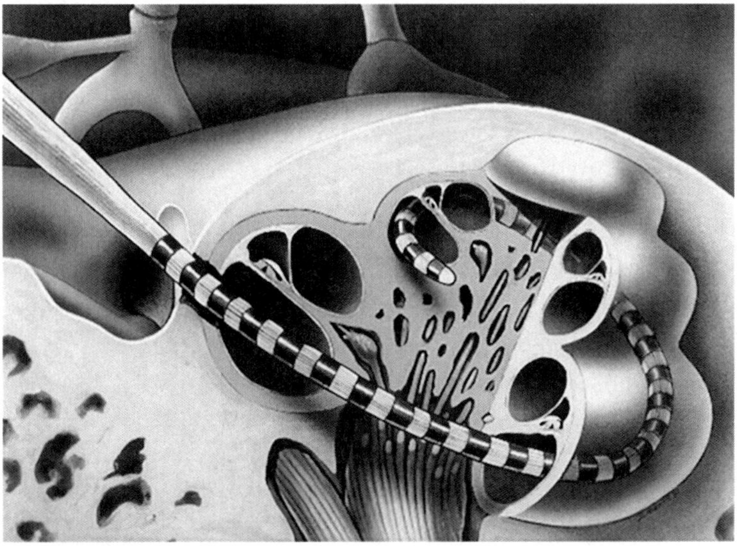

Fig. 2 Representative picture showing electrode array placement in CI [4]

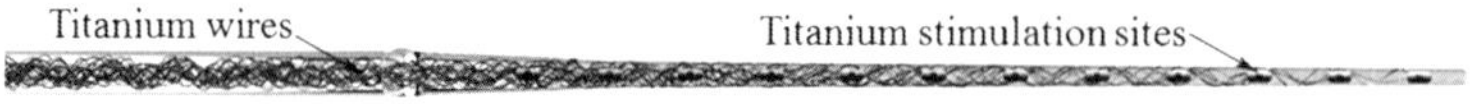

Fig. 3 An uncurled electrode array from MED-El coated with silicone, making the device flexible [5]

the design has restrictions for deeper insertion in ST thus depriving the access to the low frequency auditory neurons. To stimulate the speech regions of the cochlea having a tonotopic organisation it's necessary that the array should be inserted to a length of 14–25 mm from the stapes to achieve stimulation frequencies ranging from 500 to 3,000 Hz [6]. Over the past decades, the design of these electrodes has developed from simple single channel devices to multiple site arrays consisting of 12–22 stimulation sites. The prime goal taken into consideration while designing these electrodes includes deeper insertion into ST to match the tonotopic place of stimulation to the frequency band assigned to each electrode channel, greater operating efficiency and reducing the intra-cochlear damage caused during surgical insertion [7]. Microfabrication using photolithographic and silicon micromachining techniques in addition with micro-electro-mechanical systems (MEMS) permits high volume, batch production with microscale dimensions of reliable microelectrode arrays. Such arrays can be used for highly localized stimulation and recording of the neural tissue.

2 Electrode Materials and Methods

The requirements for microelectrode materials used in CI's and neural implants is increasing nowadays for achieving high performance and stability in their dedicated applications. The microelectrode material selection is an key factor for the success of such implants. Here in this section we talk about the microelectrode material requirements with the charge transfer techniques between the electrode material and the electrolyte.

2.1 Requirements for Stimulation Electrodes

Electrodes are the direct interface between the biological structures (auditory neurons) and the electronic system in the CI's. Stimulation electrodes inject charge into the tissue to functionally excite the nerves by electrical stimulation. In other words, electrodes measure the electric potential for charge transfer between solid metal state and electrolyte solution in liquid state inside the cochlea. For better stability implants electrode properties must be evaluated with respect to a biocompatible application for optimum stability, efficacy and life time with a minimum of toxicity. From the material point of view the requirements of an "ideal" electrode [8–10] might be summarized as follows:

- Effective charge injection from the geometrically small available electrode surface area with minimum energy consumption for stimulation.
- Low material impedance along with low frequency dependence.
- Low tendency to create irreversible reaction products caused due to high reversible charge injection limits which avoids corrosive and inflammatory products.
- Low after-potentials and low separation at the phase boundary of electrode and the electrolyte.
- Delivery of safe stimulation charge at the functional interface with stable and low electrode impedance.
- Biocompatible material coatings accepted by the body with minimum contact tissue or electrolyte reactions and its low impedance.
- Extensive performance and material stability for many years for chronic implants, i.e., negligible or no deterioration with excellent bio-stability.
- Preparation for more available electrochemical surface area by surface reconditioning methods to achieve optimum charge transfer.
- Radiographic visibility of the electrode material.

From the functioning and the biological point of view for an ideal electrode more crucial points are to be considered, however the above properties can assist us for a general selection of the electrode for a particular application.

2.2 Electrode Materials and Charge Transfer Methods

The most common noble materials used for stimulation electrodes are Platinum (Pt) [11], Platinum-Iridium (Pt-Ir) [12, 13], Titanium-Nitride (TiN) [14], Iridium (Ir) [15], Iridium Oxide (IrO) [16–18]. When a metal electrode is placed inside a biological medium for. e.g. Perilymph (equivalent to saline solution) inside the cochlea then an interface is formed between the two segments. In the first electrode metal segment which is connected to electrical circuits the electrons carry the charge. While in the second biological medium (electrolyte) charge is carried by sodium, potassium and chloride ions present in the electrolyte medium. There are two basic mechanisms of charge transfer which takes place at the electrode-electrolyte phase boundary as illustrated in Fig. 4 [19, 20] as follows:

- Non-Faradaic reactions or Capacitive type charge injection in which no electrons are transferred between the electrode surface and the electrolyte medium. The reaction includes the redistribution of charged chemical species in the conductive electrolyte medium, for e.g. such reactions are observed in Titanium Nitride (TiN) material.
- Faradaic reaction mechanism in which there is transfer of electrons between the electrode and electrolyte, which results in oxidation reduction reactions of the chemical species in the electrolyte. Faradaic reactions are further subdivided into electrochemical reversible faradaic reactions and Surface redox non-reversible faradaic reactions. In reversible process the products do not diffuse far away from

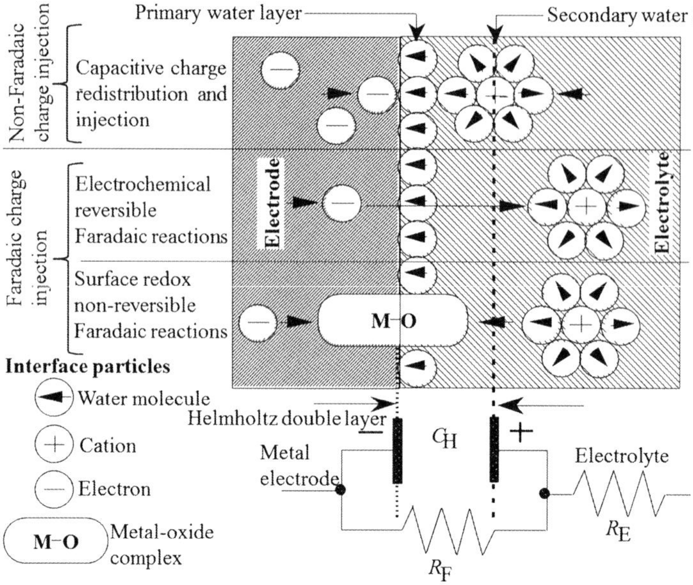

Fig. 4 Reactions at the electrode–electrolyte interface [19]

the electrode or they remain attached to the electrode surface, for e.g. Platinum (Pt), Iridium (Ir) and with the combination of Platinum-Iridium (Pt-Ir) metals.

- In the non-reversible case the products tend to diffuse away from the electrode surface by surface redox reactions resulting in hydrated oxide film with a high charge injection capacity, for e.g. these reactions occur in Iridium oxide (IrOx).

In case of CI's the auditory nerve is stimulated by the stimulation material (for e.g. Pt) used in the stimulation site. Electric charge is transferred from the stimulation site to nerve via a series of reversible electrochemical reactions at the electrode-tissue interface. This faradaic charge injection is usually capable of delivering more charge to the nerve ending than capacitive charge injection. The latter does not work with the electrochemical reactions, because it works electrostatic (charge separation or dipole orientation) or electrolytic (charge storage). It depends on the material of the electrode and the charge injection phenomenon occurring at the interface. Like stated earlier, platinum (Pt) utilizes faradaic charge injection, because it makes the use of electrochemical reactions. These reactions are:

$$Pt + H_2O \Longleftrightarrow PtO + 2H^+ + 2e^- \quad (1)$$

$$Pt + H_2O + e^- \Longleftrightarrow PtH + OH^- \quad (2)$$

Improved charge transfer capacity is commonly estimated by using a reversible charge injection process through either double layer capacitive reactions and reversible faradaic charge transfer reactions at the electrode/electrolyte interface as

Table 1 Charge-injection limits of electrodes

Material	Material properties and comments		
	Mechanism	Max. charge injection Q_{inj} (mC cm^{-2}) [20]	Comments
Pt and PtIr alloys	Faradic/Capacitive	0.05–0.15	Charge balanced biphasic pulse for Q_{inj}
Activated iridium oxide (AIROF)	Faradaic	1–5	Positive bias required for high Q_{inj}. Damaged by extreme negative potentials (< -0.6V)
Thermal iridium oxide	Faradaic	~1	Positive bias required for high Q_{inj}
Sputtered iridium oxide	Faradaic	1–5	Benefits from positive bias. Damaged by extreme negative potentials (< -0.6V)
Tantalum/ Ta_2O_5	Capacitive	~0.5	Requires large positive bias
Titanium nitride	Capacitive	~1	Oxidized at positive potentials
PEDOT	Faradaic	15	Benefits from positive bias

shown in Fig. 4. Primary and secondary water layers consist of a simple interface model known as the Helmholtz double layer. More complex models may be modelled if many primary layers are considered. From the metal electrode side electrons are forced towards interface surface by their attraction to positive ions present in the electrolyte solution for e.g., Na^+ and H^+. On the other hand electrolyte cations are also drawn to the interface surface by their attraction for electrode's electrons. An electrical filed is thus established by equal and opposite charge concentrations on each side of the electrode-electrolyte interface. In electrical analogy the Helmholtz double layer is commonly modelled as parallel RC network with plate or Helmholtz capacitance (CH) due to surface area of interface, with Faradaic resistance (RF) and electrolyte resistance (RE). Details for electrochemistry and charge transfer at electrode-electrolyte is in [21] and Table 1. Normally for electrically stimulating electrodes a faradaic process is reported due to charge availability to reach a certain activation threshold for metal electrodes exceeds from the ideal capacitive transfer [21]. However, the irreversible faradaic reactions for electrode are generally avoided.

2.3 Electrode Material Failures

Apart from the noble metals mentioned above, gold (Au), palladium (Pd) and rhodium (Rh) are also preferred for electrical stimulation [22]. However, even the working of these noble metals for electrical stimulation under saline environmental conditions cause corrosion. The corrosion is observed in terms of weight loss, metal ion dissolution, scattering of these ions in the electrolyte solution and unstable tissue layer formation. Even though the corrosion rates are minimal but considering the long term functioning of the electrode which affects not only its own deterioration but also toxicity of the electrolyte. In case of platinum metal there is a continuous deterioration/dissolution during cathodic phase and deposition during the anodic phase when a charged biphasic stimulus pulse is applied to the stimulation site. David Zhou and his colleagues found that a thick oxide layer is normally formed on Pt electrode surface due to pulse stimulation conditions for DC experiments in diluted sulphuric acid. This oxide layer shows high impedance during working which causes surface expansion, cracking of metal surface and finally leads to delamination as seen in Fig. 5 [23]. Further research and investigation is to carried out for better understanding of the electrode surface when subjected to pulse stimulation under saline electrolyte environmental conditions [24, 25].

3 Micro-Electrode Array

In view of the hurdles relating to CI electrode arrays and the challenges for material requirement different fabrication techniques have to be considered. Microfabrication offers various advantages such as batch production, micro scaling of electrode arrays, increased stimulation sites for broader frequency range etc. Over the previous years, microelectrodes fabricated by these techniques are used for extracellular neural stimulation and recording have been progressed from single electrode wires to microfabricated arrays [26]. Even though the microwires prove its robustness and functionality, one disadvantage in terms of mechanical stability is that the wires may bend during implantation causing the change pitch between two stimulation sites. Other problems associated with wire electrodes make them unsuitable for long term stability and performance. The development of the fabrication processes for the integrated-circuits

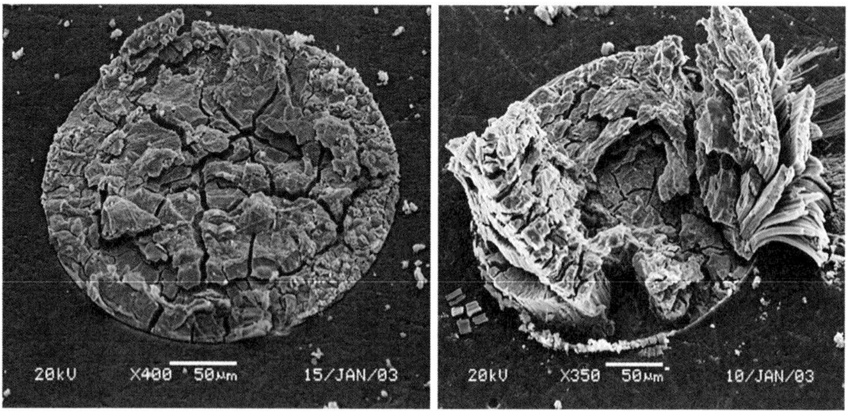

Fig. 5 Platinum Oxide surface layer expansion and cracks [23]

(IC's) consisting of photolithographic techniques with silicon etching technology enabled the path for the development and microfabrication of these microelectrode arrays. The growth of IC fabrication facilitated the earlier microfabricated thin-film microelectrode arrays in silicon. Photolithography provides the designer a room to create electrode sites of micron size to match specific application. In contrast with the wire electrodes these microelectrodes can be advantageous for highly localized stimulation and recording of the neural tissue. In addition to it, electronics and sensor integration enhances working performances with added functionality.

3.1 Stiff and Flexible Electrode Designs

In view with microelectrode arrays F. Blair Simmons in 1965 performed the first multichannel auditory prosthesis stimulation study with five stainless steel electrodes insulated with Formvar inserted into the auditory nerve (and not into the ST) and emerging out from cochlea [27]. In recent years the interest in alternative stiff electrodes, to be inserted directly into the auditory nerve, has revived. In research conducted at University of Michigan, a batch-fabricated cochlear electrode array with stacked layers of parylene and metal was fabricated by silicon micromachining techniques. The 32-site array contained IrO (Iridium Oxide) stimulation sites with a centre-to-centre site spacing of 250 μm [28]. In the following sections the stiff and flexible electrode designs from the Delft University of the Technology (TU Delft), The Netherlands are described with its micro-fabrication sequence done at The Delft Institute of Microsystems and Nanoelectronics (DIMES), TU Delft, The Netherlands.

3.1.1 Stiff Electrode Design

The stiff probe was designed to evaluate the fabrication possibilities and investigate photolithography problems encountered during the fabrication process. In the physiological perspective the intention was to penetrate and stimulate the auditory nerve bundle corresponding to their respective frequencies. The design of stiff probe, as shown in Fig. 6, contains pattered 16 Titanium Nitride stimulation sites on silicon substrate. They have diameters ranging between 75 and 250 μm with a pitch

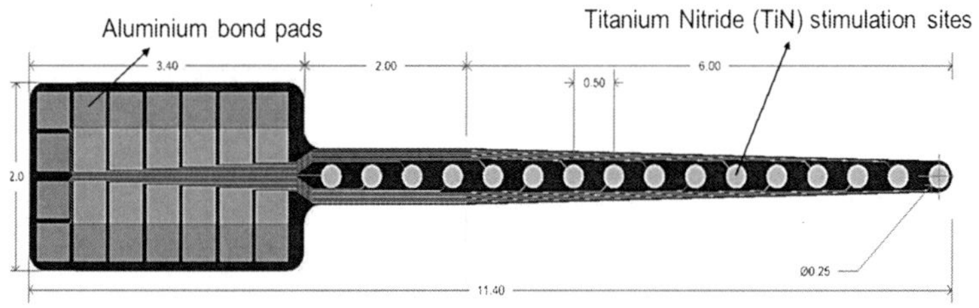

Fig. 6 Schematic view of the stiff design *(All dimensions are in mm)*

of 500 μm. These sites are connected by metal tracks to the aluminium bond pads establishing connection between the external circuitry and the sites. More details about the stiff probes are mentioned elsewhere [30, 47].

Prior to the microfabrication process, the stiff probe computer simulations were performed on the volume conduction computational model [29] developed at the medical centre in Leiden (LUMC), The Netherlands which gives the information regarding the stimulation pattern developed inside the cochlear auditory nerve bundle [31].

3.1.2 Flexible Electrode Design

The flexible electrode design is based on the stiff design, the only difference is its flexible behaviour in order to be inserted inside the curl shaped cochlea as shown in Fig. 2. The device is made flexible by replacing the silicon substrate with Polyimide (PI). PI is a polymer of imide monomers and comes in different types, changing in for example curing options [32]. It is flexible, biocompatible, and a good insulator with high chemical and electrical resistance (resistivity in the order of $10^{17}\,\Omega$. cm and dielectric constant of 3.4). Due to these properties PI is generally been used to coat medical implants. PI can be spin coated and patterned with the same techniques as used for regular chip fabrication. This makes it ideal to create micro-electrodes and combine it with metals. The flexible electrode design in our case consists of the stimulating electrode (300 nm thick of titanium nitride), sandwiched between two PI (10 μm 115 A Durimide^{TM}) layers. The upper PI layer is selectively patterned using dry etching to open the stimulating sites which will deliver the charge to stimulate the neurons. It has to be noted that the designs described here are for research purposes only, so the dimensions and layout need to be optimized for use in humans, or even for in-vivo experiments. The length of the research electrode array is 11 mm, which is smaller than the current electrode arrays (the Clarion of Advanced Bionics and the Nucleus 22 of Cochlear Ltd., both have a length of 25 mm [33]), and the electrode size needs to be verified, because the charge-injection capacity needs to be high enough. Rousche et al. [34] reported an electrode array based on polyimide for intracortical stimulation. Brain stimulation is very similar to cochlear nerve stimulation, both make use of the charge to stimulate the neurons. The process of creating such a flexible electrode array is always rather the same. It starts with a silicon wafer and an sacrificial layer like aluminium or oxide on top. Then the first layer of polyimide is applied and patterned. After that, the steps can differ somewhat, depending on the metals used. However, they all include metallization of biocompatible materials. The last step is to deposit another similar layer of PI coating the whole device and then selectively patterning the stimulation sites which would be in close proximity of the nerve endings. The detail microfabrication process and results are explained elsewhere [35]. The reported electrode arrays differ in their shape, size and materials, some of them are listed in Table 2. All of these designs make use of at least two layers of metal. Before the application of Gold there is the need for an adhesion layer. Gold has a high conductivity (4.10×10^7 S/m) [36], is corrosion resistant and

Table 2 Comparison of four papers that utilize polyimide in their electrode array design

Electrode size	Electrode pitch	Polyimide	Metals	Type of electrode array	References
Circular with a diameter of 15 μm	200 μm	10 μm of DuPont Pyralin 2525 (for the bottom) and 2555 (for the top)	Gold and titanium	Neural recordings	[38]
Square, $40 \times 40\,\mu m^2$	210 μm	2μm of Hitachi PIQ 13®	150 Å of chrome and 3000 Å of Gold	Auditory cortex measurements	[39]
Unknown	Unknown	1.3 μm of polyimide, type unknown	75 Å of chrome and 0.4 μm of gold	Neural recordings	[40]
Square, $30 \times 30\,\mu m^2$	Unknown	10–20 μm of Probimide 7520	250 Å of chrome and 200 nm of gold	Intracortical neural interface	[34]

biocompatible [37] for use in the neural environment. It does not adhere to polyimide well enough, so an adhesion layer like chromium or titanium is required.

4 Electrical Characterization Parameters for Materials

CI devices are approved by the Food and Drug Administration (FDA) for the human use for many years. The development of these devices in the form of microelectrode arrays need not only the approval from FDA but the materials been used should withstand the harsh saline environment mechanically and electrically. When the dimensions of these devices are scaled down to micron size electrical parameters such as electromigration, material impedance, Electric field distribution, Temperature Coefficient of Resistance (TCR) etc. has to be considered. Each of these parameters will be dealt in the following sections.

4.1 Electromigration

Electromigration is the phenomenon normally observed in the interconnecting metal in integrated circuits which damages the metal when the current density exceeds the safe operating limits. In CI's this parameter is important while designing the metal

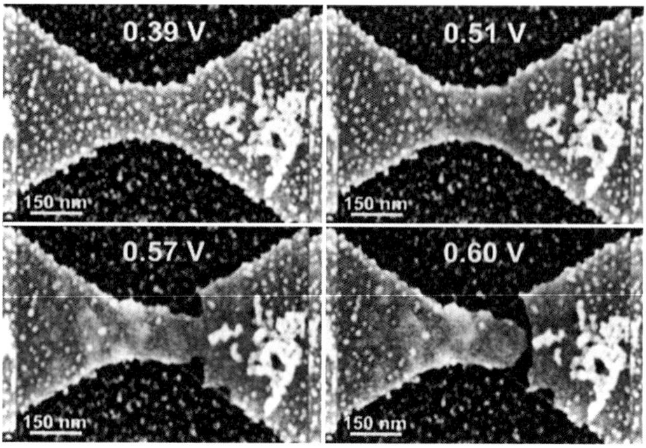

Fig. 7 Electromigration of gold (*Au*) at different voltages. Study is conducted by Thiti Taychatanapat of Cornell Center for Materials Research [41]

tracks which are connecting the bond pads with the stimulation sites in the electrode arrays.

Electromigration occurs when the dimensions shrink, but the amount of current that needs to be transported stays the same. Earlier this was only a problem in integrated circuits, but cannot be neglected for the metal tracks in CI's. Instead of the platinum wires that have a diameter of 0.025 mm (a cross sectional area of around 500 μm^2) now tracks of at most 10 μm wide and 500 nm thick (creating an area of 5 μm^2) are used in our design and generally observed in the same range as the modern microelectrode arrays. This leads to a current density increase by a factor of at least 100. It is possible to use multiple electrodes, decreasing the amount of current per channel, but there will be an increase in the current density of a single track.

Electromigration is the phenomenon which follows a forced atomic diffusion that occurs due to a large electric field [42]. Metal ions are transported through the channel due to the large amount of charge carriers. This will physically move particles from one place to another. When it has started, it will cascade until there is no metal left at the spot where it happened, because there is a local increase of current density. Fig. 7 shows what happens to gold when the current density is too high. Aluminium is a material that is widely used in integrated circuits, but is also very susceptible to electromigration. Temperature is a factor which affects this phenomenon, because the diffusion constant is largely dependant on it. Diffusion is the driving force of electromigration. Aluminium has the rule of thumb that tells that the maximum current density is roughly 1 mA/μm^2 at 125 °C.

Electromigration is the limiting factor for the maximum amount of current that can be passed through a conductor. The vacancy flux (J) due to the force induced by electromigration is related to multiple other parameters by Eq. 3 [42], c is the vacancy concentration, D the diffusivity, k Boltzmann's constant, T the absolute temperature, e the electronic charge, E the applied electric field, ρ the resistivity and

j the current density. Z^* is a dimensionless material parameter and consists of two components, as in Eq. 4.

$$J = -D\frac{\partial c}{\partial x} + \frac{Dc}{kT}Z^*eE = -D\frac{\partial c}{\partial x} + \frac{Dc}{kT}Z^*e\rho j \tag{3}$$

$$Z^* = Z_{wd} + Z_{el} \tag{4}$$

Z^* can be called the effective charge, Z_{wd} is attributed to momentum exchange between the electron current and the moving atom and Z_{el} is related to the direct electrostatic force on the moving atom. Z_{wd}, also called the electron wind, is the dominant factor, thus the atoms drift in the same directions as the electrons. The steady state solution of Eq. 3 occurs when the driving force due to generated stress equals that of electromigration, like in Eq. 5. Ω is the atomic volume and σ the stress.

$$Z^*e\rho j = \Omega\frac{\partial \sigma}{\partial x} \tag{5}$$

The diffusivity D shows a strong temperature dependence as given in Eq. 6. E_a is the activation energy for diffusion. From this equation it can be seen that higher temperatures are more susceptible to electromigration. A test at a higher temperature than body temperature (37 °C) gives the results for an accelerated test. It can be used to see if materials are affected by a certain current density at all. Ref. [44] reports that a certain amount of stress (σ_c) needs to be exceeded to start the electromigration.

$$D = D_0 e^{E_a/kT} \tag{6}$$

4.2 TCR

The Temperature Coefficient of Resistance (TCR) is a material dependent parameter that describes the change in the impedance with respect to temperature. For most materials the resistance increases with increasing temperature, some materials exhibit an opposite behaviour. These materials have a Positive Temperature Coefficient (PTC) or Negative Temperature Coefficient (NTC). In rare occasions it is possible that the resistance is close to 0 Ωwhen it reaches a very low temperature. It depends on the material what temperature this is, but the highest temperatures reported are around 125 K. The TCR of a metal can be described by a simple Eq. 7, where $\rho(T)$ is the resistivity of the material at temperature T given in (Ω. m), ρ_0 is the resistivity at reference temperature T_0 and α is the TCR [45].

$$\rho(T) = \rho_0\,(1 + \alpha(T - T_0)) \tag{7}$$

$$\alpha = \frac{1}{\rho_0}\left[\frac{\partial \rho}{\partial T}\right]_{T=T_0} \quad (8)$$

The resistivity ρ can be replaced with the resistance R in Eqs. 7 and 8, the α stays the same, because its unit is $(^{\circ}\mathrm{C})^{-1}$. These (linear) equations apply to simple materials like the metals, but semiconductor materials comply to non-linear equations.

The temperature in the ear is constant, however there might be a rise in temperature due to an illness. Though minimal, the temperature range of the human body is something to keep in mind when designing and calibrating the CI. With these measurements the resistance at every temperature can be deducted. A change in temperature can happen due to the joule heating (self heating when a current is flowing) of a track. This must be kept to a minimum, because the slightest rise in temperature in the cochlea can lead to damage of the Organs of Corti. The method we have adopted to determine the TCR is to measure the resistance at different temperatures. The results are then combined with Eq. 7 to obtain the TCR. The resistance should not be very susceptible to temperature changes, thus the TCR must be relatively low.

4.3 Impedance

Electrode impedance is an important parameter while fabricating the microelectrode array. Lower impedance is generally favoured for the stimulation material which enables high charge transfer capabilities. Looking closer to the electrode-electrolyte interface the electrode impedance is dominated by the double-layer capacitance (C_d) as explained earlier, which is in series with the resistance of the electrolyte (R_e) [43]. This acts like a high-pass filter with a cut-off frequency at:

$$f = \frac{1}{2.\pi.R_e.C_d} \quad (9)$$

At high frequencies (>10 kHz), the capacitor C_d blocks all the current, and the resulting electrode impedance is approximately equal to the electrolyte resistance R_e. At low frequencies (~0.1 Hz), C_d acts as a open circuit, resulting in a total electrode impedance of $R_e + R_i$. At intermediate frequencies, the impedance is governed by the capacitance:

$$Z = \frac{1}{j.w.C}, \ with\ C \infty A \quad (10)$$

Therefore, an electrode with rougher surface would exhibit a larger surface area, and thus resulting in higher capacitance. This would not be the case for the smoother platinium strips which are used as stimulation sites in the cochlear implant electrode arrays.

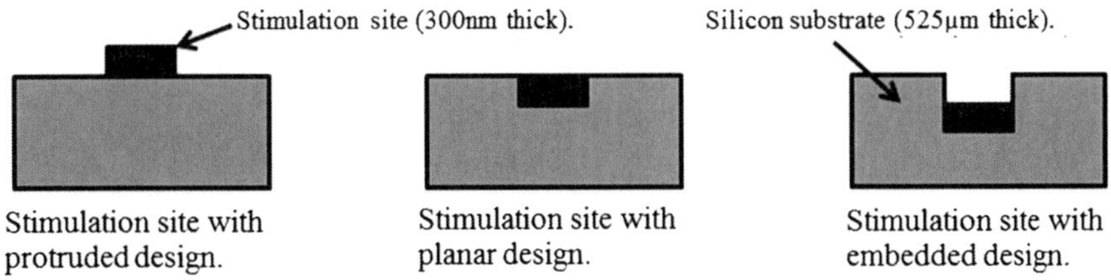

Fig. 8 Schematic view for design configurationa showing the cross-section of electrode with the stimulation site

4.4 Electric Field Distribution

When the density of the electrode increases, it is important to know the electric field distribution through the perilymph of the scala tympani. Furthermore, the effect of changing geometric parameters such as thickness of insulating layer, depth of the cavity for stimulation material and the dimensions of the electrode have to be evaluated. The electric field of one electrode should be directional and with respect to next stimulation site for localized neural stimulation. An electrode can have different geometric shapes (e.g. circular, square etc.), but to achieve a high density electrode array the stimulation site must aim to stimulate less population of neurons to avoid charge distribution in the neighbouring neurons. Furthermore, it is not desirable to have high (local) electric fields, e.g. at the corners of the stimulation site. Sharp edges in the geometry makes possible localization of the unwanted materials and organisms, like bacteria.The electric field distribution is influenced not only by the above mentioned parameters but also due to the electrolyte (perilymph in case of Cochlear Implant) contents and its electrical properties.

To understand the effect of the geometric parameter of the stimulation site on the electric field three design variations for the stimulation site were considered as illustrated in Fig. 8.

One way to calculate electric field distribution is to first calculate the electric potential distribution and then later the electric field distribution is directly obtained by minus gradient of the electric potential distribution. In a normal electrostatic field problems, electric field distribution can be written as follows [46]:

$$E = -\nabla V \tag{11}$$

From the classical Maxwell's equation

$$\nabla E = \nabla(-\nabla V) = \frac{\rho}{\varepsilon} \tag{12}$$

where, ρ is the resistivity Ω/m, ε is the material dielectric constant ($\varepsilon = \varepsilon_0\ \varepsilon_r$), ε_0 is the free space dielectric constant (8.854×10^{-12} F/m), ε_r is the relative dielectric constant of dielectric material. Placing the Eq. 11 into Eq. 12 we obtain the Poisson's equation as follows:

$$\varepsilon.\nabla(\nabla V) = -\rho \tag{13}$$

Without space charge $\rho = 0$, Poisson's equation converts to Laplace's equation.

$$\varepsilon.\nabla(\nabla V) = 0 \tag{14}$$

The Finite Element Method (FEM) study was performed to calculate the potential distribution of the electric field caused by the change in geometry of the stimualtion site. The study was simulated with COMSOL Multiphysics 4.2®. A two dimensional finite element model was created with a cross section of the electrode array in perilymph to evaluate the electric field. For simplicity purpose the three dimensional study was avoided and all the cochlear tissues were considered purely resistive. More details of the FEM analysis study can be found elsewhere [47].

5 Characterization Results and Discussions for Electrode Materials

Titanium Nitride (TiN) and Titanium (Ti) are the metals that are investigated as possible candidates for electrode arrays with Aluminium (Al) been used in our experiments for comparison purposes. Al which is being widely used in microelectronic industry as an interconnect, shows lower resistivity when combined with TiN. All of the measurements performed in the experimental setup are done on the same type of test die. There are four batches of this die, all of them are processed with different metals: Al, TiN, Ti and one has aluminium covered with titanium nitride (Al-TiN). Following sections will provide you details about the experiments performed with the experimental setup on these materials and the results thereafter. Section 5.1 goes into detail about this test die. The test die was subjected to several tests and measurements: the impedance of the different materials, the self-heating characteristics, the temperature coefficient of resistance (TCR), electromigration and endurance in a saline solution have all been measured out of which Electromigration, TCR and Impedance will be explained in detail in the later sections.

5.1 Test Die

To characterize the different metals mentioned earlier a test die (Fig. 9) was designed. The substrate of the die is a silicon (Si) wafer. This test die contains nine parallel tracks of the material that is under characterization. These tracks have different widths, the smallest is 1 µm and the largest is 5 µm. The other seven tracks are increments of 0.5 µm on the 1 µm track. They are all 10 mm long and the thickness of the material

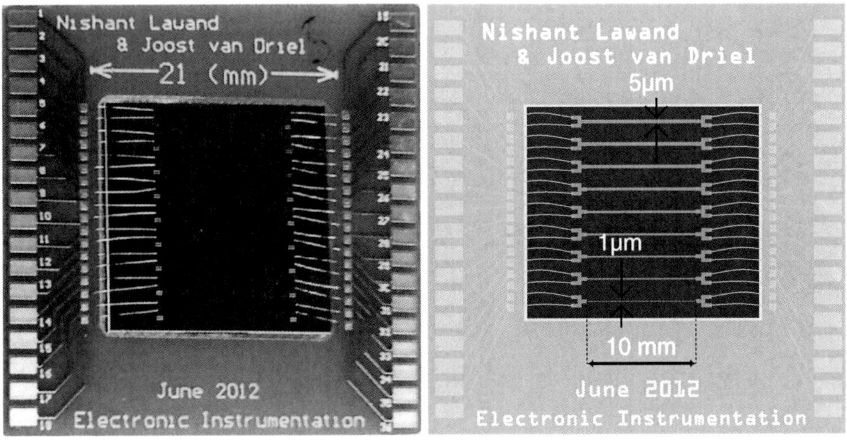

(a) (b)

Fig. 9 The test die used for measurements. **a** Photo of the test die with printed circuit board (pcb) used for measurements. **b** Schematic view of the test die showing 9 parallel lines of different widths

that needs to be characterized is 200 nm. The TiN needs an additional Ti layer of 40 nm under it for adhesion purposes. The track ends with two Al covered bonding pads at each side. Aluminium bond wires are used to connect the die to the printed circuit board (pcb). The pcb is only used as an adapter for the measurement equipment.

The processing steps of the TiN version are as follows: the substrate is a 100 p-type Si wafer of 525 +/− 15 μm thick and with a diameter of 100 mm +/− 0.2 mm. It has a resistivity of 2–5 Ω.cm. The isolation layer consists of silicon nitride (Si_3N_4) which was deposited using Low Pressure Chemical Vapour Deposition (LPCVD). This layer is 500 nm thick. The Ti and TiN layers are deposited using Physical Vapour Deposition (PVD) techniques using a Sigma 204 (Trikon) DC magnetron sputtering machine at 300 °C with a base pressure of 1.332×10^{-5} Pa and a working pressure of 0.0133×10^{-5} Pa. Both TiN and Ti use a 99.999 % high-purity Ti target for sputtering. The vacuum chamber is filled with nitrogen and argon of 99.99 % purity during deposition process. The distance between the Ti target and the wafer is 27.5 cm. The thin Ti layer under the TiN is 40 nm with TiN of 200 nm on top of it. 1.5 μm of pure Al is deposited similarly at 400° C as a bonding material. The first mask is used in the micro-fabrication process to pattern the Al layer. This patterning is done by wet etching solution (770 ml concentrated phosphoric acid (H_3PO_4, 85 %), 14 ml concentrated nitric acid (HNO_3, 65 %), 140 ml concentrated acetic acid (CH_3COOH, 100 %) and 76 ml deionized water. The remaining metal stack (Ti and TiN) is patterned using the second mask and positive photoresist and then etched using Reactive Ion Etching (RIE) in order to have the desired widths. The resulting layers are shown in Fig. 10a. The Al, Ti and Al-TiN versions differ from the TiN version process. The build-up of the layers of these test dies are shown

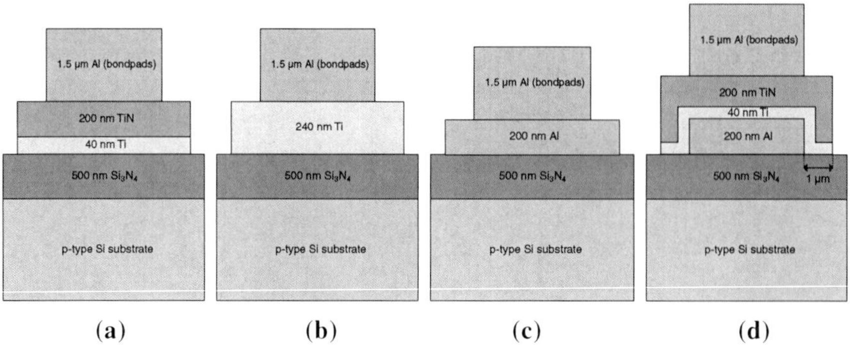

Fig. 10 Schematic view of metal layers used for different test dies. **a** The layers of the TiN test die. **b** The layers of the Ti test die. **c** The layers of the Al test die. **d** The layers of the Al-TiN test die

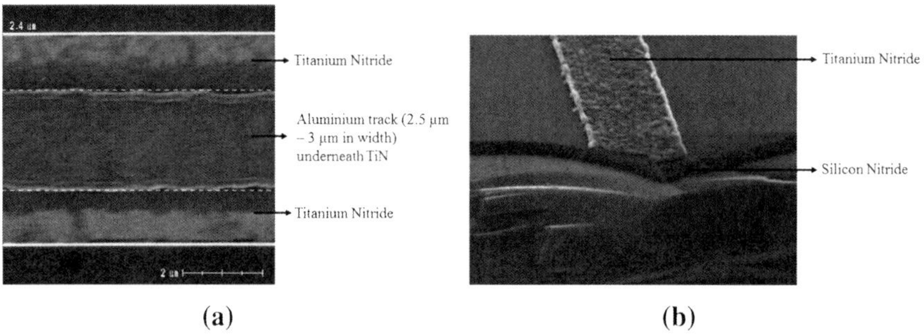

Fig. 11 SEM images of the fabricated tracks. **a** Cross-section SEM image of the Al-TiN track. **b** SEM image showing TiN with Si_3N_4 on silicon

in Fig. 10. The techniques described in the previous paragraph are also used to create the other versions as shown in Fig.10d.

Wafers with all the above mentioned versions were fabricated in the DIMES cleanroom of TU Delft. Scanning Electron Microscope (SEM) images of the micro-fabricated tracks are as shown in Fig. 11. A cross-sectional SEM image of a single Al-TiN diced through the centre of die in y-direction is as shown in Fig. 11. In this cross-section image Al (2.5 μm in width) can be seen in between the TiN (5 μm in width).

5.2 Electromigration

Electromigration phenomenon is tested by driving current through the smallest (the 1 μm) track. The smallest track is chosen because of the high current density even though the heat dissipation stays low. This is relevant, because a failure due to a too

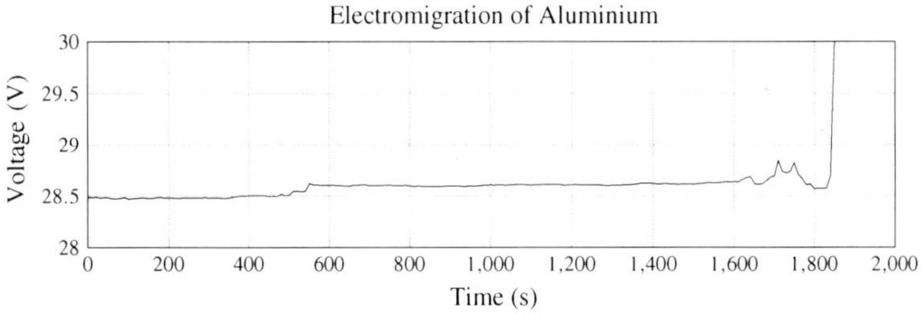

Fig. 12 Last 30 m of the electromigration test on aluminium, please note that the measurement was running for 3, 5 days, so there is an offset of 3×10^5 s

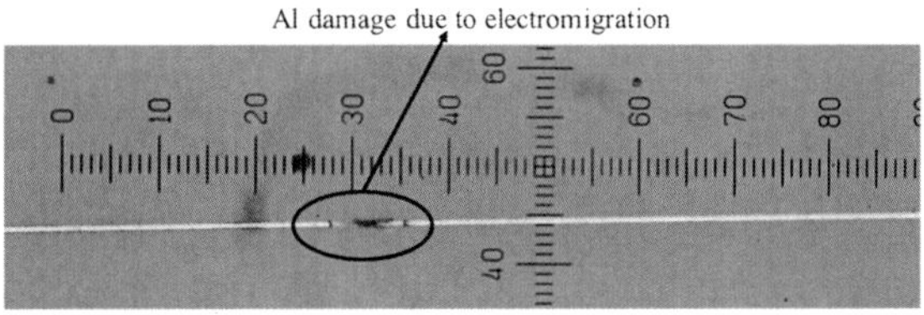

Fig. 13 Microscope image of the damage due to electromigration in the Aluminum track

high temperature is not the purpose of this test. The test has been done with the three materials (Al, Ti and TiN) using the *Yokogawa GS200 DC voltage/current source* and *Binder FD53 Drying/heating oven*. The GS200 is controlled and monitored using Labview® through a GPIB connection. The test die is put in the oven at a constant temperature and current. The measurement stops when the voltage hits 30 V, the maximum of the GS200. At this point the track will be interrupted. Both titanium and titanium nitride did not show any problems after a week of applying approximately 28 V (due to a constant current) and 37 °C. The titanium nitride was also exposed to higher temperatures, but apart from the higher potentials due to the TCR, there was no effect. Aluminium did break down after 3.5 days of passing 1.1 mA (current density of 5.5 $\frac{mA}{\mu m^2}$) and the last 30 minutes as shown in Fig. 12. The first damage to the track was done at 500 seconds, a slight increase in the potential was the result of that. A part of the track is missing, but there is still a direct connection. However, the current density at the location of the damage is higher than before, so a fatal damage will occur faster. This can be observed between 1,600 and 1,900 s: the line breaks down after some spikes. These spikes show the physical movement of the track due to the electric field, but the first two movements did not result in the fatal breakdown. A image using a regular light microscope has been made to verify the breakdown. This image is shown in Fig. 13.

Table 3 The TCRs of the different test dies ($°C^{-1}$)

Track width (μm)	Aluminium	Titanium	Titanium Nitride
1.0	$3.845 \cdot 10^{-3}$	$4.651 \cdot 10^{-3}$	$5.826 \cdot 10^{-4}$
1.5	$3.656 \cdot 10^{-3}$	$4.559 \cdot 10^{-3}$	$5.818 \cdot 10^{-4}$
2.0	$3.572 \cdot 10^{-3}$	$4.504 \cdot 10^{-3}$	$5.840 \cdot 10^{-4}$
2.5	$3.588 \cdot 10^{-3}$	$4.457 \cdot 10^{-3}$	$5.852 \cdot 10^{-4}$
3.0	$3.493 \cdot 10^{-3}$	$4.417 \cdot 10^{-3}$	$5.867 \cdot 10^{-4}$
3.5	$3.311 \cdot 10^{-3}$	$4.385 \cdot 10^{-3}$	$5.871 \cdot 10^{-4}$
4.0	$3.3 \cdot 10^{-3}$	$4.352 \cdot 10^{-3}$	$5.915 \cdot 10^{-4}$
4.5	$3.223 \cdot 10^{-3}$	$4.321 \cdot 10^{-3}$	$5.967 \cdot 10^{-4}$
5.0	$3.333 \cdot 10^{-3}$	$4.302 \cdot 10^{-3}$	$6.050 \cdot 10^{-4}$
Average	$3.5 \cdot 10^{-3}$	$4.4 \cdot 10^{-3}$	$5.9 \cdot 10^{-4}$

5.3 *TCR*

The Temperature Coefficient of Resistance (TCR) determines the relation between resistance and temperature. Three different test dies with Al, Ti and TiN were investigated to determine the coefficient for each material. A *Yokogawa GS200 DC voltage/current source* with a GPIB link to Labview® and a *Binder FD53 Drying/heating oven* have been used to obtain reliable results. A current sweep from zero to 30 V is done at different temperatures. The current step size can be adjusted in Labview® front panel. Materials with a higher resistance have a smaller step size to achieve the same amount of steps per material.

The current is applied through two of the four connections of a track at the test die. The voltage is measured at the other two connections of the track. This four point measurement makes sure that the measured voltage is only the potential difference over the track. The measurements are done per test die and it is possible to switch between the tracks with the use of a knob. The applied current is plotted against the measured voltage. Ohm's Law tells us that: $V = I \times R$, so the slope of the (linear) graph is the resistance. The slope is obtained using the fit function of Matlab. The TCR is then calculated using Eq. 7 by plotting the calculated resistances against the temperatures. The results from the plotted graphs can be then summarized as shown here in Table 3.

The results should all be the same per material, but both Ti and Al show relative large differences between the different track widths. Titanium nitride is showing linear results and has a relatively low TCR, which is favourable for a lot of applications. Aluminium changes its resistance with an increase of the current, probably due to self-heating. This is as shown in Fig. 14. The notable increase in resistance of Al is probably due to self-heating, because the current is much higher than supplied to Ti and TiN and each measurement took more time.

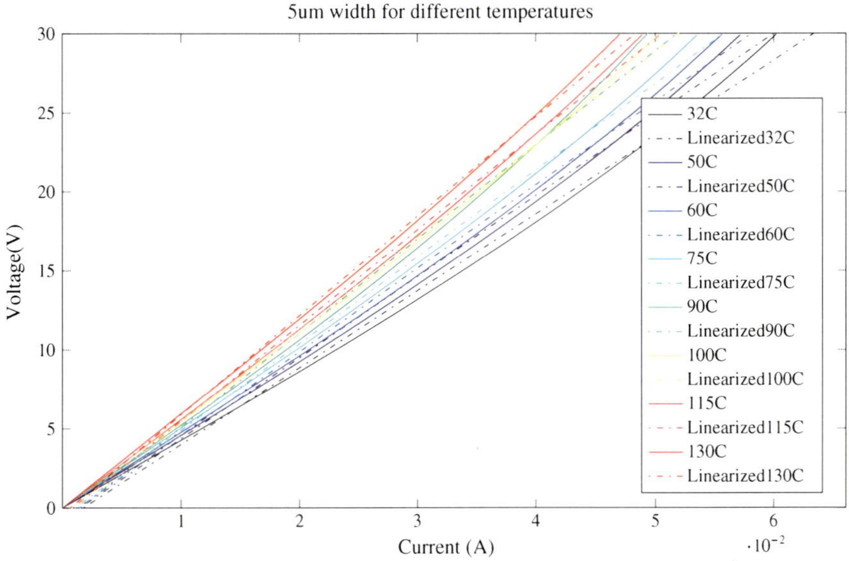

Fig. 14 The current-voltage relationship of the 5 μm track of an Aluminium test die is not linear and can be a cause of the differences in TCR in Table 3

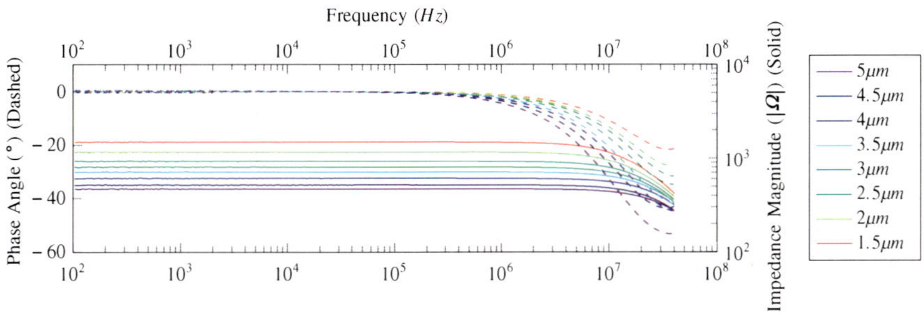

Fig. 15 Bode plot of the different tracks on the **Aluminium** test die

5.4 Impedance

An impedance measurement is done to see if the metal is purely resistive. If this is not the case, then it could indicate that signals (especially square waves) are distorted due to capacitive or inductive behaviour. It is also interesting to know the sheet resistance of the material to use in further calculations. The impedance characteristics are shown as a Bode plot, including both the impedance and the phase angle. The measurements have been done with a *HP 4194A Impedance/Gain Phase Analyzer* at room temperature, using a short integration time (500 μs), no averaging and a sweep from 100 to 40 MHz. This is the complete range of the device. A GPIB link from the device to a PC with Labview® has been used to transfer the data.

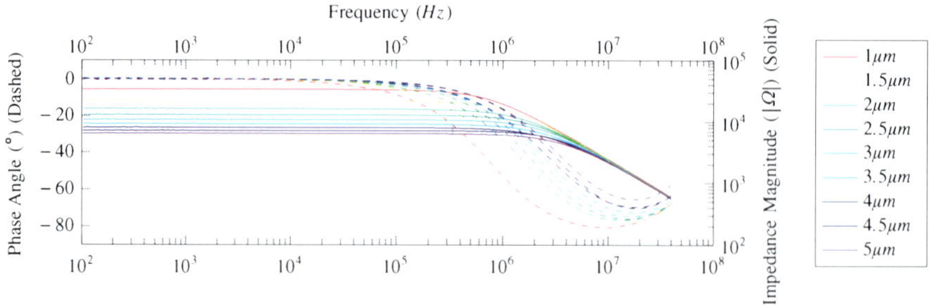

Fig. 16 Bode plot of the different tracks on the **Titanium** test die

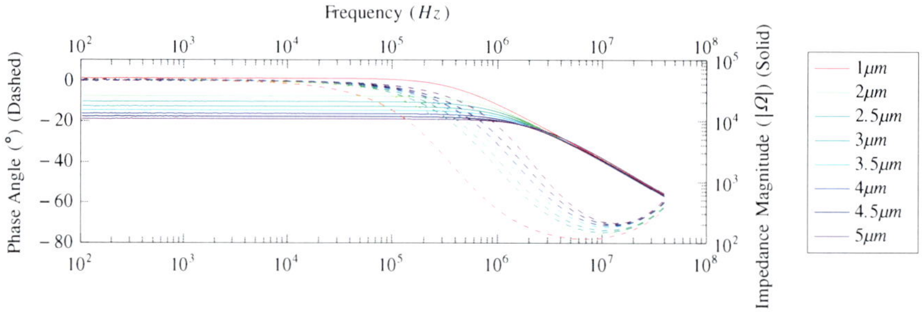

Fig. 17 Bode plot of the different tracks on the **Titanium Nitride** test die

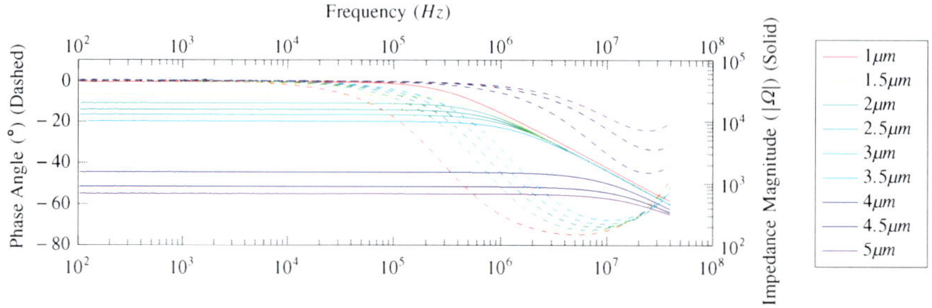

Fig. 18 Bode plot of the different tracks on the **Aluminium covered with Titanium Nitride**

The Bode plots of the four different test dies are shown in Figs. 15, 16, 17 and 18. The characteristics can be modelled as a resistor with a capacitor in parallel. The value of the resistor is the horizontal part of the impedance magnitude in the Bode plot and the capacitor defines the cut-off frequency. The capacitor acts as an open circuit at lower frequencies, but its impedance magnitude decreases with higher frequencies. The cut-off frequency is taken at the point where the phase is at -45°. There was a problem when determining the cut-off frequency for some of the aluminium tracks, because the phase did not get to -45°. The slope downwards was extrapolated at its linear piece to obtain a 'virtual cut-off frequency'. This gives an indication, but is

Table 4 The calculated values of the resistances and capacitances of the tracks of the different materials

	Al		Ti		TiN	
Track width (μm)	R(Ω)	C (F)	R (Ω)	C (F)	R (Ω)	C (F)
1.0			$3.43 \cdot 10^4$	$3.85 \cdot 10^{-11}$	$5.04 \cdot 10^4$	$4.53 \cdot 10^{-11}$
1.5	1488	$3.89 \cdot 10^{-11}$	$2.22 \cdot 10^4$	$3.90 \cdot 10^{-11}$		
2.0	1165	$2.25 \cdot 10^{-11}$	$1.64 \cdot 10^4$	$3.84 \cdot 10^{-11}$	$2.58 \cdot 10^4$	$4.50 \cdot 10^{-11}$
2.5	923	$3.62 \cdot 10^{-11}$	$1.32 \cdot 10^4$	$3.94 \cdot 10^{-11}$	$2.06 \cdot 10^4$	$4.50 \cdot 10^{-11}$
3.0	811	$3.48 \cdot 10^{-11}$	$1.09 \cdot 10^4$	$3.92 \cdot 10^{-11}$	$1.74 \cdot 10^4$	$4.53 \cdot 10^{-11}$
3.5	705	$2.55 \cdot 10^{-11}$	$9.32 \cdot 10^3$	$3.90 \cdot 10^{-11}$	$1.49 \cdot 10^4$	$4.36 \cdot 10^{-11}$
4.0	610	$2.46 \cdot 10^{-11}$	$8.11 \cdot 10^3$	$4.07 \cdot 10^{-11}$	$1.33 \cdot 10^4$	$4.28 \cdot 10^{-11}$
4.5	518	$2.32 \cdot 10^{-11}$	$7.35 \cdot 10^3$	$4.06 \cdot 10^{-11}$	$1.20 \cdot 10^4$	$3.91 \cdot 10^{-11}$
5.0	470	$1.14 \cdot 10^{-11}$	$6.53 \cdot 10^3$	$3.78 \cdot 10^{-11}$	$1.08 \cdot 10^4$	$3.59 \cdot 10^{-11}$

not as reliable as a real measurement. The modelled resistor and capacitor values of aluminium, titanium and titanium nitride are shown in Table 4.

The Bode plot of the Al-TiN test die (Fig. 18) shows that 3 of the 9 tracks have a combination of aluminium and titanium nitride. This can be concluded from the fact that 3 of the tracks have a much lower resistance than the regular titanium nitride test die. The (parasitic) capacitance is mostly from the track to substrate with silicon nitride in between. The change in impedance will not affect the working of the CI, because the cutoff frequency is much higher than the spectrum that CI's use.

6 Conclusions

The microfabrication of electrode arrays built with silicon micromachining techniques illustrates an positive approach towards future CI electrode array development in respect to the traditional manufacturing method used now days. Also lithography and MEMS technology facilitates the addition of enhanced functionality to the microelectrode arrays. There is, however, still a long way to go until these devices can be used in real Cochlear Implants. The fabrication possibilities and characterization of different CMOS compatible metals (Ti, TiN and Al) provides a strong base to go ahead with further research in this direction. In our electrical tests done we conclude that TiN is able to withstand a high current density $\left(2.8\frac{\text{mA}}{\mu\text{m}^2}\right)$, while aluminium failed due to electromigration; even when coated with TiN. The resistance of a 10 mm long and 5 μm wide track decreases from $1.08 \times 10^4\ \Omega$ to $6.9 \times 10^2\ \Omega$ when a combination of Al and TiN is made. This solves reasonably the high resistance of TiN. The self-heating and the change in resistance due to temperature changes (TCR) are measured, because the amount of dissipated heat should stay as low as possible. TiN has a low TCR, $5.9 \cdot 10^{-4}$. TiN was able to withstand the harsh current acceleration tests at high temperatures between 60 to 70 °C. Long term stability of these devices is a prime importance for the leading CI manufacturers. In that concern these microfabricated devices coated with bio-compatible insulation for e.g. Parylene are to be tested for long term electrochemical experiments in saline solution. Surface enhancement and coating with bioactive conducting polymers can improve the life time of the implants but its in-vivo and in-vitro capabilities have yet to be proven for this type of application in future.

Acknowledgments The authors gratefully acknowledge the Dutch Technical Foundation (STW) for their financial support in this project (Project no. 10056) and The Delft Institute of Microsystems and Nanoelectronics (DIMES) especially Mr. Johannes M. W. Laros (Mario) for the processing support. The authors would also like to thanks Advanced Bionics^{TM}, USA for their support in this project.

References

1. Hair cell regeneration and hearing loss, Research portfolio Online Reporting Tools (RePORT) http://health.nih.gov/,(March, 2012)
2. H.R. Cooper, L.C. Craddock, *Cochlear Implants—A Practical Guide* (Whurr Publishers, London, 2006), pp. 1–20
3. A.R. Moller, *Cochlear and Brainstem Implants*, 1st edn. (S. Karger AG, Richardson, 2006)
4. G. Clark, *Cochlear Implants: Fundamentals and Applications*, 1st edn. (Springer, New York, 2003)
5. Cochlear Implants for Hearing Loss-MED-EL http://www.medel.com/ (March, 2013)
6. D.D. Greenwood, J. Acoust. Soc. Am. 2592–2605 (1990)
7. D. Zhou, E. Greenbaum, *Implantable Neural Prostheses 1-Devices and Applications* (Springer, Berlin, 2009), pp. 87
8. A. Bolz, *Die Bedeutung der Phasengrenze zwischen alloplastischen Festkrpern und biologischen Geweben fr die Elektrostimulation* (Fachverlag Schiele und Schn, Berlin, 1995)
9. G.E. Loeb, R.A. Peck, J. Martyniuk, J. Neurosci. Methods **63**, 175–183 (1995)
10. T. Stieglitz, J. -U. Meyer, Microtechnical interfaces to neurons, in *Microsystem Technology in Chemistry and Life Science*, vol. 194, ed. by A. Manz, H. Becker (Springer, Berlin, 1998), pp. 131–162
11. T.L. Rose, L.S. Robblee, IEEE Trans. Biomed. Eng. **37**, 1118–1120 (1990)
12. S.F. Cogan, P.R. Tryok, J. Ehrlich, T.D. Plante, IEEE Trans. Biomed. Eng. **52**, 1612 (2005)
13. K. Yoshida, K. Horch, IEEE Trans. Biomed. Eng. **40**, 492–4 (1993)
14. F. Yuan, J. Wiler, K. Wise, D. Anderson, in *Proceedings of the IEEE BMES/EMBS Conference* (1999)
15. B. Wessling, W. Mokwa, U. Schnakenberg, J. Electrochem. Soc. **155**(5), F61–F65 (2008)
16. S.F. Cogan, T.D. Plante, J. Ehrlich, in *Proceedings of the 26th Annual International Conference of the IEEE Engineering in Medicine and Biology Society*, vol. 2, pp. 4153 (2004)
17. S.F. Cogan, J. Ehrlich, T.D. Plante, A. Smirnov, D.B. Shire, M. Gingerich, in *Annual International Conference of the IEEE Engineering in Medicine and Biology Society*, vol. 6, pp. 4153–4156 (2004)
18. N.S. Dias, J.P. Carmo, A. Ferreira da Silva, P.M. Mendes, J.H. Correia, Sens. Actuators, A **164**(1—2), 28–34 (2010)
19. M.M. Schaldach, *Electrotherapy of the Heart* (Springer, Berlin, 1992)
20. S.F. Cogan, Annu. Rev. Biomed. Eng. **10**, 275–309 (2008)
21. D.R. Merrill, *Implantable Neural Prostheses 2* (Springer, New York, 2010), pp. 85
22. P.F. Johnson, L.L. Hench, Brain Behav. Evol. **14**, 23–45 (1977)

23. D. Zhou, A. Chu, A. Agazaryan, *Proceedings of the 207th Meeting of the Electrochemical Society*, Canada, 2005, p. 275
24. E.M. Hudak, J.T. Mortimer, J. Neural Eng. **7**, 026005 (2010)
25. A. Hung, D. Zhou, R. Greenberg, J. Electrochem. Soc. **154**, C479–C486 (2007)
26. K.C. Cheung, Biomed. Microdevices **9**, 923–938 (2007)
27. F. Blair, Science, **148**(2666), 104–106 (1965)
28. A.C. Johnson, K.D. Wise, IEEE MEMS, pp. 1007–1010 (2010)
29. J.J. Briaire, Phd thesis, Cochlear Implants: from models to patients (2008)
30. N.S. Lawand, P.J. French, J.J. Briaire, J.H.M Frijns, in *Proceedings of IEEE Sensors*, pp. 1827–1830 (2011)
31. N.S. Lawand, P.J. French, J.J. Briaire, J.H.M. Frijns, *Advanced Materials Research*, vol. 254, pp. 82–85 (2011)
32. Fujifilm Electronic Materials (2012) Polyamic Acid. Durimide.
33. J.T.J. Roland, T.C. Huang, A.J. Fishman, Cochlear implant electrode history, in *Cochlear Implants*, ed. by J. Roush, 2nd edn. (Thieme Medical Publishers, New York, 2006), pp. 110–125
34. P.J. Rousche, D.S. Pellinen, D.P. Pivin, J.C. Williams, R.J. Vetter, D.R. Kipke, IEEE Trans. Biomed. Eng. **48**, 361–371 (2001)
35. N.S. Lawand, W. Ngamkham, G. Nazarian, P.J. French, W.A. Serdijn, G.N. Gaydadjiev, J.J. Briaire, J.H.M. Frijns, IEEE EMBS Yet to be published (July 2013)
36. R.A. Serway, *Principles of Physics Fort Worth*, 2nd edn. (Saunders College Pub, London, 1998)
37. M. Marelli, G. Divitini, C. Collini, L. Ravagnan, G. Corbelli, C. Ghisleri, A. Gianfelice, C. Lenardi, P. Milani, L. Lorenzelli, J. Micromech. Microeng. **21**, 045013 (2011)
38. S.A. Boppart, B.C. Wheeler, C.S. Wallace, IEEE Trans. Biomed. Eng. **39**, 37–42 (1992)
39. A.L. Owens, T.J. Denison, H. Versnel, M. Rebbert, M. Peckerar, S.A. Shamma, J. Neurosci. methods **58**, 209–220 (1995)
40. M. Peckerar, S.A. Shamma, M. Rebbert, J. Kosakowski, P. Isaacson, Rev. Sci. Instrum. **62**(9), 2276 (1991)
41. T. Taychatanapat (CMMR), (2007) http://www.ccmr.cornell.edu/
42. D. Pierce, P. Brusius, Microelectron. Reliab. **37**, 1053–1072 (1997)
43. D. Zhou, E. Greenbaum, *Implantable Neural Prostheses 2-Techniques and Engineering Approaches* (Springer, New York, 2010), p. 201
44. I.A. Blech, J. Appl. Phys. **47**(4), 1203 (1976)
45. A. Kusy, J. Appl. Phys. **62**(4), 1324 (1987)
46. S. Sangkhasaad, High Voltage Engineering, Thailand, 3rd edn. Printed in Bangkok, pp. 121–123 (2006)
47. N.S. Lawand, J. van Driel, P.J. French, Bioscience and Bioengineering, COMSOL 2012 (2012)
48. N.S. Lawand, P.J. French, J.J. Briaire, J.H.M Frijns, Sensing Technology (ICST), pp. 533–537 (2012)

Chapter 2
Machine Vision Based Techniques for Automatic Mango Fruit Sorting and Grading Based on Maturity Level and Size

C. S. Nandi, B. Tudu and C. Koley

Abstract In recent years automatic vision based technology has become more potential and more important to many areas including agricultural fields and food industry. An automatic electronic vision based system for sorting and grading of fruit like Mango (Mangifera indica L.) based on their maturity level and size is discussed here. The application of automatic vision based system, aimed to replace manual based technique for sorting and grading of fruit as the manual inspection poses problems in maintaining consistency in grading and uniformity in sorting. To speed up the process as well as maintain the consistency, uniformity and accuracy, a prototype electronic vision based automatic mango sorting and grading system using fuzzy logic is discussed. The automated system collects video image from the CCD camera placed on the top of a conveyer belt carrying mangoes, then it process the images in order to collect several relevant features which are sensitive to the maturity level and size of the mango. Gaussian Mixture Model (GMM) is used to estimate the parameters of the individual classes for prediction of maturity. Size of the mango is calculated from the binary image of the fruit. Finally the fuzzy logic techniques is used for automatic sorting and grading of mango fruit.

Keywords Electronic vision · Fruit sorting and grading · Video image · Maturity prediction · Gaussian mixture model (GMM) · Fuzzy logic

---

C. S. Nandi (✉)
Applied Electronics and Instrumentation Engineering Dept. University Institute of Technology, The University of Burdwan, Burdwan, India
e-mail: chandrasekharnandi@gmail.com

B. Tudu
Instrumentation and Electronics Engineering Department, Jadavpur University, SaltLake Campus, Kolkata, India

C. Koley
Electrical Engineering Department, National Institute of Technology, Durgapur, India

A. Mason et al. (eds.), *Sensing Technology: Current Status and Future Trends II*, Smart Sensors, Measurement and Instrumentation 8, DOI: 10.1007/978-3-319-02315-1_2,

1 Introduction

Machine vision and image processing techniques have been found increasingly useful in the fruit industry, especially for applications in quality inspection and defect sorting applications. Thus fruit produced in the garden are sorted according to quality and maturity level and then transported to different standard markets at different distances based on the quality and maturity level. Sorting of fruits according to maturity level is most important in deciding the market it can be sent on the basis of transportation delay.

In present common scenario, sorting and grading of fruit according to maturity level are performed manually before transportation. This manual sorting by visual inspection is labor intensive, time consuming and suffers from the problem of inconsistency and inaccuracy in judgment by different human. Which creates a demand for low cost exponential reduction in the price of camera and computational facility adds an opportunity to apply machine vision based system to assess this problem.

The manual sorting of fruits replaced by machine vision with the advantages of high accuracy, uniformity and processing speed and more over non-contact detection is an inevitable trend of the development of automatic sorting and grading systems [1]. The exploration and development of some fundamental theories and methods of machine vision for pear quality detection and sorting operations has been accelerate the application of new techniques to the estimation of agricultural products' quality [2].

Many color vision systems have been developed for agricultural grading applications which include direct color mapping system to evaluate the quality of tomatoes and dates [3], automated inspection of golden delicious apples using color computer vision [4]. Color image processing techniques, presented in [5] was proposed to judge maturity levels or growing of the agricultural products.

In recent years, machine vision based systems has been used in many applications requiring visual inspection. As examples, a color vision system for peach grading [6], computer vision based date fruit grading system [7], machine vision for color inspection of potatoes and apples [8], and sorting of bell peppers using machine vision [9]. Some machine vision systems are also designed specifically for factory automation tasks such as intelligent system for packing 2-D irregular shapes [10], versatile online visual inspections [11, 12], automated planning and optimization of lumber production using machine vision and computer tomography [13], camera image contrast enhancement for surveillance and inspection tasks [14], patterned texture material inspection [15], and vision based closed-loop online process control in manufacturing applications [16].

Here the proposed technique applies machine vision based system to predict the maturity level and size of mango from its RGB image frame, collected with the help of a CCD camera. In additions, the method can be extended and applied to analyze and extract the visual properties of other agricultural products as well. The materials and methods, details preprocessing of image, different feature extraction methods and the basic theory of GMM are discussed in the following sections. Finally the results, discussions and conclusions are summarized at the last section.

2 Materials and Methods

2.1 Sample Collection

For the experimental works total 750 number of unsorted mangoes of five varieties locally termed as "Kumrapali" (KU), "Amrapali" (AM), "Sori" (SO),"Langra" (LA) and "Himsagar" (HI) were collected from three gardens, located at different places of West Bengal, India. Collection of mangoes were performed in three batches with an interval of one week in between batches and in each batch 250 numbers of mango were collected, having 50 numbers of each variety i.e. KU, AM, SO, LA and HI. Steps were taken to ensure randomness in mango collection process from the gardens in each batch. After collection of mangoes each mango were tagged with some unique number generated on the basis of variety, name of the origin garden, batch number and serial number etc. Three independent human experts work in the relevant field were selected for manual identification of maturity.

Each mango was used to pass through a conveyer belt every day until it rotten and was presented to the experts (after removing tags) for recording of human expert predicted maturity level. Then the mangoes were stored in a manner as used during transportation.

2.2 Experimental Procedure

A schematic diagram of electronic vision based system for automatic mango fruit sorting and grading is shown in Fig. 1. The camera used in the study was a 10 megapixel CCD (charge-coupled device) camera for capturing the video image. Then the still frame are extracted using MATLAB software from the video images with frame rate of 30 frames/sec. The camera was interfaced with a computer through USB port. The proposed algorithm was implemented in Lab VIEW®Real Time Environment for automatic sorting and grading. Light intensity inside the closed image capturing chamber was kept at 120 lux, measured with the help of lux meter (Instek-GLS-301) and was controlled automatically by the light sensor along with light intensity controller and light source supplier.

The automated fruit sorting and grading system consists of a motor driven conveyer belt to carry the fruits serially. The fruit placer places one fruit at a time on the conveyer belt and the belt carries it to the imaging chamber where the video image of fruit is captured by the computer through CCD camera. The proposed algorithm runs into the computer automatically to classify the fruit on the basis of four different maturity level like raw (M1), semi-matured (M2), matured (M3) and over matured (M4) and then give a direction to the sorting unit to place the mango in appropriate bin. The sorting unit consists of four solenoid valves driven by respective drive units, which are controlled by the computer. The time delay in between image capturing

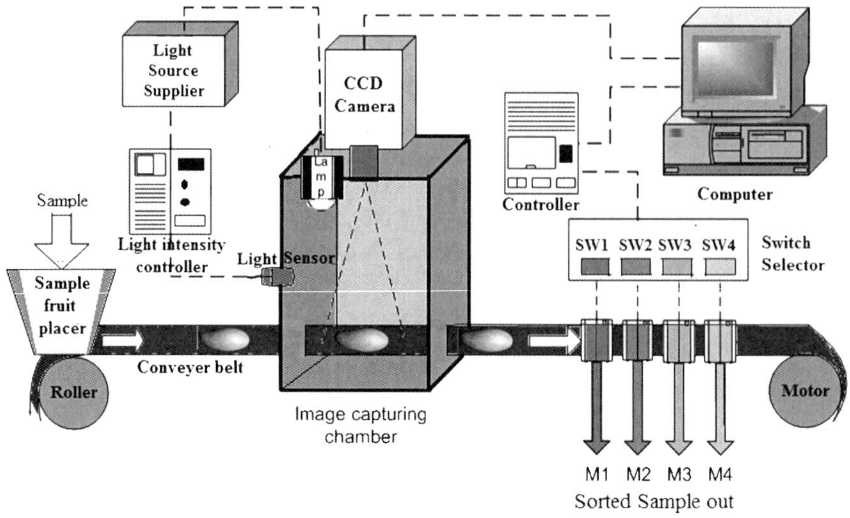

Fig. 1 Proposed model of machine vision based automated mango fruit sorting and grading system

of a fruit and the triggering the solenoid valve is estimated by the computer on the basis of conveyer belt speed.

The color of the conveyer belt was chosen blue for two reasons. First, blue does not occur naturally in mangoes. Second, blue is one of the three channels in the RGB color space, making it easier to separate the background from the image of mango.

The image capturing chamber is a wooden box and the ceiling of the chamber is quoted with reflective material to reduce the shading effect. A CCD camera is mounted in the top center of the image capturing chamber. One fluorescent lamp is mounted at the top of the chamber. The camera is mounted right side the light source for the best imaging. The distance between the CCD camera and the fruit on the conveyer belt is kept 20 cm.

The light intensity inside the imaging chamber is measured and consequently controlled by a separate light intensity controller, which keep the light intensity constant irrespective of power supply voltage and any variation of the filament characteristics and changes in ambient environment. However, even with the lamp current being constant, the light output of the lamp still varies resulting from the lamp ageing, filament or electrode erosion, gas adsorption or desorption, and ambient temperature. These effect cause changes in the RGB values of the images. The light intensity controller corrects for the lamp output changes, maintaining a constant short and long-term output from the lamp. The light output regulating unit is made up of a light sensing head and a controller. The (silicon based) light sensor is also mounted near to the sample fruit inside the chamber monitors part of the light source output; the controller constantly compares the recorded signal to the pre-set level and changes the power supply output to keep the measured signal at the set level i.e. 120 lux.

In this system the motor speed and distance between the two consecutive mangoes are taken as input. If the motor speed and distance between two consecutive mangoes are known then we can find a frame that will be the best still image of full mango within the imaging chamber. In our system the speed of the conveyer belt was 2 ft/sec, length of the imaging chamber was 1 ft and the distance between two consecutive mangoes on the conveyer belt was 1 ft. So 7,200 samples/hour can be sorted by our system. This rate can be increased by increasing the speed of the conveyer belt and reducing the distance between two consecutive samples in the conveyer belt. But if we increase the speed of the conveyer belt the motion blur effect will occur. So this can't be increased very much. We have to trade off between this two. The size of the still frame in this system was 480 × 640 pixels.

2.3 Color Calibration of CCD Camera

The value of RGB is device dependent. So camera calibration [17] is essential for color inspection systems based upon machine vision to get the intrinsic and extrinsic parameters for providing accurate and consistent color measurements. When a camera creates an image, that image does not represent a single instant of time. Because of technological constraints or artistic requirements, the image may represent a scene over a period of time. Most often this exposure time is brief enough that the image captured by the camera appears to capture an instantaneous moment, but this is not always so, and a fast moving object or a longer exposure time may result in blurring artifact which makes this apparent. As object in a scene movie, an image of that scene must represent an integration of all positions of those objects, as well as the camera viewpoints, over the period of exposure determine by the shutter speed. Motion blur error during extraction of still frame from video image can be reduced by increasing the shutter speed of the camera and can be eliminated by using camera having global shutter technology. Many software products (e.g. Adobe Photoshop or GIMP) offer simple motion blur filters. However, for advanced motion blur filtering including curves or non-uniform speed adjustment, specialized software products are necessary.

2.4 Images of Different Category and Different Maturity Level of Mango

An image matrix for five varieties ("KU", "SO", "LA", "HI" and "AM") of mango having four different maturity levels (M1, M2, M3 and M4) are shown in Fig. 2.

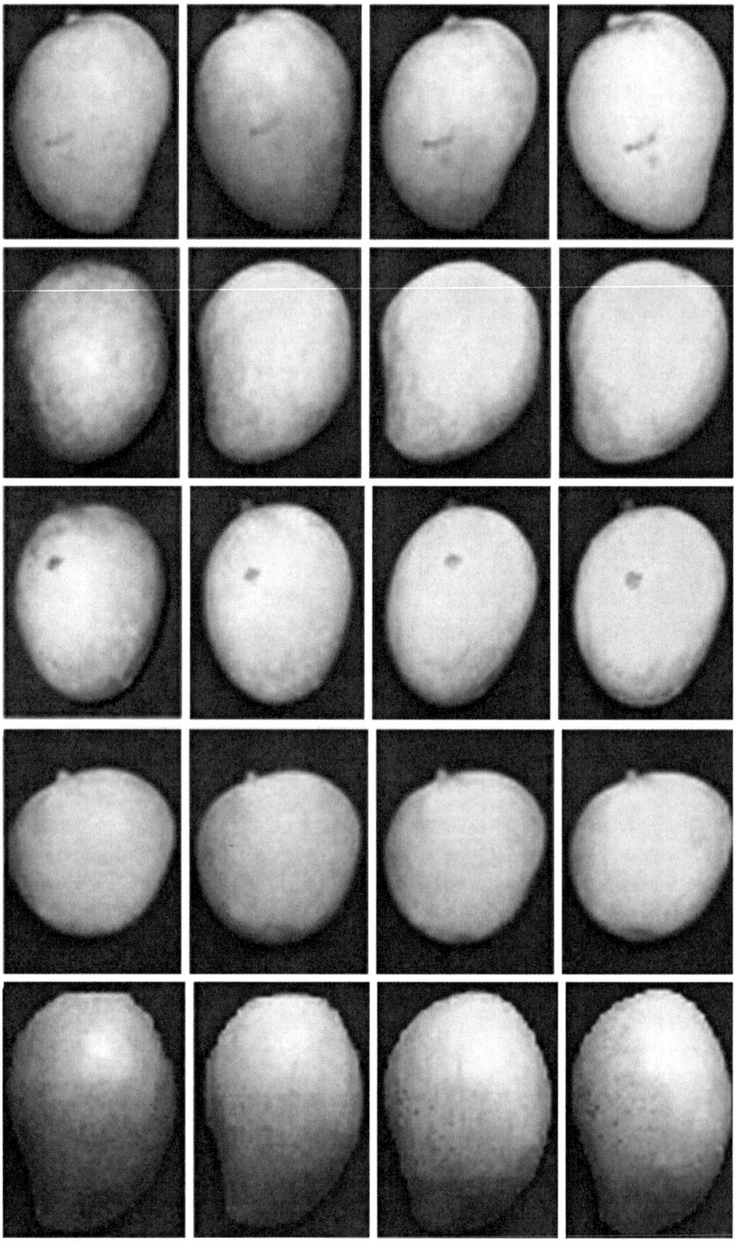

(a) (b) (c) (d)

Fig. 2 Images of five varieties mango having different maturity level. 1st row: "KU", 2nd row: "SO", 3rd row :"LA" and 4th row: "HI" and 5th row: "AM". Images are taken with an interval of 2 days, shown in (**a**) 1st colomn: raw (M1) (**b**) 2nd colomn: semimatured (M2) (**c**) 3rd colomn: matured (M3) (**d**) 4th column: over-matured (M4)

3 Pre-processing of Images

The performance of the sorting and grading system depends on the quality of the images captured by the video camera, since various measures/features calculated from the images of the mangoes will be used for sorting and grading. Video signal collected from camera found to be contaminated with motion blur artifact and noise, thus proper extraction of images from video frames and then filtering is essential. On the other hand mangoes moving through the conveyer belt, can be at any position along with the background thus in order to reduce computation, removal of background by detecting the edges of the image and alignment of the mango images in axial position is necessary. The present section discuss about all these preprocessing issues, in brief.

3.1 Still Frame Extraction from Video Image

At first the video streams acquired by the CCD camera are separated into sequence of images and then to remove motion debluring wiener filtering method [18] was used. Motion compensated four frames are shown in Fig. 3.

3.2 Filtering of Mango Image

Though the images were taken in controlled environment under fixed illumination of light of 120 lux with the help of tungsten filament lamp, but there were some noises in the picture. To remove these noises a simple median filter found to provide reasonable good performance but it is computationally intensive. For that pseudo-median filter [19] was used in the work, as it is computationally simpler and possesses many of the properties of the median filter. This filtering process often helped to obtain smooth continuous boundary of the mango.

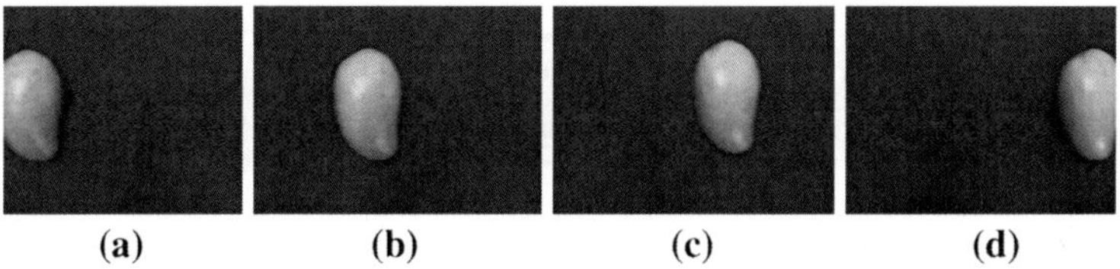

Fig. 3 Extracted still frames from the video image with an interval of 5 frames, (**a**) frame no.10, sample entering into the imaging chamber (**b**) frame no.15, sample near the middle of the imaging chamber (**c**) frame no. 20, sample crossing the middle position of imaging chamber (**d**) frame no. 25, sample going out from imaging chamber

3.3 Edge Detection and Boundary Tracing

Experimentally it was observed that for all the set of the RGB mango images the G value is always greater than the B value for that reason the color of the background was kept blue with R, G and B value of close to 0, 0 and 255 respectively, as much as possible. So with the help of simple comparison the back ground was eliminated, and the image was converted to binary image (BW). Small patches containing number of pixel less than 800 were removed. This cutoff number was determined experimentally, by studying all the images of the mango.

In order to find the boundary or the contour of the mango, a graph contour tracking method based on chain-code was adopted. This algorithm found to work reasonable fast, as the boundary of the mango is not so complex. The details of algorithm can be found in [20]. Here a brief description is given. The algorithm first detects every run at each row and records every single run's serial number and the corresponding start-pixel coordinates and end-pixel coordinates are stored in a table named as "ABS". Through this method, the run-length code of the image is obtained. Then, a 3 × 3 pixel box is adopted to detect the relationship between the objective pixel and its eight-connected surrounding pixels. All the runs are categorized into 5 classes, and then each run's serial number along with the corresponding class label was recorded in another table named as "COD". Then, the algorithm searches the ABS-table and COD-table sequentially. Based on the coordinates and class of each run recorded on the table, the starting pixel of the contour is recognized and the contour is successfully followed. And the chain code is generated while following the contour. An image of mango along with the traced boundary is shown in Fig. 4.

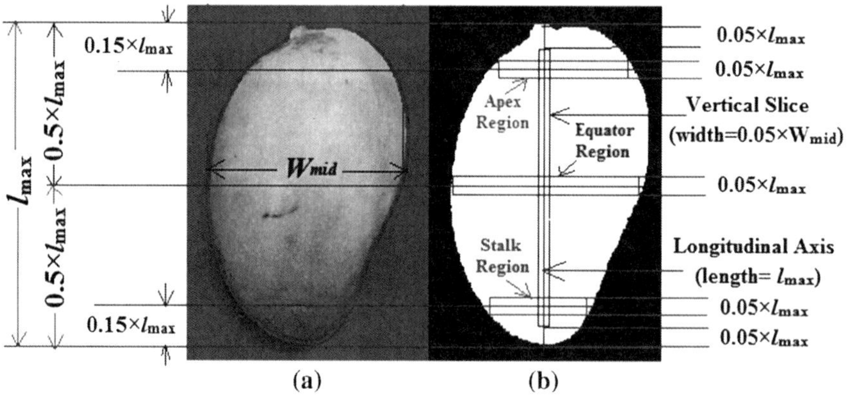

Fig. 4 Filtered images of Mango, **a** raw mango image along with the obtained contour, **b** binary image of the same mango after removing the small patches, also shows the different positions and their name as used in the work

3.4 Alignment of the Mango Image

After getting the contour of the mango, search to find the longitudinal axis of the mango is obtained. Let $C\ (x_i, y_i),\ i = 1, 2, \ldots.., N(N$ is the total number of points in the contour) represent the obtained contour of a mango, then two end points $\left(x_1^l, y_1^l\right)$ and $\left(x_2^l, y_2^l\right)$ are the points for which $l = \sqrt{\left(x_i^l - x_j^l\right) + \left(y_i^l - y_j^l\right)}$ is maximum, where $i, j = 1, 2, \ldots..N$. After getting the coordinates of the two boundary points along the longitudinal axis the image was rotated by an angle determined by $\theta = \tan^{-1}\left[\left(y_2^l - y_1^l\right)/\left(x_2^l - x_1^l\right)\right]$, this rotation will align the mango vertically but may not able to place the "apex" region at the top, it may be at the bottom position. To fix the problem another rotation of 180° was made if the "apex" region is in bottom position. The detection of the "apex" or the "stalk" was made on the basis of geometrical properties of the varieties of the mango under test. For all the varieties of the mango the width of the apex is always higher than the stalk, this properties was utilized to place the apex region at the top of the image. The center point of the apex/stalk region is the point lies on the longitudinal axis at a distance of $0.15 \times l_{\max}$ from the two end points of the longitudinal axis. This relation was determined experimentally, and found true for all the varieties of the mango under test.

4 Extraction of Features

In order to predict the maturity level with the help of computer, some suitable measures collected from the images of the mangoes need to be investigated, which are most correlated with the maturity level. This section discuss about the various features, selection of the features are mainly based on the experienced gain by the authors, while discussing this issue with the experts involves in manual sorting and grading process. Total 27(F1, F2, F3, ...F27) numbers of features are extracted, out of these 15(F1, F2, F3 ....F15) numbers are main features and rest 12(F16, F17 ....F27) numbers are derived features. Different main features and the derived features are discussed in the following sections.

4.1 Main Features

4.1.1 Average R, G and B Value of the Entire Mango

This represent the average R, G and B value of the entire mango and was calculated from the following equation:

$$A_{k=R,G,B} = \frac{1}{rc}\sum_{i=1}^{r}\sum_{j=1}^{c}(I_k \times BW) \quad (1)$$

where, BW is the binary image acting as a mask set the region outside the contour of the mango to 0, and I_k is the captured RGB image, r and c represent the total number of rows and columns of the image.

4.1.2 Gradient of R, G and B Value Along the Longitudinal Axis

Due to the fact that the mango start ripe from the apex region, so the slope of the R, G and B varies from apex region to stalk region (shown in Fig. 5), and this variation found to be different under different maturity level of the mango. The slope of the R, G and B were determined by the following equation. First by taking a slice image along the longitudinal axis, the width of the sliced image is the 5 % of the width of the mango at the middle position of the longitudinal axis and length in between 5 % below and above of the two end points of the longitudinal axis.

$$S_{k=R,G,B} = slope\left(\sum_{i=-1}^{p} s_k(i,j)\right) \quad (2)$$

where p, is the width of the slice. The slope was determined by searching the best fit straight line by least mean square sense.

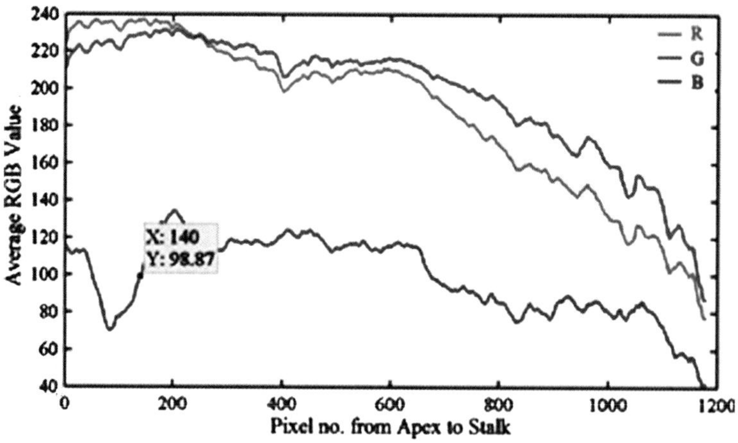

Fig. 5 Variation of R, G and B value along the longitudinal axis

4.1.3 Average R, G and B value of the Apex, Equator and Stalk region

For collecting the average R, G and B value for these three regions, slice images along the horizontal axis were extracted from the RGB image of the mango, the width of the each slice (along the longitudinal axis) is $0.05 \times l_{\max}$ and length is in between the end points of boundary along the horizontal axis cutting the center point of each region as shown in Fig. 4b. The average values were calculated according to:

$$As_k = \frac{1}{rc}\sum_{i=1}^{r}\sum_{j=1}^{c} Is_k \tag{3}$$

where Is_k, is the sliced RGB image, $k = R,\ G\ and\ B$ and $s = Apex,\ Equator\ and\ Stalk$.

4.2 Derived Features

From the above discuss main features some other derived features were calculated, these are as follows:

4.2.1 Difference of Average R, G and B Value of the Entire Mango

Differences of average R, G and B value of the entire mango i.e.

$$(A_R - A_G),\ (A_G - A_B)\ and\ (A_R - A_B) \tag{4}$$

4.2.2 Differences of Average R, G and B Value for the Corresponding Apex, Equator and Stalk Region

Differences of corresponding average R, G and B value for the apex, equator and stalk region, i.e.

$$\begin{aligned}&(Aapex_R - Aequator_R),\ (Aequator_R - Astalk_R)\ and\\&(Aapex_R - Astalk_R)\end{aligned} \tag{5}$$

similarly for G and B also.

In the present work parameters of the individual classes are estimated from the above features using Gaussian Mixture Model. There are several techniques available for estimating the parameters of a GMM [21]. A brief theory of GMM is presented in next section, the details and the methods adopted in the present work for estimation of individual classes can be found in [22].

5 Gaussian Mixture Model

A Gaussian mixture density is a weighted sum of mixture component densities. The Gaussian mixture density can be described as:

$$p(\vec{x}|\theta) = \sum_{k=1}^{M} \omega_k p_k(\vec{x}) \tag{6}$$

where, M is the number of mixture components and ω_k, $k = 1, \ldots\ldots, M$, are the mixture weights, subject to $\omega_k > 0$ and $\sum_{k=1}^{M} \omega_k = 1$, $p_k(\vec{x})$ are the component densities and $\vec{x}$ be a d-dimensional feature vector, $\vec{x} \in \Re^d$. With $d \times 1$ mean vector μ_k and $d \times d$ covariance matrix S_k, each component density is a d-variate Gaussian density function given by:

$$\begin{aligned} p_k(\vec{x}) &\approx N(\mu_k, S_k) \\ &= \frac{1}{(2\pi)^{\frac{d}{2}} |S_k|^{\frac{1}{2}}} \exp\left\{-\frac{1}{2}(\vec{x} - \mu_k)^T S_k^{-1} (\vec{x} - \mu_k)\right\} \end{aligned} \tag{7}$$

For the classification of different category of mango, one needs to find the proper values of $\omega_{k,}\mu_k$ and s_k, so that the GMM provide the best representation for distribution of the training feature vectors.

6 Maturity Prediction Using Gaussian Mixture Model

The variation of the three measures i.e. average R of entire mango, average R of apex region and difference between average R to G value for the three mangoes (two from KU variety and one from HI variety) with respect to different maturity level(day) are shown in Fig. 6a, b and c. From the Fig. 6a, b and c, it can be observed that these features are correlated with the maturity level, on the other hand the nature of variation of these measures for different variety of mangoes are different. Prediction of maturity level of mango using Support Vector Machine based Regression analysis can be found in [23]. The variation of average R with respect to average G for the four different maturity levels (i.e. M1, M2, M3 and M4 of KU) is shown in Fig. 6d. From this figure it can be observed that these two measures are not sufficient to classify the mangoes into four different classes accurately, due to overlapping of the features, which is due to variation of color texture of different samples.

The Probability Density Function (PDF) estimated using only two features i.e. average values of total G and B values is shown in Fig. 7a. When the same GMM was used to find the PDF of raw mangoes came from different gardens over three batches, it was observed that there are strong correlations of the mangoes in a batch originated from a specific garden, but PDF distribution of different batches and different gardens

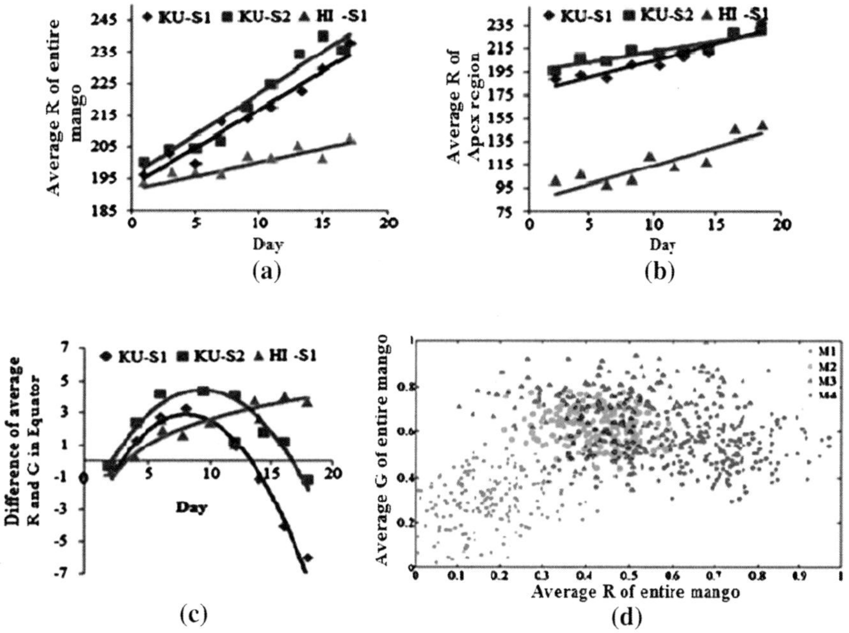

Fig. 6 **a** Variation of average R of entire mango with maturity level(day), **b** variation of average R in apex region with maturity level(day), **c** variation of difference of average R and G in equator with maturity level(day), **d** variation of average R to average G with four different maturity level(i.e. M1, M2, M3 and M4 of KU)

are often found to be different. This fact can be observed from the Fig. 7b, where it is seen that the several GMM components has been formed for the mangoes originated from different gardens and in different batches.

The summary statistics for some of the most correlated features is represented by box–whiskers plots, shown in Fig. 7c. The box corresponds to the inter-quartile range where top and bottom bound indicate 25 and 75th percentiles of the samples respectively, the line inside the box represents median, and the whiskers extend to the minimum and maximum values. The details of the box–whiskers plot can be found in [24]. After estimation, the evaluation of the classification performance was performed on a test data set, in which the objective was to find the class model which has the maximum a posteriori probability for a given observation sequence. The classification accuracy obtained using GMM. Misclassification may occur when different maturity level mangoes having similar color pattern, but it was observed that extraction of multiple features particularly gradient based features helped to correctly identify those mangoes, as some raw mangoes having color pattern of matured mango particularly in the apex region, but those mangoes shows high gradient value of 'R' along the longitudinal axis.

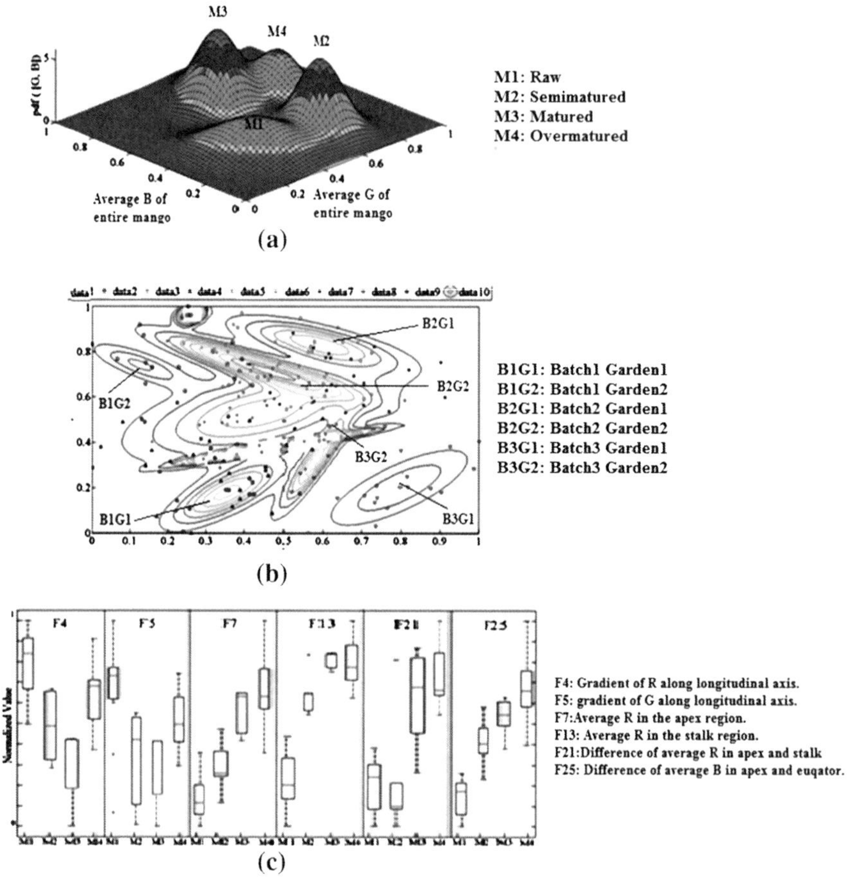

Fig. 7 **a** PDF distribution with average value of B and G of entire mango, **b** PDF distribution with the data set of mango collected from different batches and different gardens, and **c** box-whiskers plots for most corelated features

7 Size Calculation

The fruit size is another quality attribute used by farmers, the bigger size fruit is considered of better quality. The size is estimated by calculating the area covered by the fruit image. First the fruit image is binaries shown in Fig. 4b to separate the fruit image from its background. The number of pixels that cover the fruit image is counted and estimated the size. The fruits are categorized as small, medium, big and very big depending on the number of pixel of the binary image of the mango. According to the variety the range is converted in normalized domain and the corresponding membership function is shown in Fig. 11.

Table 1 Fuzzy if-then rules for classification

M/ /S	M1	M2	M3	M4
SS	Q1	Q2	Q2	Q2
MS	Q1	Q2	Q3	Q3
BS	Q2	Q3	Q4	Q3
VB	Q2	Q4	Q4	Q3

M1: Raw
M2: Semi-mature
M3: Mature
M4: Over mature
SS: Small size
MS: Medium size
BS: Big size
VB: Very big size
Q1: Poor quality
Q2: Medium quality
Q3: Good quality
Q4: Verygood quality

8 Sorting and Grading Methodology Using Fuzzy Logic Technique

The sorting system depends on the prediction of maturity level and calculation of size. The flowchart of sorting and grading process is shown in Fig. 8.

Figure 9 shows the complete process of developing a fuzzy inference system (FIS) for sorting and grading process using MATLAB [25]. The process consist of three main steps: defining the input and output in Membership Function Editor, set the fuzzy rule in Rule Editor, and obtaining the output for each rule in Rule and Surface Viewer.

Fuzzy rule based algorithm is used to sort the mango fruit into four quality grades like poor (Q1), medium (Q2), good (Q3) and very good (Q4) based on their maturity level and size. The grading system has two inputs (maturity level and size) and one output (quality). Total 16 fuzzy if-then rules are created in order to sort the mango fruits into four different quality levels (Q1, Q2, Q3, and Q4). The fuzzy rule base is shown in Table 1. Examples of some rules are illustrated as in Table 2. The fuzzy inference system is shown in Fig. 10.

The rule viewer shown in Fig. 13 consists of the system's inputs and output column. The first and second column is the inputs which are maturity level and size while third column is the quality level output (defuzzification output). Based on the defuzzification results from the rule viewer in Fig. 13 where the maturity (= 0.66) is in matured set with membership grade 1 and size (= 0.5) is in big and medium set both with 0.5 membership grade. The defuzzification value of quality is calculated by using centroid method. Then the classification of fruit is determined based on crisp logic given in Table 3.

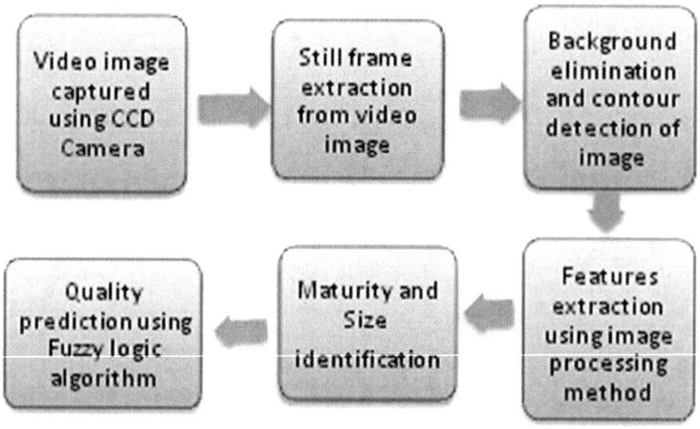

Fig. 8 Flowchart of sorting and grading process

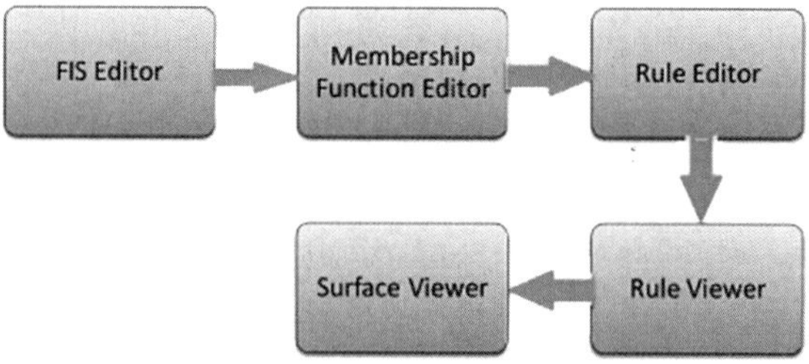

Fig. 9 Flowchart of Fuzzy Inference System

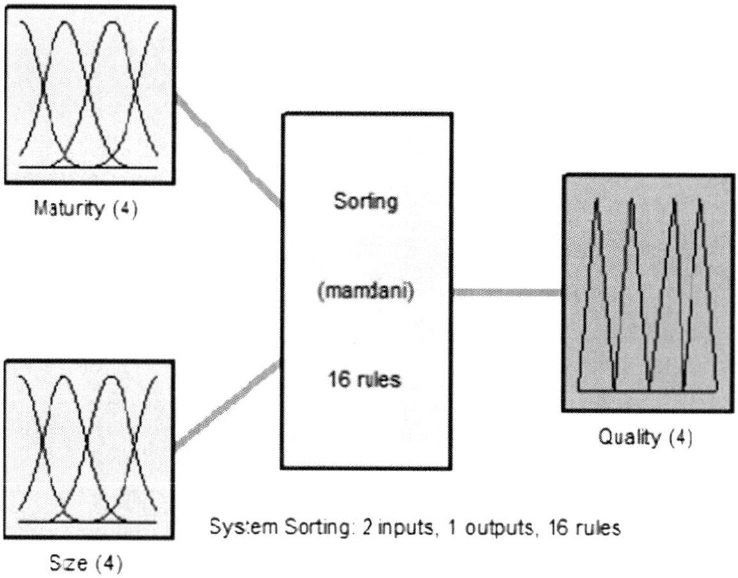

Fig. 10 Fuzzy inference systems

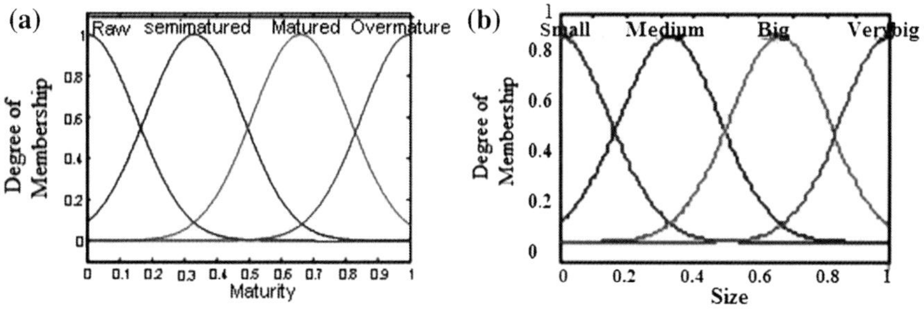

Fig. 11 Input membership functions representing maturity level (a) and size (b)

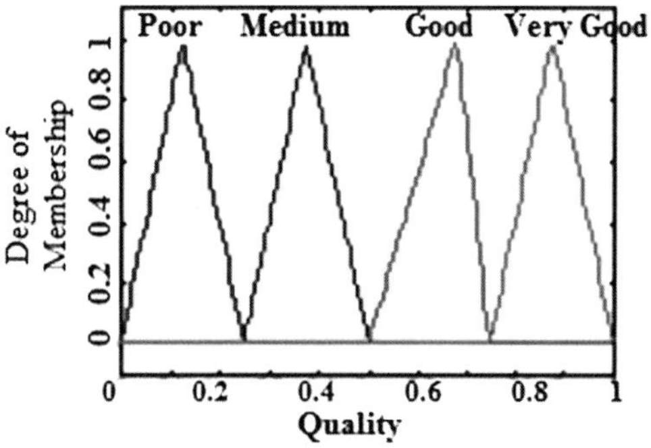

Fig. 12 Membership function representing quality output

Table 2 Fuzzy if-then rules

Rule1	Rule 6
If maturity level is raw and size is small then quality is poor	If maturity level is semimature and size is medium then quality is medium
Rule 11	Rule 16
If maturity level is mature and size is big then quality is very good	If maturity level is overmatured and size is very big then quality is good

The quality value is 0.626. Therefore, the mango used in experiment is graded as good quality.

9 Results and Discussion

The classification accuracy obtained using fuzzy logic algorithm based system and the average classification accuracy by the three experts for the five varieties of mango is presented in Table 4.

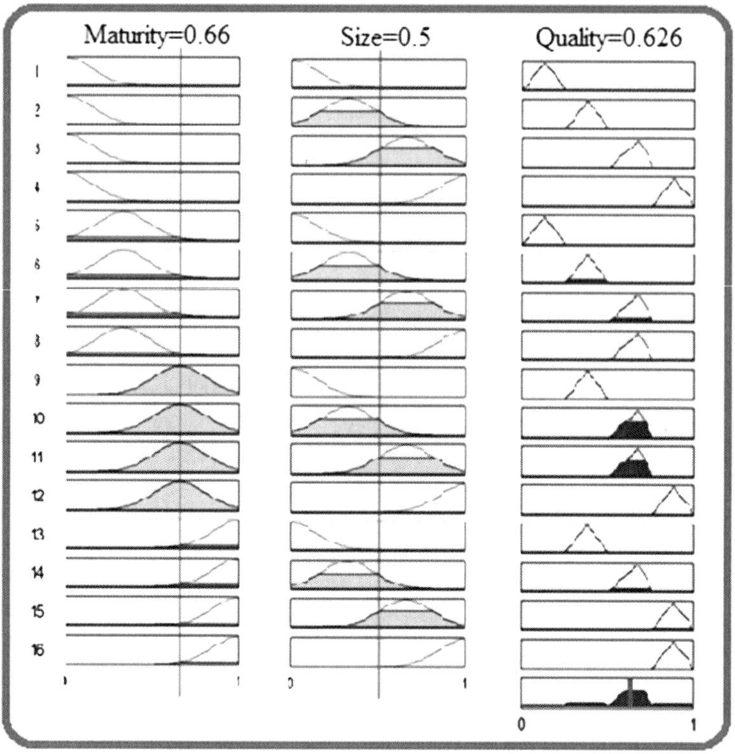

Fig. 13 Defuzzification results from rule viewer

Table 3 Fuzzification algorithm

Defuzzification output	Quality grade
(Quality value<0.25)	Poor (Q1)
(Quality value>=0.25) && (Quality value<0.5)	Medium (Q2)
(Quality value >=0.5) && (Quality value <0.75)	Good (Q3)
(Quality value >=0.75)	Very good (Q4)

10 Conclusions

In this chapter an application of machine vision based technique for automatic sorting and grading of mango according to the maturity level and size using fuzzy logic is discussed. Different image processing techniques were evaluated, to extract different features from the images of mango. Gaussian mixture model is used to estimate the parameters of individual classes to predict the maturity level. This technique found to be low cost effective and moreover intelligent. The speed of sorting system is limited by the conveyer belt speed and the gap maintained in between two mangoes

Table 4 Performance analysis

Variety (Local name)	Q1		Q2		Q3		Q4	
	Experts	System	Experts	System	Experts	System	Experts	System
KU	91.2	90.4	89.4	88.7	90.3	89.4	90.2	89.4
AM	90.7	90.1	90.0	89.1	88.9	88.2	90.1	89.4
SO	90.2	89.7	90.7	89.1	90.0	88.3	89.8	89.5
LA	91.1	90.5	90.9	88.6	89.2	88.8	91.0	89.7
HI	90.1	89.4	90.6	90.0	90.4	89.1	91.3	90.0

rather than response time of the computerized vision based system, which is on the order of $\sim$50 ms.

Test has been conducted for the five varieties of mango fruit only, but can be extended for other fruits where there are reasonable changes in skin color texture occur with maturity. The variations of classification performances with the variation of other factors like changes in ambient light, camera resolution, and distance of the camera were not studied. The study shows that, machine vision based system performance, is closer to the manual experts, where experts judge the mangoes maturity level not only by the skin color but also with firmness and smell.

References

1. B. Jarimopas, N. Jaisin, An experimental machine vision system for sorting sweet tamarind. J. Food Eng. **89**(3), 291–297 (2008)
2. Y. Zhao, D. Wang, D. Qian, Machine vision based image analysis for the estimation of Pear external quality, in *Second International Conference on Intelligent Computation Technology and Automation*, 2009, pp 629–632
3. D.J. Lee, J.K. Archibald, G. Xiong, Rapid color grading for fruit quality evaluation using direct color mapping, IEEE Trans. Autom. Sci. Eng. **8**(2), 292–302 (2011)
4. Z. Varghese, C.T. Morrow, P.H. Heinemann, H.J. Sommer, Y. Tao, R.W. Crassweller, Automated inspection of golden delicious apples using color computer vision, Am. Soc. Agric. Eng. **16**, 1682–1689 (1991)
5. Y. Gejima, H. Zhang, M. Nagata, Judgement on level of maturity for tomato quality using L*a*b color image processing, in *Proceedings 2003 IEEE/ASME International Conference on Advanced Intelligent Mechatronics (AIM 2003)*, 2003
6. B.K. Miller, M.J. Delwiche, A color vision system for peach grading. Trans. ASAE **32**(4), 1484–1490 (1989)
7. A. Janobi, Color line scan system for grading date fruits, in *ASAE Annual International Meeting*, Orlando, Florida, USA, 12–16 July, 1998
8. Y. Tao, P.H. Heinemann, Z. Varghese, C.T. Morrow, H.J. Sommer, Machine vision for color inspection of potatoes and apples. Trans. ASAE **38**(5), 1555–1561 (1995)
9. S.A. Shearer, F.A. Payne, Color and defect sorting of bell peppers using machine vision. Trans. ASAE **33**(6), 2045–2050 (1990)
10. A. Bouganis, M. Shanahan, A vision-based intelligent system for packing 2-D irregular shapes. IEEE Trans. Autom. Sci. Eng. **4**(3), 382–394 (2007)
11. H.C. Garcia, J.R. Villalobos, Automated refinement of automated visual inspection algorithms. IEEE Trans. Autom Sci. Eng. **6**(3), 514–524 (2009)

12. H.C. Garcia, J.R. Villalobos, G.C. Runger, An automated feature selection method for visual inspection systems. IEEE Trans. Autom. Sci. Eng. **3**(4), 394–406 (2006)
13. S.M. Bhandarkar, X. Luo, R.F. Daniels, E.W. Tollner, Automated planning and optimization of lumber production using machine vision and computed tomography. IEEE Trans. Autom. Sci. Eng. **5**(4), 677–695 (2008)
14. N.M. Kwok, Q.P. Ha, D. Liu, G. Fang, Contrast enhancement and intensity preservation for gray-level images using multiobjective particle swarm optimization. IEEE Trans. Autom. Sci. Eng. **6**(1), 145–155 (2009)
15. H.Y.T. Ngan, G.K.H. Pang, Regularity analysis for patterned texture inspection. IEEE Trans. Autom. Sci. Eng. **6**(1), 131–144 (2009)
16. Y. Cheng, M.A. Jafari, Vision-based online process control in manufacturing applications. IEEE Trans. Autom. Sci. Eng. **5**(1), 140–153 (2008)
17. W. Qi, F. Li, L. Zhenzhong, Review on camera calibration IEEE, in *Chinese Control and Decesion Conference*, 2010, pp 3354–3358
18. X. Jiang, C. Cheng, S. Wachenfeld, K. Rothaus, in *Motion Deblurring Seminar: Image Processing and Pattern Recognition*, Winter Semester 2004 / 2005
19. A. Rosenfeld, A.C. Kak, *Digital Image Processing*, (Academic Press, New York, 1982), cap 11
20. S.D. Kim, J.H. Lee, J.K. Kim, A new chain-coding algorithm for binary images using run-length codes. CVGIP **41**, 114–128 (1988)
21. S. Biswas, C. Koley, B. Chatterjee, S. Chakravorty, A methodology for identification and localization of partial discharge sources using optical sensors, IEEE Trans. Dielectr. Electr. Insulation **19**(1), 18–28 (2012)
22. E. S. Gopi, *Algorithm Collections for Digital Signal Processing Applications using Matla* (Springer, New York, 2007)
23. C. S. Nandi, B. Tudu, C. Koley, Support vector Machine based maturity prediction, in *WASET, International Conference ICCESSE-2012*, 2012, pp. 1811–1815
24. K. Fukunaga, *Introduction to Statistical Pattern Recognition* (Academic Press, Boston, 1990)
25. The Mathworks, Fuzzy Logic Toolbox User's Guide, (Release 2009b)

Chapter 3
Region Adaptive, Unsharp Masking Based Lanczos-3 Interpolation for 2-D Up-Sampling: Crisp-Rule Versus Fuzzy-Rule Based Approach

A. Acharya and S. Meher

Abstract Up-sampling problem is very crucial in image communication as it plays an important role in restoring the low resolution 2-D signals at the receiver. Generally, at the transmitting end, a video intra frame is sub-sampled to lessen the bandwidth required for transmission. At the receiver, the resolution of the sub sampled intra frame is improved to the original by a suitable interpolation technique. This process lessens the signal bandwidth for transmission through a communication link and hence avoids channel congestion. Most of the up-sampling scheme based on interpolation generates undesirable blurring artefacts in the up-sampled video intra frame. This results in signal deterioration in terms of loss of fine details and critical edge information. Such problems occur due to the resemblance of the up-sampling process with the low pass filtering operation. In order to resolve this problem, both crisp-rule and fuzzy-rule based hybrid interpolation techniques are proposed here. A crisp-rule based technique (Proposed-1) although gives better results in certain cases but fails to provide considerable performance under varying constraints such as variation in zoom in or zoom out conditions, change in compression ratio and video characteristics. This is basically due to improper mapping by crisp-rule based technique between input and output values. Such problems can be avoided by fuzzy-rule based technique (Proposed-2) which employs a proper and precise mapping between input and output values using fuzzy inference system. This book chapter critically compares the capabilities and limitations of crisp-rule and fuzzy-rule based up-sampling techniques and their relevance in perspective of video scalability, compatibility, complexities, quality enhancement and real time applications.

Keywords Image and video processing · Unsharp masking · Lanczos-3 interpolation · Variance · Up-sampling · Fuzzy logic

---

A. Acharya (✉) · S. Meher
National Institute of Technology, Rourkela, Odisha, India
e-mail: aditya.acharya2011@gmail.com

A. Mason et al. (eds.), *Sensing Technology: Current Status and Future Trends II*,
Smart Sensors, Measurement and Instrumentation 8, DOI: 10.1007/978-3-319-02315-1_3,

1 Introduction

Video frame resizing has gained much importance in the contemporary video communication because of its potential features like scalability and compatibility with various receiving devices with different display resolutions. This scalable feature is because of the interpolation technique which makes the video compatible over a wide range of display devices starting from mobile phones to HDTV. Frame resizing also plays a key role in reducing the transmission bandwidth requirement which consequently avoids channel congestion. For instance, at the transmitter, the original video is sub-sampled for efficient use of transmission channel bandwidth whereas at the receiver, the resolution of the sub-sampled video is improved to its original size by making use of a suitable interpolation scheme. Up-sampled high resolution video not only gives a better visual quality to a viewer but also provides additional information for various post processing applications such as inspection or recognition. In medical imaging, remote sensing and video surveillance applications, very often it is desired to improve the native resolution offered by imaging hardware for subsequent analysis and interpretation. Video interpolation aims to generate high resolution video from the associated low resolution capture and hence is very essential for the aforesaid applications.

Interpolation plays a significant role in video quality improvement which may deteriorate due to optical distortion, aliasing, motion blur or noise contamination during capturing of a video sequence. In the process of interpolation the intermediate values of a continuous event are estimated from the discrete samples in order to resize the image and to correct spatial distortion meant for a better video quality. Interpolation is also used in discrete image manipulation, such as geometric alignment and registration to improve the image quality of display devices. Furthermore, it plays a key role in the field of lossy image compression wherein some pixels or some frames are discarded during the encoding process and must be regenerated from the remaining information for decoding.

In medical image processing, image interpolation methods have taken a vital role for image generation and post-processing. In computed tomography (CT) or magnetic resonance imaging (MRI), image reconstruction requires interpolation to approximate the discrete function to be back projected for inverse Rondon transform. In modern X-ray imaging system such as digital subtraction angiography (DSA), interpolation is used to enable the computer-assisted alignment of the current radiograph and the mask image. Moreover, zooming or rotating medical images after their acquisition often is used in diagnosis and treatment. In order to achieve this, interpolation methods are incorporated into systems for computer aided diagnosis (CAD), computer assisted surgery (CAS), and picture archiving and communication systems (PACS) [1, 2].

Scalability is one of the key features of video interpolation which is exploited in internet technology and consumer electronics applications. For instance, while remote browsing a video database, it would be more convenient and economical to send a low resolution version of a video clip to the user. If the user shows interest the resolution can be progressively enhanced using interpolation. Similarly HDTV exploits the scalable feature of video interpolation for its compatibility with most of the existing video compression standards such as H.263 and H.264. In addition, the video is made adaptive to variable bit rate and computational capacities of different receiving devices by utilizing the scalable feature of interpolation. Thus the analysis and exploitation of video interpolation are quite essential to improve the performance of contemporary video communication in terms of quality, scalability and compatibility.

There are several interpolation techniques are used for the up-sampling process. One of the simplest interpolation technique is a nearest neighbour interpolation. In this case, the value of a new point in the interpolated image is taken as the value of old coordinate which is located nearest to the new point. Although it is a simple technique, it suffers through blocking artefacts. Another simple interpolation technique is bilinear interpolation where the value of a new point is computed using linear interpolation of four pixels surrounding the new point [3]. Bilinear interpolation though is simple and less complex, it has undesirable blurring artefacts. There are widely used interpolation techniques [4, 5] such as Bicubic and B-spline which consider sixteen pixels for determining a new interpolated point. These techniques provide better performance in terms of quality at the cost of computational complexities. Bicubic and B-spline [6, 7] interpolation techniques provide a less degree of blurring in comparison to bilinear interpolation. Lanczos is another spatial domain interpolation technique which is implemented by multiplying a sinc function with a sinc window which is scaled to be wider and truncated to zero outside of a range [8, 9]. Even if Lanczos interpolation [10, 11] gives good results, it is slower than other approaches and provides a blurring effect in the reconstructed image.

Many approaches for image resizing have been developed in transform domain [12, 13]. Up-sampling in DCT domain is implemented by padding zero coefficient to the high frequency side. Image resizing in DCT domain shows very good result in terms of scalability and image quality. However, this technique suffers through undesirable blurring and ringing artefacts [14, 15]. Several hybrid interpolation techniques have been developed in order to reduce the blurring artefacts. However they do have certain limitations. The adaptive unsharp masking based sharpening is a spatial domain pre-processing approach which extracts high frequency details in the spatial domain, sharpens the sub-sampled video to a certain degree through an adaptive selection of weight factor by referring the original frame after a regular interval so as to compensate the blurring caused by the subsequent discrete DCT based interpolation technique. The main drawback of this technique is the requirement of the corresponding original frame after a regular frame interval and the undesirable ringing artefacts [16].

There are so many no reference hybrid interpolation techniques which are giving better results. No reference, region adaptive unsharp masking based Lanczos-3 interpolation technique [17] is based on both local and global processing as per the region statistics of a video intra frame. This method makes use of a combination of anticipatory, spatial domain, region adaptive, unsharp masking operation coupled with Lanczos-3 interpolation for retaining some of the fine details and critical edge information in the reconstructed video frame. The region adaptive unsharp masking is a preprocessing approach which sharpens the intra frame regions locally as per their statistical local variance so as to compensate the blurring caused by the subsequent Lanczos-3 interpolation technique. The degree of sharpening is increased as per the rise in the statistical local variance of a neighbourhood and vice versa. Furthermore, the unsharp masking operation is made globally adaptive by multiplying the unsharp mask with a global scaling factor which is obtained by adding one to the global variance of an intra frame. Although this technique provides good results with reduced blurring artefacts, it lacks in several aspects. Since these techniques are developed using a crisp rule, they are unable to adapt with varying constraints and thereby unable to provide better performance for different types of images and video sequences. In case of video sequences subjected to zoom in or zoom out conditions, these techniques fail drastically. Their performance also deteriorates if subjected to variation in compression ratio and video characteristics. It is because, there are only few output values for a large variation in the input values so the problems lies in the in-proper mapping between input and output values using crisp rule base.

Such problems can be resolved by using the proposed fuzzy based mapping technique in which there will be as a de-fuzzified crisp output value corresponding to each crisp input value which may vary over a large range. According to this fuzzy rule based technique, the regions with high variance are sharpened more than the regions with low variance based on the Fuzzy rule base. This consequently results in the restoration of fine details and edge information in the reconstructed up-sampled video intra frame with improved objective and subjective quality. Thus there is a precise and accurate mapping between input and output by using fuzzy inference system. Furthermore, this consequently improves the adaptability of the proposed fuzzy based technique with the varying conditions. Therefore, the proposed technique aims to produce very least degree of blurring and at the same time flexible enough to provide considerable performance under varying constraints such as compression ratio, video characteristics and zooming conditions.

2 Crisp-Rule Based approach: Region Adaptive, Unsharp Masking Based Lanczos-3 Interpolation (Proposed Method-1)

Generally in a transmitter, a sub-sampled video is produced by alternate deletion of rows and columns for effective use of transmission channel bandwidth where as at the receiver, the resolution of the sub-sampled video is enhanced using the

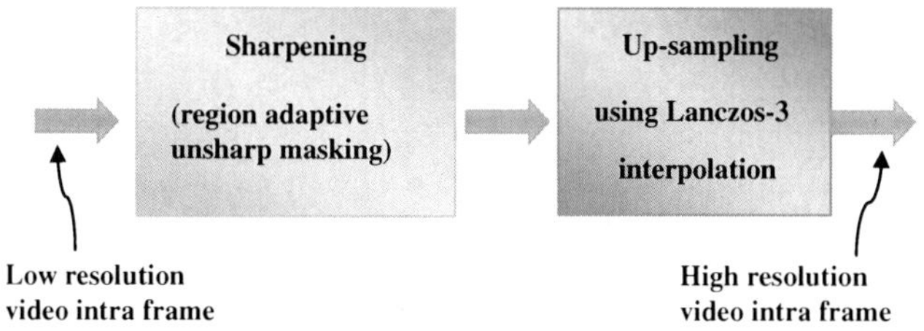

Fig. 1 Region adaptive unsharp masking based Lanczos-3 interpolation using crisp-rule based algorithm

suitable interpolation technique. The proposed method as shown in Fig. 1 is an anticipatory spatial domain preprocessing step coupled with the Lanczos-3 interpolation scheme in order to retain some of the fine details and critical edge information which may be lost while converting a low resolution video to it's high resolution counterpart.

Lanczos-3 based up-sampling scheme is analogous to a low pass filtering operation. This consequently results in the loss of high frequency details which leads to blurring in the up-sampled video intra frame. To overcome this problem, we make use of a sharpening technique in spatial domain to compensate the high frequency loss. Therefore, a hybrid interpolation technique is proposed here which exploits the advantage of the spatial domain, region adaptive unsharp masking operation prior to Lanczos-3 interpolation for preserving the fine details and critical edge information. The region adaptive unsharp masking operation is a preprocessing step that sharpens the sub-sampled video intra frame to a certain degree depending on its statistical local variety. The local regions with high local variance are proportionately sharpened more than the regions with less local variance by the proposed adaptive algorithm. During this operation, an unsharp mask obtained by subtracting the unsharp or smooth version of the video frame from the original. The smooth version of the video frame is obtained by an adaptive low pass filtering operation using a region adaptive Gaussian mask whose centre pixel weight is made adaptive as per the statistical local variance of a neighbourhood.

Furthermore, the unsharp mask is enhanced by a global scaling factor which is obtained by adding one to the global variance of the intra frame for better objective and subjective video quality. The unsharp mask thus obtained is added to the original video frame so as to obtain the sharpened video frame. The degree of sharpening obtained using the aforesaid operation compensates the extent of blurring caused by the subsequent Lanczos-3 interpolation technique and hence enhances the objective and subjective quality. The proposed method basically consists of two steps. They are namely region adaptive unsharp masking and Lanczos-3 interpolation which are described in subsequent subsections in detail.

2.1 Region Adaptive Unsharp Masking

The region adaptive unsharp masking process is used to sharpen a video frame by subtracting an unsharp or smoothed version of it from the original. The smooth version of a video frame is obtained by blurring it using a region adaptive Gaussian mask whose centre pixel weight is varied depending on the statistical local variance of a neighbourhood. If the local variance of the neighbourhood is more, the weight of the centre pixel is reduced proportionately as per the region adaptive blurring algorithm in order to provide more blurring to the high variance regions. This consequently results in more degree of sharpening in the high variance regions. Similarly, the reverse operation is performed in the case of low variance regions. Subsequently, the region adaptive unsharp mask is obtained by subtracting the blurred video frame from the original.

In order to improve the sharpening performance, the region adaptive unsharp mask is made globally adaptive by multiplying it with a global scaling factor. The global scaling factor is obtained by adding one to the global variance of the corresponding video intra frame. This modified mask is then added to the original video frame to form a sharpened video intra frame. Here, both local and global pre-processing techniques are employed to substantially improve the objective and subjective quality of the up-sampled video intra frame. The local pre-processing tends to improve the objective quality where as the global pre-processing technique tends to improve the subjective quality and the overall human visual system (HVS) performance. In brief, the region adaptive unsharp masking consists of the following steps [18].

- Blur the original video frame using region adaptive Gaussian low pass filtering operation.
- Subtract the blurred video frame from the original to generate the region adaptive unsharp mask.
- Modify the mask by multiplying with a global scaling factor which is obtained by adding one to the global variance of an intra frame.
- Add the modified mask to the original and repeat this operation for all the frames.

Let $g_1(x, y, n)$ and $f(x, y, n)$ denote the blurred video sequence using region adaptive Gaussian mask and the original video sequence respectively. Now the region adaptive unsharp mask is given by:

$$g_{mask}(x, y, n) = f(x, y, n) - g_1(x, y, n) \tag{1}$$

Then the mask is modified by a global scaling factor and is added back to the original video frame for sharpening and is given by:

$$g_2(x, y, n) = f(x, y, n) + (V(n) + 1)g_{mask}(x, y, n) \tag{2}$$

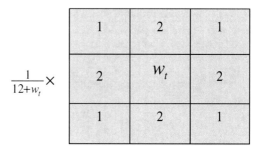

Fig. 2 Region adaptive Gaussian mask

$$\frac{1}{12+w_t} \times \begin{bmatrix} 1 & 2 & 1 \\ 2 & w_t & 2 \\ 1 & 2 & 1 \end{bmatrix}$$

where $g_2(x, y, n)$ and $V(n)$ denote the sharpened video sequence and global variance of an intra frame respectively. n is the frame number that represents discrete time. $g_1(x, y, n)$ is the blurred video sequence which is obtained by using the region adaptive algorithm given the Sect. 2.2.

The global mean $M(n)$ and global variance $V(n)$ of a video intra frame are represented in (3) and (4) respectively.

$$M(n) = \frac{1}{PQ} \sum_{x=1}^{P} \sum_{y=1}^{Q} f(x, y, n) \tag{3}$$

$$V(n) = \frac{1}{PQ} \sum_{x=1}^{P} \sum_{y=1}^{Q} [f(x, y, n) - M(n)]^2 \tag{4}$$

where P and Q represents the number of rows and columns of a video intra frame. The region adaptive Gaussian mask for sharpening operation is shown in Fig. 2 where w_t represents the central weight of the Gaussian mask which is made adaptive as per the statistical local variance of a 3×3 neighbourhood.

2.2 Region Adaptive Algorithm

2.3 Up-Sampling Using Lanczos-3 Interpolation

Lanczos is a spatial domain interpolation technique which is implemented by multiplying a sinc function with a sinc window which is scaled to be wider and truncated to zero outside of the main lobe. In case of Lanczos-3 interpolation, the main lobe of the sinc function along with the two subsequent side lobes on either side is used as a sinc window. The Lanczos window is a product of sinc functions $\sin c(x)$ with the scaled

Region adaptive algorithm to obtain blurred video sequence $g_1(x, y, n)$

$for \quad n = 1$ to frame number do
Find the maximum local variance $\upsilon_{\max}$ and minimum local variance $\upsilon_{\min}$ for each frame.
Find the difference d between the maximum and minimum local variance.
$d \leftarrow \upsilon_{\max} - \upsilon_{\min}$
Find the step size s by dividing the difference by 6
$s \leftarrow \frac{\upsilon_{\max} - \upsilon_{\min}}{6}$

$for \quad x = 1 \quad to \quad P$
$for \quad y = 1 \quad to \quad Q$
Find local mean m and local variance υ for each pixel in a neighbourhood

$$m \leftarrow \frac{1}{9} \sum_{s=-1}^{1} \sum_{t=-1}^{1} w(x+s, y+t)$$

$$\upsilon \leftarrow \frac{1}{9} \sum_{s=-1}^{1} \sum_{t=-1}^{1} [f(x+s, y+t) - m]^2$$

$if \quad \upsilon > \upsilon_{\max} - s$

$w_t = 2$

$elseif \ \upsilon > \upsilon_{\max} - 2s \ and \ \upsilon \leq \upsilon_{\max} - s$

$w_t = 3$

$elseif \ \upsilon > \upsilon_{\max} - 3s \ and \ \upsilon \leq \upsilon_{\max} - 2s$

$w_t = 4$

$elseif \ \upsilon > \upsilon_{\max} - 4s \ and \ \upsilon \leq \upsilon_{\max} - 3s$

$w_t = 8$

$elseif \ \upsilon > \upsilon_{\max} - 5s \ and \ \upsilon \leq \upsilon_{\max} - 4s$

$w_t = 16$

$else$

$w_t = 32$

$endif$

$$h \leftarrow \frac{1}{w_t + 12} \times \begin{bmatrix} 1 & 2 & 1 \\ 2 & w_t & 2 \\ 1 & 2 & 1 \end{bmatrix}$$

$$g_1(x, y) \leftarrow \frac{1}{w_t + 12} \sum_{s=-1}^{1} \sum_{t=-1}^{1} h(s, t) f(x+s, y+t)$$

$endfor$
$endfor$
$endfor$

version of the sinc function $\sin c(x/a)$ restricted to the main period $-a \leq x \leq a$ to form a convolution kernel for re-sampling the input field [8]. In one dimension, the Lanczos interpolation formula is given by

$$L(x) = \begin{cases} \sin c(x)\ \sin c(x/a), & -a \le x \le a \\ 0, & otherwise \end{cases} \quad (5)$$

where 'a' is a positive integer, typically 2 or 3, is used for controlling the size of the kernel. The parameter 'a' corresponds to the number of lobes of the sinc function. The three lobed Lanczos windowed sinc function (Lanczos-3) is given by:

$$Lanczos\ 3(x) = \begin{cases} \frac{\sin(\pi x)}{\pi x} \frac{\sin(\pi x/3)}{\pi x/3}, & -3 \le x \le 3 \\ 0, & otherwise \end{cases} \quad (6)$$

For a two dimensional function such as an image $g_2(x, y)$, an interpolated value at an arbitrary point (x_0, y_0) using Lanczos-3 interpolation is given by:

$$\hat{g}_2(x_0, y_0) = \sum_{i=\lfloor x_0 \rfloor - a + 1}^{\lfloor x_0 \rfloor + a} \sum_{j=\lfloor y_0 \rfloor - a + 1}^{\lfloor y_0 \rfloor + a} g_2(i, j)\, L(x_0 - i)\, L(y_0 - j) \quad (7)$$

where $a = 3$ for Lanczos-3 kernel which denotes the size of the kernel. The Lanczos-3 interpolation in 2-D uses a support region of $6 \times 6 = 36$ pixels from the original image [9]. In case of a 3-D signal such as video, the above 2-D interpolation is operated in discrete time. The final equation for video interpolation using Lanczos-3 interpolation, upon substituting $a = 3$ is given by:

$$\hat{g}_2(x, y, n) = \sum_{i=\lfloor x \rfloor - 2}^{\lfloor x \rfloor + 3} \sum_{j=\lfloor y \rfloor - 2}^{\lfloor y \rfloor + 3} g_2(i, j, n)\, L(x - i)\, L(y - j) \quad (8)$$

where $\hat{g}_2(x, y, n)$ denotes the interpolated up-sampled video. x, y represents spatial co-ordinates and n is the frame number that represents discrete time.

2.4 Experimental Results and Discussions

To demonstrate the performance of the proposed hybrid technique, the input video sequences are down-sampled in the spatial domain by deleting the alternate rows and columns at (4:1) compression ratio. Then we interpolate the frames back to their original size to allow the comparison with the original video frame. Experimental results show the proposed hybrid interpolation technique outperforms DCT and other spatial domain interpolation scheme in terms of objective and subjective measures with reduced ringing artefacts. In Table 1, we have illustrated the average PSNR (dB) comparison of various existing techniques such as DCT, Bicubic and Lanczos-3 with the proposed-1 interpolation technique in 4:1 compression ratios for different CIF and QCIF sequences. The results show that the proposed technique shows up to 0.362 dB average PSNR improvement than DCT at 4:1 compression ratio in the case

Table 1 Average PSNR comparison of different CIF and QCIF sequence sequences at 4:1 compression ratio

Sequences	Average PSNR (dB)			
	Bicubic	Lanczos3	DCT	Proposed-1
Flower_sif	67.4448	67.6323	67.6192	67.7124
Salesman_cif	77.1100	77.4577	77.5707	77.7681
Bus_cif	73.3932	73.8475	73.9253	74.2872
Tennis_cif	73.3649	73.5604	73.5845	73.6115
News_cif	76.4648	77.0882	77.2933	77.3543
City_cif	75.7227	76.0100	75.9829	76.2195
Football_cif	76.6988	77.4973	77.8178	78.0112
Mobile_cif	69.3281	69.7275	69.8889	70.1455
Coastguard_cif	74.6310	75.0699	75.2115	75.4961
News_qcif	73.1883	73.5988	73.7390	74.0117
Salesman_qcif	77.1254	77.3786	77.4667	77.5604

of bus sequence. In Figs. 7 and 8, we have shown the variation of PSNR with respect to the frame index at 4:1 compression ratio for different sequences. In Figs. 10 and 11, the subjective performances of different interpolation techniques are shown for the 20th frame of akiyo and the football sequence respectively in 4:1 compression ratio. Experimental results show, the blurring is much reduced and the edges are more pronounced with fine detail preservation in comparison to other existing techniques irrespective of the video types.

In addition, This region adaptive technique provides considerable objective performance improvement irrespective of the variation in resolution and characteristics of a video sequence than the other mentioned techniques.

2.5 Advantages and Limitations of Crisp-Rule Based Approach

The crisp-rule based, no reference hybrid interpolation technique not only restores a sub-sampled video with high precision but also yields a very low degree of blurring and ringing effect with fine detail preservation. It provides better performance and flexibility under the variation in resolution and the video characteristics because of the coarse variation of the adaptive weight factor by making use of crisp-rule based algorithm. Since the proposed technique is based on spatial domain processing, it is faster, computationally less complex and more efficient than the transform domain techniques such as DCT. In addition, since this technique is a pre-processing approach, it imparts more computational burden on the transmitting side than the receiver and thus eases the computational burden on the receiver. This makes the receiver less complex, faster, cost effective and suitable for various real time applications.

A crisp-rule based technique although gives better results in certain cases but fails to provide considerable performance under varying constraints such as variation in zoom in or zoom out conditions, change in compression ratio and video characteristics. This is basically due to improper mapping by crisp-rule based technique between input and output values. In case of video sequences subjected to zoom in or zoom out conditions, this technique fails drastically. It is because, there are only few output values to map for a large variation in the input values. So, the problem lies in the in-proper mapping between input and output values. The crisp-rule based algorithm is unable to perform fine and precise mapping between input and output because of the coarse variation of adaptive weight factor with respect to continuous variation of the local variance as input parameter. Such problems can be avoided by fuzzy-rule based technique which employs a proper and precise mapping between input and output values using fuzzy inference system.

3 Fuzzy-Rule Based Approach: Fuzzy Weighted Unsharp Masking Based Lanczos-3 Interpolation Technique (Proposed Method-2)

Fuzzy weighted unsharp masking based Lanczos-3 interpolation technique is an anticipatory, spatial domain, fuzzy logic based preprocessing approach which sharpens the sub-sampled or low resolution video intra frames depending on their region statistics in order to compensate the blurring caused by the subsequent Lanczos-3 interpolation technique. According to this method, the regions with high variance are sharpened more than the regions with low variance based on the Fuzzy rule base. The fuzzy based sharpened output is further sharpened by Laplacian for further improvement. This consequently results in the restoration of fine details and edge information in the reconstructed up-sampled video intra frame with improved objective and subjective quality. In this proposed technique, there is a precise and accurate mapping between input and output by using fuzzy inference system. This consequently improves the adaptability of the proposed fuzzy based technique with the varying conditions. Therefore, this technique aims to produce very least degree of blurring and at the same time flexible enough to provide considerable performance under varying constraints such as compression ratio, video characteristics and zooming conditions.

Generally, at the transmitting end, an image or a video intra frame is sub-sampled by alternate deletion of rows and columns in order to save the bandwidth required to transmit the signal. At the receiver, the sub-sampled video intra frame is restored to its original size by using a suitable interpolation technique. In a communication system, the 2-D signal sub-sampling and up-sampling process can be modeled as a low pass filtering operation which results in loss of high frequency information of an image such as fine details and edge information. However, the low frequency information such as smooth and slowly varying region details are retained due to such operations. For instance, a Lanczos-3 based up-sampling scheme has an important property of preserving the low frequency components generated by smooth and fast changing area in a video intra frame because of its similarity with the low pass filtering (LPF) operation. Thus, in an image or video intra frame, the high frequency details are more degraded than the low frequency details. Keeping this thing in view, an efficient, hybrid interpolation technique is proposed here which will perform the inverse operation so as to restore the high frequency details more than low frequency details based on fuzzy rule base. Thus, the proposed method utilizes a preprocessing, fuzzy based, region adaptive unsharp masking technique which sharpens the video intra frame locally as per the statistical local variance so as to alleviate the blurring caused by 2-D sub-sampling and up-sampling operation. The fuzzy based sharpened output is further sharpened by the 2-D Laplacian operator for much improved performance prior to up-sampling operation.

As per the fuzzy knowledge base, the region with higher variance are sharpened more than the region with lower variance so as to perform the inverse operation of high frequency degradation caused by 2-D sub-sampling and up-sampling process. It is due to the precise and accurate mapping between input and output variable of the fuzzy inference system, the proposed system is made adaptive to varying constraints such as zooming conditions, compression ratios and video characteristics. In addition, to restore the high frequency details, the proposed sharpening technique is performed prior to the Lanczos-3 based up-sampling such that a precise amount of sharpening will compensate the degree of blurring caused by the subsequent Lanczos-3 based interpolation technique. Consequently, this results in reduced blurring in the up-sampled video intra frame with better visual quality.

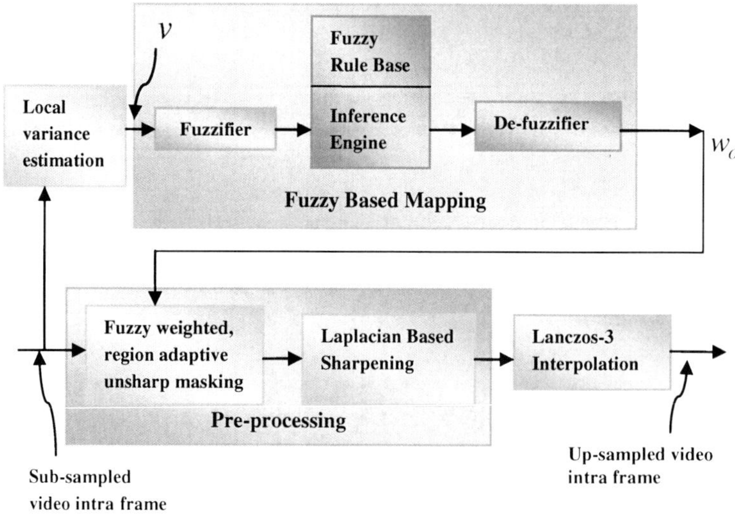

Fig. 3 Block diagram of the proposed fuzzy weighted, region adaptive unsharp masking based Lanczos-3 interpolation technique

The proposed fuzzy-rule based method as shown in Fig. 3 basically comprises of five steps.

- Local variance estimation.
- Fuzzy based mapping.
- Fuzzy weighted region adaptive unshap masking.
- Laplacian based sharpening.
- Up-sampling using Lanczos-3 interpolation.

3.1 Local Variance Estimation

Local variance is the measure of high frequency details in a neighbourhood of an intra frame. Since the proposed method is a region based technique, the local variance is calculated at each pixel of an intra frame in order to show the level of high frequency details in a neighbourhood. The local variance is used as the input variable to the Fuzzy inference system which generates a defuzzified crisp output as an adaptive central weight of the Gaussian mask based on Fuzzy rule base. This technique will be fully explained in the subsequent step. Let m and υ represent the local mean and local variance of a 3×3 neighbourhood of an intra frame $f(x, y)$ respectively. The local variance υ is given by:

$$m = \frac{1}{9} \sum_{s=-1}^{1} \sum_{t=-1}^{1} W_{window}(x + s, y + t) \tag{9}$$

$$\upsilon = \frac{1}{9} \sum_{s=-1}^{1} \sum_{t=-1}^{1} [f(x + s, y + t) - m]^2 \tag{10}$$

3.2 Fuzzy Based Mapping

Fuzzy based techniques are designed to handle various nonlinear problems. The blurring caused by the up-sampling operation is nonlinear since more blurring takes place in the high variance regions than the low variance regions. In order to resolve this nonlinear problem of nonuniform blurring, fuzzy based mapping technique is used which maps the central weight of the Gaussian mask w with respect to the local variance υ as per the fuzzy rule base. υ and w are denoted as input and output variable of the fuzzy inference system respectively.

Fuzzy logic controllers are governed by a set of if-then rule known as a knowledge base or rule base. The fuzzy rule base drives the inference engine to produce the output in response to one or a set of inputs. The inputs are in general real world analog signals are termed as crisp input. These inputs are converted to fuzzy variable in fuzzifier. The fuzzifier inputs are sent to the inference engine to generate a controller response in fuzzy environment with the help of the fuzzy rule base. The inference engine operates with various fuzzy based operators such as min-max or product–max etc. These responses are then defuzzified to real world analog signal in the De-fuzzifier [19].

The fuzzy logic controller maps a crisp input to a crisp output through four blocks. The fuizzifier changes the crisp input based on membership function into fuzzy values or fuzzy sets. These fuzzy values are mapped in the inference engine to another set which is derived from the knowledge base. The knowledge base consists of rules provided by an expert or even obtained from numerical data. These rules are expressed as a set of "if-then" explicitly defining the nature of control action to be achieved for a certain input or a set of inputs. These rules decide the quality of the control action. The inference engine assigns a weight to the rule based on inputs, implications and aggregation operators. This caters to the quantitative nature of the control action. Hence the designing of fuzzy logic controller for a specific application deals with

- Deciding the membership function for fuzzifier and De-fuzzifier.
- Deciding the input and output ranges based on heuristic knowledge.
- Creation of knowledge base established on expert operator strategies.
- Deciding the inference operator or the De-fuzzification method.

Deciding the membership function plays an important role in solving a specific problem. In this method, the triangular membership function is selected for input variable while a combination of triangular and trapezoidal membership function is taken for output to combat the nonuniform blurring problem. This is a single input single output fuzzy system where the statistical local variance is taken as input parameter while the central weight of the Gaussian kernel is taken as the output parameter. The input parameter (local variance) is normalized to (0–100) scale whereas the range of the output parameter variable w is set within 0 to 100 based on heuristic knowledge. The input membership function is shown in Fig. 4a whereas the output membership function is shown in Fig. 4b. Both of the figure represent the plot of membership grade with respect to the variation in the input and output parameter respectively. Now the expressions for the input and output membership functions are given by

$$\mu_{AL}(v;0,\ 0,\ 50)=\begin{cases}0, & v\le 0\\ \frac{50-v}{50}, & 0\le v\le 50\\ 0, & 50\le v\end{cases} \tag{11a}$$

$$\mu_{AM}(v;\ 25,\ 50,\ 75)=\begin{cases}0, & v\le 25\\ \frac{v-25}{50-25}, & 25\le v\le 50\\ \frac{75-v}{75-50}, & 50\le v\le 75\\ 0, & 75\le v\end{cases} \tag{11b}$$

$$\mu_{AH}(v;\ 50,\ 100,\ 100)=\begin{cases}0, & v\le 50\\ \frac{v-50}{100-50}, & 50\le v\le 100\\ 0, & 100\le v\end{cases} \tag{11c}$$

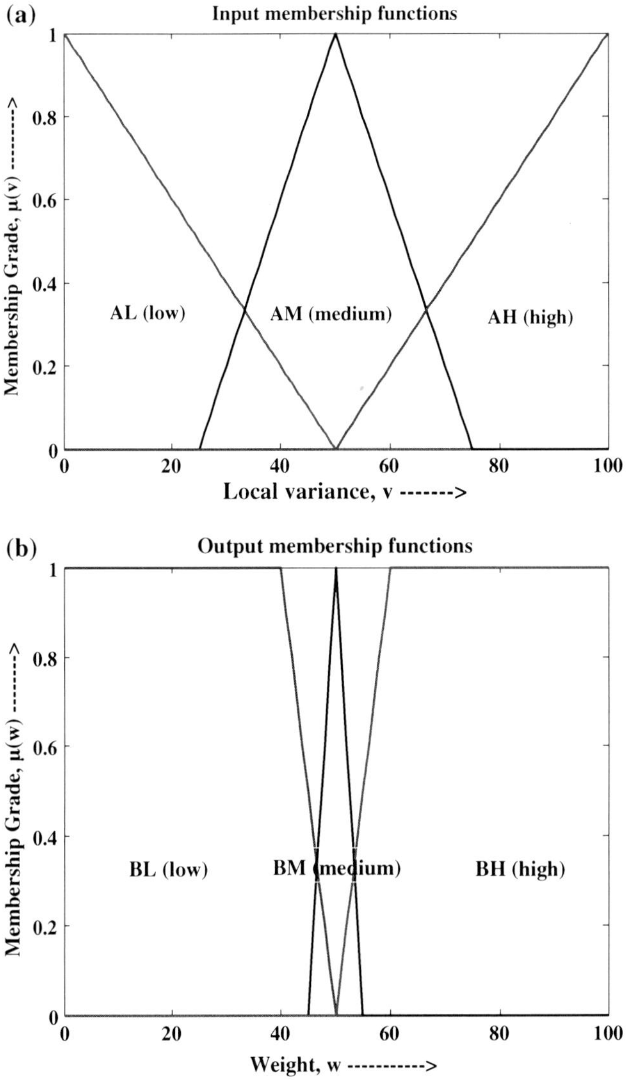

Fig. 4 Membership function plot: **a** input membership function, **b** output membership function

where $\mu_{AL}(v)$, $\mu_{AM}(v)$ and $\mu_{AH}(v)$ denote the low, medium and high membership function for the input variable respectively. Similarly the low, medium and high membership function of the output variable is denoted by $\mu_{BL}(w)$, $\mu_{BM}(w)$ and $\mu_{BH}(w)$ respectively. Input and output membership functions are shown in Fig. 4. The input variable is represented as the local variance, v. Similarly the output variable is represented as the central weight of the adaptive Gaussian mask, w. The expressions for output membership function are given by

$$\mu_{BL}(w;\ 0,\ 0,\ 40,\ 50) = \begin{cases} 1, & 0 \le w \le 40 \\ \frac{50-w}{50-40}, & 40 \le w \le 50 \\ 0, & w \ge 50 \end{cases} \quad (12a)$$

$$\mu_{BM}(w;\ 45,\ 50,\ 55) = \begin{cases} 0, & w \le 45 \\ \frac{w-45}{50-45}, & 45 \le w \le 50 \\ \frac{55-w}{55-50}, & 50 \le w \le 55 \\ 0, & 55 \le w \end{cases} \quad (12b)$$

$$\mu_{BH}(w;\ 50,\ 60,\ 100,\ 100) = \begin{cases} 0, & w \le 50 \\ \frac{w-50}{60-50}, & 50 \le w \le 60 \\ 1, & 60 \le w \le 100 \end{cases} \quad (12c)$$

Fuzzy logic controllers are governed by a set of if-then rule known as a knowledge base or rule base. The fuzzy rule base drives the inference engine to produce the output in response to one or a set of inputs. The knowledge base established on a heuristic approach is given as follows.

FUZZY IF – THEN RULES :
RULE I : if local variance(v) is low then weight (w) is high
OR
RULE II : if local variance is medium then weight is medium
OR
RULE III : if local variance is high then weight is low

The rule base contains all the information required to relate the inputs and outputs. In this case, the independent variables of the membership functions of input and output are different, so the result will be two dimensional. Hence the minimum of the input and the output membership function is performed as per the fuzzy if-then rules. Subsequently, the inference engine operates with the min-max operator to generate the output responses. The output responses are then de-fuzzified to produce a crisp output using the center of gravity method. Now the minimum of the two membership function is given by:

$$\mu_{AL \cap BH}(v, w) = \min\{\ \mu_{AL}(v),\ \mu_{BH}(w)\ \} \quad (13a)$$

$$\mu_{AM \cap BM}(v, w) = \min\{\ \mu_{AM}(v),\ \mu_{BM}(w)\ \} \quad (13b)$$

$$\mu_{AH \cap BL}(v, w) = \min\{\ \mu_{AH}(v),\ \mu_{BL}(w)\ \} \quad (13c)$$

Now to determine the output weight w_o for a specific input v_o, AND operation is performed between $\mu_{AL}(v_o)$ and the general result $\mu_{AL \cap BH}(v, w)$ evaluated at v_o according to fuzzy if-then rule-1. Let it be $Q_1(w)$. Similarly $Q_2(w)$ an $Q_3(w)$ are calculated for rule-2 and rule-3 respectively.

$$Q_1(w) = \min\{\mu_{AL}\ (v_o), \mu_{AL\cap BH}(v_o, w)\ \} \tag{14a}$$

$$Q_2(w) = \min\{\mu_{AM}\ (v_o), \mu_{AM\cap BM}(v_o, w)\ \} \tag{14b}$$

$$Q_3(w) = \min\{\mu_{AH}\ (v_o), \mu_{AH\cap BL}(v_o, w)\ \} \tag{14c}$$

In order to obtain the overall response, the individual responses are aggregated by OR operation. Thus the overall response $Q(w)$ is given by:

$$\begin{aligned} & Q(w) = Q_1(w) \cup Q_2(w) \cup Q_3(w) \\ \Rightarrow\ & Q(w) = \max\ \{\ Q_1\ (w),\ \max\ \{\ Q_2(w),\ Q_3(w)\ \}\ \} \end{aligned} \tag{15}$$

The crisp output w_o from the fuzzy set Q is obtained by center of gravity de-fuzzification method. Since $Q(w)$ can have K possible values, $Q(1), Q(2), \ldots,$ $Q(K)$, its centre of gravity is given by:

$$w_o = \frac{\sum_{w=1}^{K} wQ(w)}{\sum_{w=1}^{K} Q(w)} \tag{16}$$

Now, the de-fuzzified crisp output $w_o$is used to update the central weight of the Gaussian mask for the subsequent region adaptive unsharp masking operation.

3.3 Fuzzy Weighted Region Adaptive Unsharp Masking

The Fuzzy weighted, region adaptive unsharp masking operation is used to sharpen a video intra frame by subtracting the unsharp or smoothed version of it from the original. The smooth version of a video intra frame is obtained by blurring it using

Fig. 5 Two dimensional Laplacian kernel

0	-1	0
-1	4	-1
0	-1	0

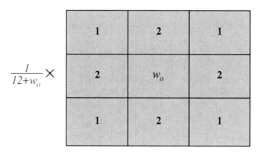

Fig. 6 Fuzzy weighted, region adaptive Gaussian mask

$$\frac{1}{12+w_o} \times \begin{bmatrix} 1 & 2 & 1 \\ 2 & w_o & 2 \\ 1 & 2 & 1 \end{bmatrix}$$

a Fuzzy weighted region adaptive Gaussian mask whose central weight is varied depending on the statistical local variance of a neighbourhood. The central weight is updated by the de-fuzzyfied crisp output w_o As per the fuzzy rule base, the central weight of Gaussian mask will be more if the local variance is less and vice versa. This results in proportionately low degree of blurring in low variance region and high degree of blurring in the high variance regions. Subsequently, the blurred video intra frame is subtracted from the original to produce the unsharp mask. In our proposed method, the unsharp mask is further sharpened using 2-D Laplacian kernel [18] as illustrated in Fig. 5. It is then by adding the modified unsharp mask to the original, a sharpened video frame is obtained in which the regions with higher local variance are sharpened more than the regions with lower local variance in order to compensate the blurring caused by the subsequent Lanczos-3 interpolation technique. Thus the proposed technique lessens the blurring effect by Lanczos-3 interpolation and improves the PSNR (dB) gain of the reconstructed video intra frame. In brief, the Fuzzy weighted region adaptive unsharp masking consists of the following steps.

- Determination of local variance of an intra frame as input variable.
- Fuzzy based mapping between local variance υ and central weight of the Gaussian mask w_o by using the knowledge base.
- Central weight updation of the Gaussian mask by the de-fuzzyfied crisp output w_o as shown in Fig. 6.
- The original video frame is blurred by using the fuzzy weighted region adaptive Gaussian mask.
- The blurred video frame is subtracted from the original. The resulting difference is called as the unsharp mask. This operation is repeated for all the frames.
- The resulting unsharp mask is further enhanced by Laplacian operator.
- The modified unsharp mask is then added to the original sub-sampled intra frame to obtain the sub-sampled sharpened sequence for more improved performance.

Let $g_1(x, y, n)$, $f(x, y, n)$ denote the blurred and the original video sequence respectively. Therefore, the fuzzy weighted, region adaptive unsharp mask is given by:

$$g_{mask}(x, y, n) = f(x, y, n) - g_1(x, y, n) \tag{17}$$

$g_1(x, y, n)$ is obtained by using the fuzzy weighted, region adaptive blurring technique.$n$represents the frame number that represents discrete time. The fuzzy weighted, region adaptive Gaussian mask is illustrated in Fig. 6. In this figure, w_o represents the center pixel weight of the mask which is made adaptive as per the statistical local variance of a 3×3 neighborhood.

3.4 Laplacian Based Sharpening

Laplacian based sharpening is used to sharpen the fuzzy weighted unsharp mask for further enhancement of high frequency information. Finally, this sharpened unsharp mask is added to the original sub-sampled video intra frames to obtain the sharpened video sequence $g_2(x, y, n)$ which is given in the subsequent equation.

$$g_2(x, y, n) = f(x, y, n) + c\ [\ \nabla^2\ g_{mask}(x, y, n)] \tag{18}$$

where ∇^2 denotes the Laplacian operator and c is a constant which varies according to the compression ratio. $c = 0.25$ at 4:1 compression ratio and $c = 0.5$ at 16:1 compression ratio. It is because, the degree of blurring at 16:1 compression ratio is more than the blurring at 4:1 compression ratio. Therefore, in order to compensate the high frequency loss in the subsequent Lanczos-3 interpolation, the degree of sharpening at 16:1 compression ratio is kept more than 4:1 compression ratio. Laplacian based sharpening is responsible for further improvement of the objective and subjective quality of the up-sampled video sequence.

The sharpened sub-sampled sequence is then finally up-sampled using Lanczos-3 interpolation technique to generate the high resolution video sequence. The Lanczos-3 interpolation technique is discussed in detail in Sect. 2.3.

3.5 Experimental Results and Analysis

To demonstrate the performance of the proposed hybrid technique, the input video sequences are down-sampled in the spatial domain by deleting alternate rows and columns at (4:1) and (16:1) compression ratio respectively. Then for each scheme, we interpolate the frames back to their original size to allow the comparison with the original video frame. Tables 2 and 3 illustrate theaverage PSNR comparison of DCT,

Table 2 Average PSNR comparison of different sequences at 4:1 compression ratio

Sequences	Average PSNR (dB)				
	Bicubic	Lanczos3	DCT	Proposed-1	Proposed-2
Flower	67.4448	67.6323	67.6192	67.7124	**67.8417**
Salesman	77.1100	77.4577	77.5707	**77.7681**	77.7558
Tennis	73.3649	73.5604	73.5845	73.6115	**73.7800**
News	76.4648	77.0882	77.2933	77.3543	**77.5418**
Akiyo	81.0421	81.5810	81.7782	81.7035	**82.0062**
City	75.7227	76.0100	75.9829	76.2195	**76.2912**
Football	76.6988	77.4973	77.8178	78.0112	**78.2981**
Ice	80.7227	81.2724	81.3500	80.8460	**81.7192**
Mobile	69.3281	69.7275	69.8889	70.1455	**70.1853**
Soccer	78.3849	78.7956	78.8791	78.7989	**79.1657**
Stefan	70.8642	71.1441	71.1566	71.1057	**71.4423**
Coastguard	74.6310	75.0699	75.2115	**75.4961**	75.4490
Room	78.2512	78.6650	78.7058	78.1700	**79.0335**
Highway 1	85.7051	86.3905	86.5953	85.3643	**86.8565**
Container	73.6984	74.1394	74.3871	74.4511	**74.4936**
Miss America	85.0846	85.2259	84.9645	84.8288	**85.3536**
Bus	73.3932	73.8475	73.9253	74.2872	**74.2903**

The bold values signifies the highest PSNR gain or the highest objective performance of a particular method among all the methods

Table 3 Average PSNR comparison of different sequences at 16:1 compression ratio

Sequences	Average PSNR (dB)				
	Bicubic	Lanczos3	DCT	Proposed-1	Proposed-2
Flower	64.9665	65.0323	65.0454	65.0814	**65.1355**
Salesman	73.2679	73.3983	73.3981	73.5152	**73.6042**
Tennis	70.8640	70.9054	70.8735	70.8026	**70.9548**
News	70.7512	70.9707	70.9456	71.2020	**71.3128**
Akiyo	76.2144	76.4112	76.4478	76.4182	**76.6514**
City	72.3711	72.4679	72.4509	72.4545	**72.5861**
Football	71.3216	71.5169	71.6358	71.5535	**71.7570**
Ice	75.5415	75.8449	**75.9274**	75.5418	75.9073
Mobile	65.8343	65.9303	65.9213	66.0015	**66.0741**
Soccer	74.2612	74.4484	74.4829	74.3310	**74.5544**
Stefan	67.6652	67.7742	67.8474	67.7539	**67.9395**
Coastguard	70.6563	70.7569	70.7682	70.3467	**70.8854**
Room	73.4384	73.6894	**73.7722**	73.3312	73.6787
Highway 1	78.4452	78.8337	78.9462	78.6124	**79.0227**
Container	69.6644	69.7546	69.7354	69.7261	**69.8580**
Miss America	81.2423	81.4991	**81.7596**	79.1835	81.3263
Bus	69.0830	69.1872	69.1570	69.2723	**69.3525**

The bold values signifies the highest PSNR gain or the highest objective performance of a particular method among all the methods

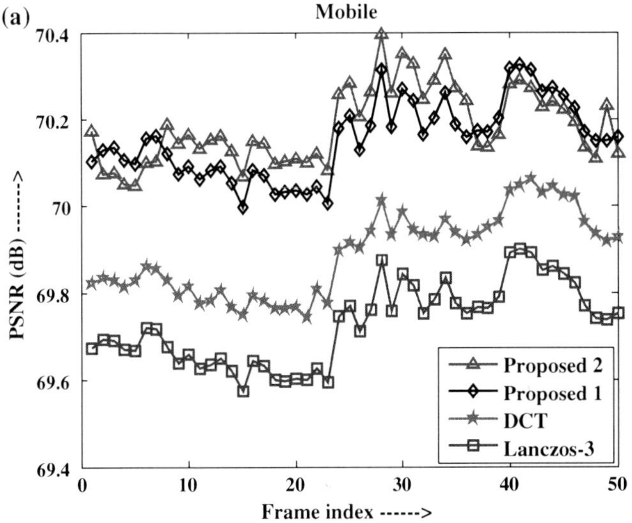

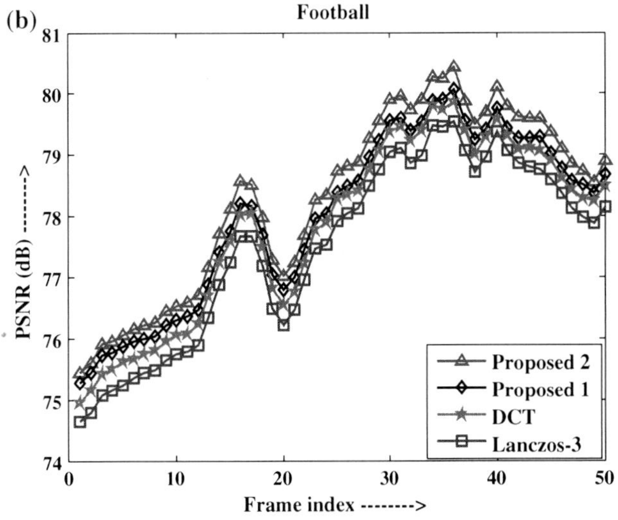

Fig. 7 PSNR (dB) comparison of different interpolation techniques at 4:1 compression ratio for different video sequence: **a** mobile, **b** football

Bicubic, Lanczos-3, proposed-1 and proposed-2 interpolation techniques at 4:1 and 16:1 compression ratios respectively. The bold values signify the maximum PSNR (dB) gain among all the methods. Experimental results reveal, at 4:1 compression ratio the proposed technique shows the average PSNR improvement up to 0.48 dB than DCT and an improvement up to 1.6 dB than the popular Bicubic interpolation technique particularly in the case of football sequence. Similarly the proposed

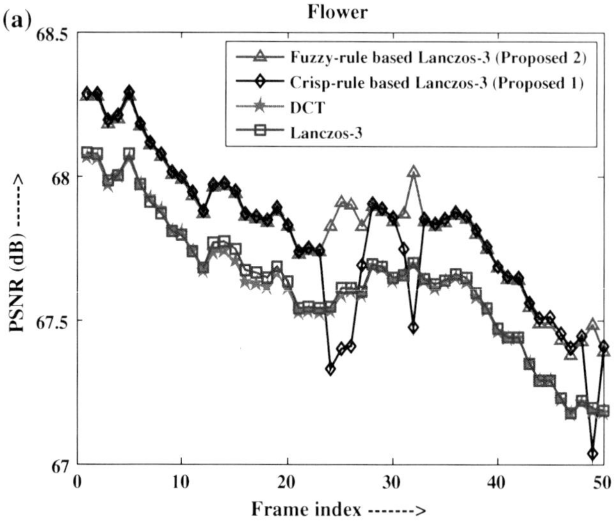

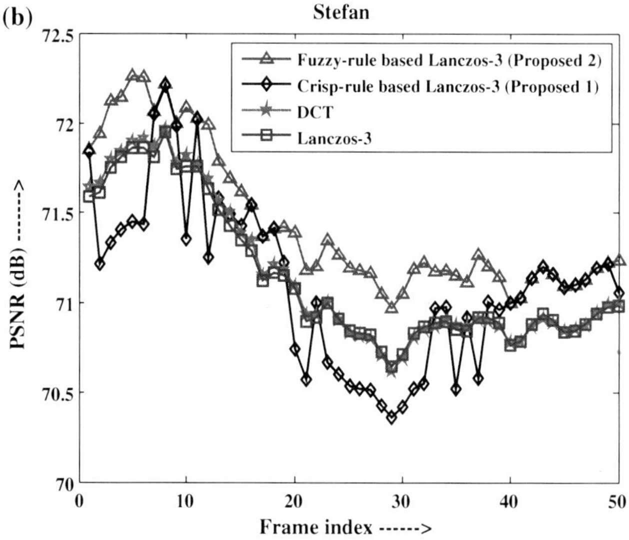

Fig. 8 Performance comparison between the proposed fuzzy based technique and the crisp rule based technique under varying zooming conditions at 4:1 compression ratio: **a** flower, **b** stefan

technique achieves a gain up to 0.367 dB than DCT and an improvement up to 0.56 dB than the popular Bicubic interpolation particularly in case of news sequence at 16:1 compression ratio. The average PSNR gain at 4:1 compression ratio is more than the gain at 16:1 compression ratio. It is because, at a high compression ratio, most of the high frequency details are lost,finally giving a flat and blurred output. Since the proposed method employs the high frequency details of the sub-sampled

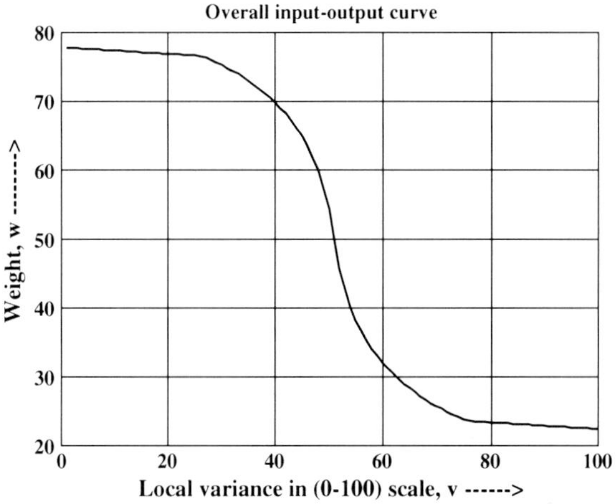

Fig. 9 Overall input-output curve between local variance (v) and central weight (w) of 11th frame of football sequence

intra frame for sharpening it so as to reduce the blurring caused by the subsequent Lanczos-3 interpolation, the PSNR gain is less at a higher compression ratio than the low compression counterpart. In Figs. 7 and 8, the variations of PSNR with respect the frame index are shown at 4:1 compression ratio. In either of the case, the proposed method-2 yields better PSNR gains than the other widely used interpolation techniques for different types of sequences.

In Fig. 8, the performance comparison between the proposed fuzzy based technique and the crisp rule weighted unsharp masking based Lanczos-3 interpolation is shown under varying zooming conditions. Experimental results reveal that the crisp-rule based technique is inconsistent in providing better results under the variation in zooming conditions. Although, the crisp rule based method gives better performance in some of the frames, due to lack of adaptability, the performance deteriorates for the remaining frames. This problem is due to improper mapping between input and output variable. In the crisp rule based method, a wide range of input values is mapped to only few (six) output values. This leads to loss of intermediate values to map the exact output value as required. On the other hand, in the proposed method, a wide variation in the input values is mapped to as many numbers of output values based on fuzzy rule base. Thus, there is an accurate and precise mapping between input and output as shown in Fig. 9 which not only provide better performance but also makes the proposed method adaptive to varying constraints such as zooming conditions, compression ratio and the video characteristics. The subjective performances of different interpolation techniques are illustrated in Figs. 10 and 11 for the 20th frame of akiyo and football sequence respectively at 4:1 compression ratio. Experimental results show, the blurring is much reduced and the edges are more pronounced with fine detail preservation in comparison to other existing interpolation techniques irrespective of the video types. Besides this, since the proposed method employs Lanczos-3 interpolation, it is free from ringing artifacts unlike DCT interpolation. It may be observed from all the result that, unlike crisp-rule based technique, the

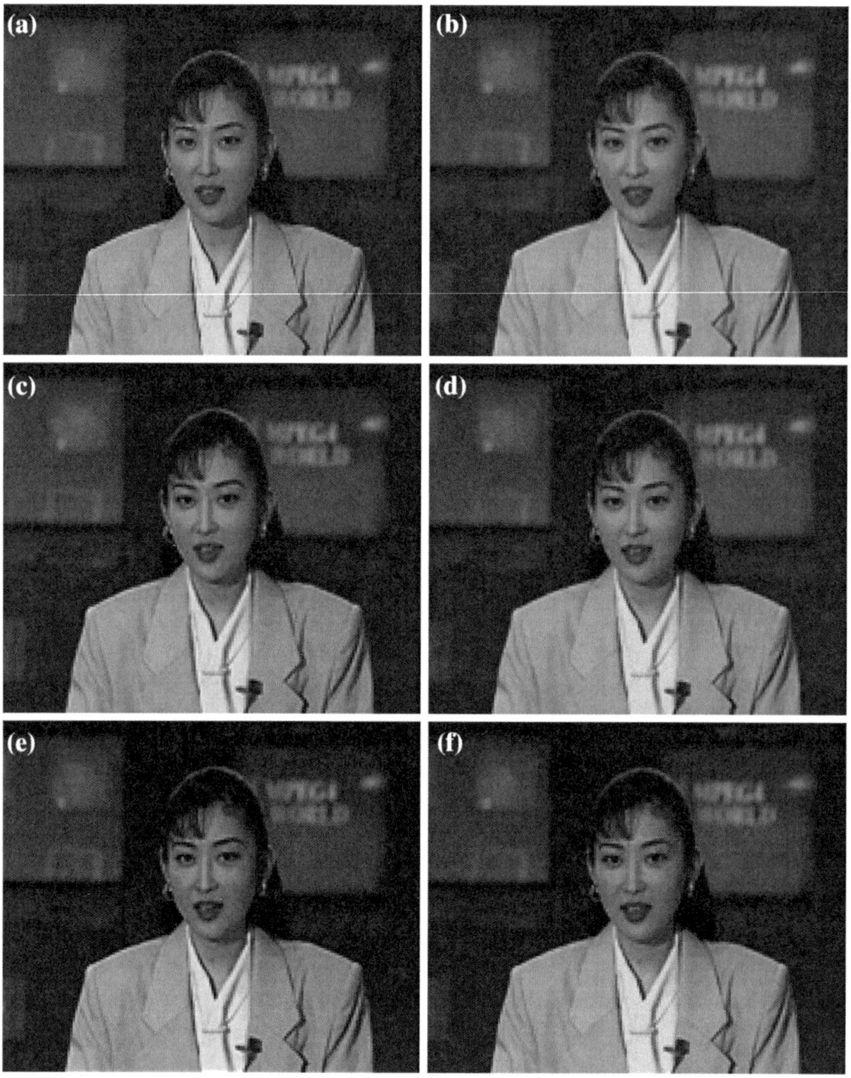

Fig. 10 Subjective performance of the 20th frame of the akiyo sequence at 4:1 compression ratio using different interpolation techniques: **a** Original, **b** Bicubic, **c** Lanczos-3, **d** DCT, **e** Crisp-rule based Lanczos-3 (Proposed-1), **f** Fuzzy-rule based Lanczos-3 (Proposed-2)

fuzzy-rule based technique achieves significant improvement over the existing techniques for all types of sequence irrespective of change in compression ratio, zooming conditions and video types. Hence, the proposed fuzzy-rule based technique is flexible enough to provide considerable performance under the wide variation in different constraints.

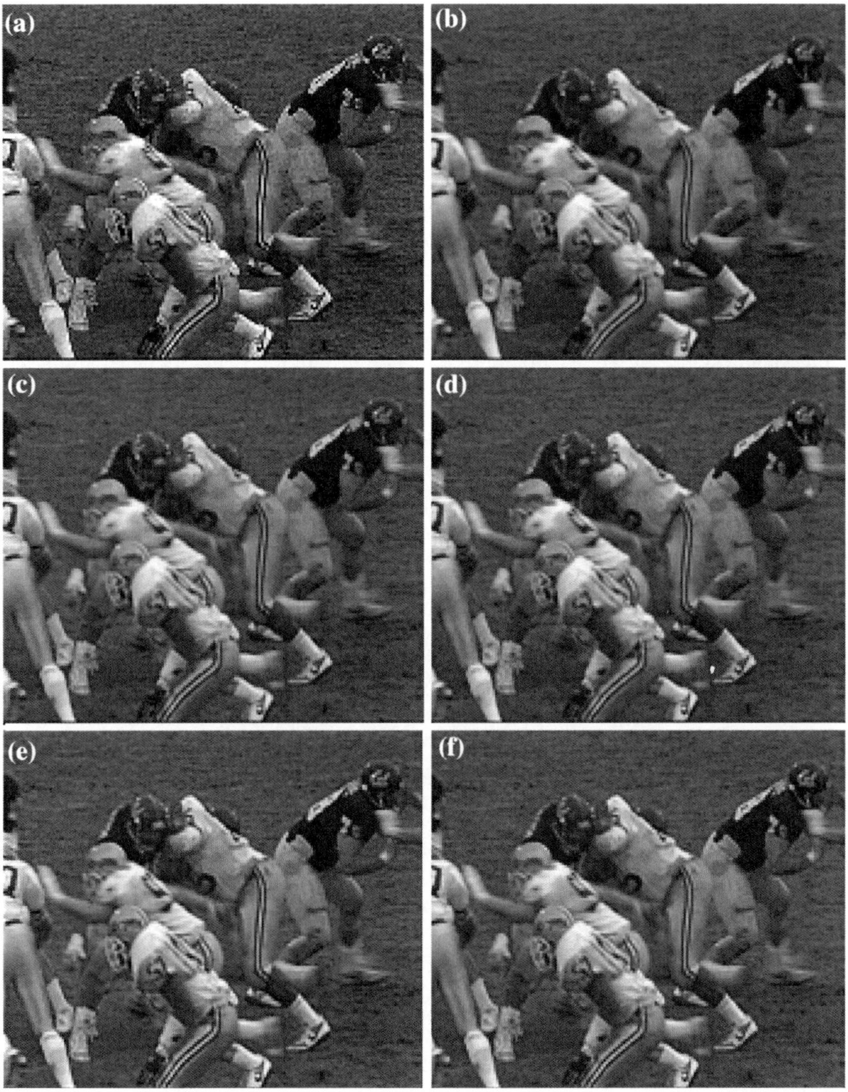

Fig. 11 Subjective performance of the 20th frame of the football sequence at 4:1 compression ratio using different interpolation techniques: **a** Original, **b** Bicubic, **c** Lanczos-3, **d** DCT, **e** Crisp-rule based Lanczos-3 (Proposed-1), **f** Fuzzy-rule based Lanczos-3 (Proposed-2)

4 Conclusions

This book chapter critically compares the capabilities and limitations of crisp-rule and fuzzy-rule based up-sampling techniques and their relevance in view of robustness, adaptability to varying constraints, complexities, quality enhancement and their effectiveness in real time applications.

The proposed crisp-rule based (Proposed-1) technique although gives better results in certain cases but fails to provide considerable performance under varying constraints such as variation in zoom in or zoom out conditions, change in compression ratio and video characteristics. This is basically due to improper mapping by crisp-rule based technique between input and output values. In case of video sequences subjected to zoom in or zoom out conditions, this technique fails drastically. It is because, there are only few output values to map for a large variation in the input values. So, the crisp-rule based algorithm is unable to perform fine and precise mapping between input and output because of the coarse variation of an adaptive weight factor with respect to continuous variation of the local variance as input parameter. Such problems can be avoided by fuzzy-rule based technique which employs a proper and precise mapping between input and output values using fuzzy inference system.

The proposed fuzzy-rule based technique not only restores a sub-sampled video with high precision but also yields a very low degree of blurring with fine detail preservation. It delivers superior performance and high degree of flexibility under a variety of constraints such as change in compression ratio and the video types. It achieves better performance of video reconstruction by exploiting the advantages of fuzzy weighted, region adaptive unsharp masking and 2-D Laplacian operation. The incorporation of fuzzy based preprocessing technique makes the proposed method highly adaptive to varying constraints and hence it works fine with different types of video subjected to change in compression ratio. The 2-D Laplacian operation enhances the high frequency details so as to considerably improve the objective performance of the proposed algorithm. In addition, by making use of fuzzy based mapping in between local variance and central weight of the Gaussian mask, the proposed method achieves much better objective and subjective performance under the change in zooming conditions and compression ratio.

Generally, in a terrestrial communication system, a transmitter possesses more processing ability than the receiver. Therefore, the major computational burden is easily taken up by the transmitter and less computational burden is left for the receiver. Since the proposed method is based on a preprocessing approach, it imparts more computational burden on the transmitting side than the receiving end and thus makes the receiver computationally less complex, fast and suitable for various real time applications. In addition, since this method is a spatial domain approach, it is computationally less complex than transform domain techniques such as DCT and wavelet. The proposed fuzzy-rule based method is a low complex, highly flexible and efficient technique that works fine with all types of video data.

References

1. T.M. Lehmann, C. Gonner, K. spitzer, Survey: Interpolation methods in medical image processing. IEEE Trans. Med. Imaging **18**(1), 1049–1075 (1999)
2. A. Goshtasby, D.A. Turner, L.V. Ackerman, Matching of tomographic slices for interpolation. IEEE Trans. Med. Imaging **11**, 507–516 (1992)

3. J. Lu, X. Si, S. Wu, An improved bilinear interpolation algorithm of converting standard definition images to high definition images, in *WASE International Conference on Information Engineering*, pp. 441–444 (2009)
4. R.G. Keys, Cubic convolution interpolation for digital image processing. IEEE Trans. Acoust. Speech Signal Process **ASSP-29**(6), 1153–1160 (1981)
5. H.S. Hou, H.C. Andrews, Cubic splines for image interpolation amd digital filtering. IEEE Trans. Acoust. Speech Signal Processsng, vol. ASSP-26 (1978)
6. S.K. Park, R.A. Schowengerdt, Image reconstruction by parametric cubic convolution. Comput. Vis. Graphics Image Processing **23**(3), 258–272 (1983)
7. L.A. Ferrari, J.H. Park, A. Healey, S. Leeman, Interpolation using a fast spline to transform (FST). IEEE Trans. Circuits Syst. I **46**, 891–906 (1999)
8. W. Ye, A. Entezari, A geometric construction of multivariate sinc functions. IEEE Trans. Image Processing **19**(12) (2011)
9. W. Burger, M.J. Burge, Principles of Digital Image Processing: Core Algorithms (Springer, Berlin, 2009), pp. 231–232
10. L.P. Yaroslavsky, Signal sinc-interpolation: a fast computer algorithm. Bioimaging **4**, 225–231 (1996)
11. S.R. Dooley, A.K. Nandi, Notes on the interpolation of discrete periodic signals using sinc function related approaches. IEEE Trans. Signal Processing **48**, 1201–1203 (2000)
12. T. Blu, M. Unser, Quantitative Fourier analysis of approximation techniques: Part I. Interpolators and projectors. IEEE Trans. Signal Processing **47**, 2783–2795 (1999)
13. I. Daubechies, *Ten Lectures on Wavelets* (SIAM, Philadelphia, 1992)
14. R. Dugad, N. Ahuja, A fast scheme for image size change in the compressed domain. IEEE Trans. Circuit Syst. Video Technol. **11**, 461–474 (2001)
15. J.I. Agbinya, Two dimensional interpolation of real sequences using the DCT. Electron. Lett., **29**(2), 204–205 (1993)
16. A. Acharya, S. Meher, An efficient, adaptive unsharp masking based interpolation for video intra frame upsampling, in *Proceedings of IEEE Asia Pacific Conference on Postgraduate Research in Microelectronics and Electronics*, Prime Asia, Dec. 2012, pp. 100–105
17. A. Acharya, S. Meher, Region adaptive unsharp masking based Lanczos3 interpolation for video intra frame up-sampling, in *Proceedings of IEEE 6th International Conference on Sensing Technology, ICST*, Dec. 2012
18. R. Gonzalez, R. Woods, *Digital Image Processing* (Pearson Publications, Upper Saddle River, 2008)
19. L.A. Zadeh, The concept of a linguistic variable and its application to approximate reasoning (III). Inf. Sci. **9**, 43–80 (1975)

Chapter 4
Gaze-Controlled Stereo Vision to Measure Position and Track a Moving Object: Machine Vision for Crane Control

Yasuo Yoshida

Abstract Stereo vision using parallel optical axes is used as a machine vision, but gaze-controlled stereo vision may be not used in practical use. But gaze-controlled stereo vision can always track a moving object with catching in the center of camera image. Boom cranes are used in many fields, in which a suspended load must be transported and positioned at the desired position without any swing. Therefore positioning and swing suppression of a boom crane using visual feedback control are studied. The boom crane has broad working space, so we use a stereo vision installed on the boom's rotary axis that has gazing motion to track the suspended load. The three dimensional position and the swing angle of the load are measured using this tracking stereo vision to control the crane. The experimental control results are presented concerning with the tracking of the stereo vision, the positioning of the boom and the swing suppression of the suspended load.

Keywords Stereo vision · Gazing · Tracking · Visual feedback control · Boom crane · Swing suppression · Positioning

1 Introduction

When working by visual information, it is necessary to grasp what you have in the outside world at sight, and a result of recognition must be reflected by an action. A two-dimensional moving point with one camera mounted on a robot was chased [1–3]. But, two cameras are necessary in the case of three dimensions. Binocular disparity of a base-line stereo that both optical axes are parallel is usually used [4]. Gaze control is a method to determine an object on a intersection point of optical axes of both cameras and makes it possible to always catch an observation object at the center of a field of vision [5, 6]. This method has concentration of gaze

Y. Yoshida (✉)
Department of Mechanical Engineering, Chubu University, 1200 Matsumoto-cho, Kasugai, Aichi 487-8501, Japan
e-mail: yyoshida@isc.chubu.ac.jp

A. Mason et al. (eds.), *Sensing Technology: Current Status and Future Trends II*, Smart Sensors, Measurement and Instrumentation 8, DOI: 10.1007/978-3-319-02315-1_4,

domain image processing and an advantage of a movement body chase. A suspended load of a crane must be transported and positioned at the desired location without any swing. Dynamics and control of cranes was reviewed [7]. A boom crane has derricking, rotary and hoisting motions. Therefore, it has broad working space and is used in many fields. On the other hand, visual servo which uses visual information for motion control is being utilized in fields such as a robot [8–10], a car [11–13], medical care [14]. However, there are a few applications of visual servo to a crane, but visual controls of an overhead crane using separate type stereo vision or installed 3D-camera are shown in from [14] to [18]. And a stereo vision installed on the rotary axis of a boom crane is used to have gazing motion to track the suspended load. The three dimensional position and the swing angle of the load are measured using this tracking stereo vision. The experimental results are presented concerning with stereo vision gazing tracking control, boom derricking/rotary positioning control, and load swing suppression control.

1.1 Gaze-Controlled Stereo Vision

Gaze-controlled stereo vision is shown in Fig. 1. This stereo vision has four-degree-of-freedom, θ_1, θ_2 to rotate and derrick two cameras, and θ_3, θ_4 to move each CCD camera's optical axis.

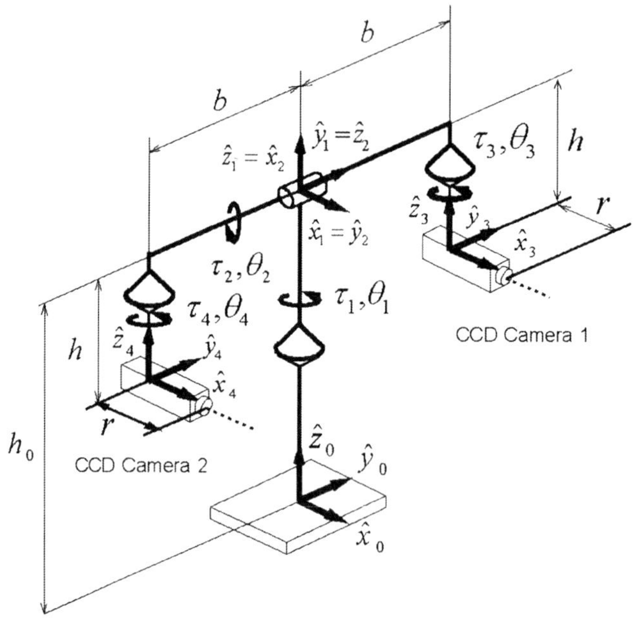

Fig. 1 Gaze-controlled stereo vision

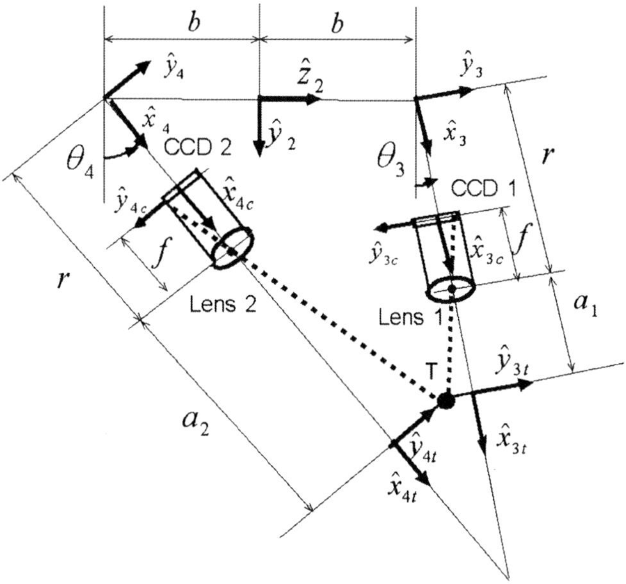

Fig. 2 Camera Coordinates

Camera coordinates are shown in Fig. 2, where f, r are focal distance and lens position each on optical axis. Camera image positions of moving point T indicated by camera coordinate 3c, 4c are expressed as:

$$^{3c}P_T = \begin{bmatrix} 0 \ y_{3c} \ z_{3c} \end{bmatrix}^T, {}^{4c}P_T = \begin{bmatrix} 0 \ y_{4c} \ z_{4c} \end{bmatrix}^T. \tag{1}$$

Therefore, point T indicated by target coordinate 3t, 4t can be expressed with image data:

$$^{3t}P_T = \frac{a_1}{f}\begin{bmatrix} 0 \ y_{3c} \ z_{3c} \end{bmatrix}^T, {}^{4t}P_T = \frac{a_2}{f}\begin{bmatrix} 0 \ y_{4c} \ z_{4c} \end{bmatrix}^T, \tag{2}$$

a_1, a_2 are distances from the lens center to the target-point along optical axes using $c_i \equiv \cos\theta_i$, $s_i \equiv \sin\theta_i (for\, i = 1 \sim 4)$ and $c_{34} \equiv \cos(\theta_3 - \theta_4)$, $s_{34} \equiv \sin(\theta_3 - \theta_4)$:

$$a_1 = \frac{2b\left(\frac{y_{4c}}{f}s_4 - c_4\right) - r\left\{\frac{y_{4c}}{f}(1 - c_{34}) + s_{34}\right\}}{\frac{y_{3c}-y_{4c}}{f}c_{34} + \left(1 + \frac{y_{3c}y_{4c}}{f^2}\right)s_{34}},$$

$$a_2 = \frac{2b\left(\frac{y_{3c}}{f}s_3 - c_3\right) + r\left\{\frac{y_{3c}}{f}(1 - c_{34}) - s_{34}\right\}}{\frac{y_{3c}-y_{4c}}{f}c_{34} + \left(1 + \frac{y_{3c}y_{4c}}{f^2}\right)s_{34}} \tag{3}$$

Positions by joint coordinate 3, 4 can be expressed:

$$
{}^{3}P_T \equiv \begin{bmatrix} {}^{3}x_T \\ {}^{3}y_T \\ {}^{3}z_T \end{bmatrix} = \begin{bmatrix} r + a_1 \\ (a_1/f)\, y_{3c} \\ (a_1/f)\, z_{3c} \end{bmatrix},\ {}^{4}P_T \equiv \begin{bmatrix} {}^{4}x_T \\ {}^{4}y_T \\ {}^{4}z_T \end{bmatrix} = \begin{bmatrix} r + a_2 \\ (a_2/f)\, y_{4c} \\ (a_2/f)\, z_{4c} \end{bmatrix} \tag{4}
$$

From ${}^{3}P_T$, the position expressed with base coordinate 0 is given by:

$$
{}^{0}P_T \equiv \begin{bmatrix} {}^{0}x_T \\ {}^{0}y_T \\ {}^{0}z_T \end{bmatrix} = \begin{bmatrix} c_1 s_2 \left({}^{3}z_T - h\right) + c_1 c_2 \left(c_3{}^{3}x_T - s_3{}^{3}y_T\right) - s_1 \left(s_3{}^{3}x_T + c_3{}^{3}y_T + b\right) \\ s_1 s_2 \left({}^{3}z_T - h\right) + s_1 c_2 \left(c_3{}^{3}x_T - s_3{}^{3}y_T\right) + c_1 \left(s_3{}^{3}x_T + c_3{}^{3}y_T + b\right) \\ c_2 \left({}^{3}z_T - h\right) - s_2 \left(c_3{}^{3}x_T - s_3{}^{3}y_T\right) + h_0 \end{bmatrix} \tag{5}
$$

Similarly, the position of the target can be obtained from ${}^{4}P_T$. Figure 3a shows the gazing state where the target is on a point of intersection of optical axes of both cameras where the position is given by putting $y_{3c} = y_{4c} = 0,\quad z_{3c} = z_{4c} = 0$ into eq. (3) and (4). Figure 3b shows the looking state of parallel optical axes of both cameras by putting $\theta_3 = \theta_4 = 0$ into eq. (3) and (4). Therefore, expressions of (3 and 4) include not only gaze but also parallel stereo.

We can see that the three-dimensional position of a moving object is the function with camera angles, image data, the focus distance and stereovision dimensions. It is gaze control that there is a moving point in the center of camera image, and tracking control that always catches a moving point to camera image. When an intersection point of two cameras optical axes accords with a moving point, stereovision angles become target desired angles for the gaze/tracking control.

Fig. 3 Looking states using gaze and parallel optical axes of stereo cameras

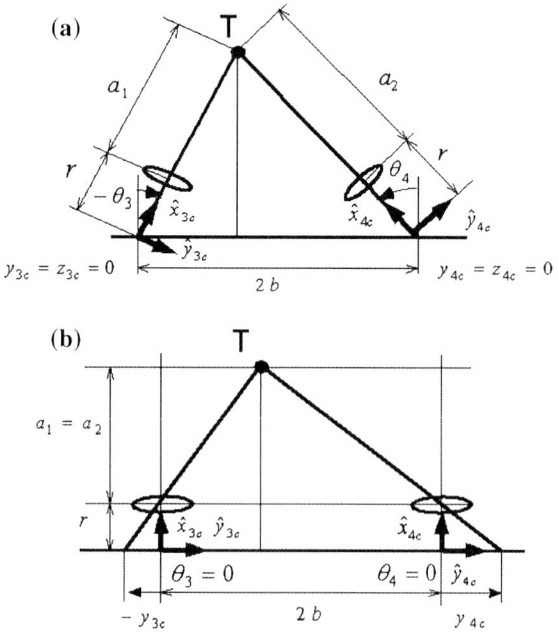

Target desired angles as shown in Figure 4 using measured position of eq.(5) are provided by inverse kinematics:

$$\theta_{1d} = A\tan 2\left({}^{0}y_T, {}^{0}z_T\right),$$

$$\theta_{2d} = -\frac{\pi}{2} - A\tan 2\left({}^{0}z_T - h_0, \sqrt{\left({}^{0}x_T\right)^2 + \left({}^{0}y_T\right)^2}\right)$$
$$+ A\tan 2\left(\sqrt{\left({}^{0}x_T\right)^2 + \left({}^{0}y_T\right)^2 + \left({}^{0}z_T - h_0\right)^2 - h^2}, h\right), \quad (6)$$

$$\theta_{3d} = -\theta_{4d},$$

$$\theta_{4d} = A\tan 2\left(b, \sqrt{\left({}^{0}x_T\right)^2 + \left({}^{0}y_T\right)^2 + \left({}^{0}z_T - h_0\right)^2 - h^2}\right)$$

Digital controller by which the controlled angle θ_i follows the desired angle θ_{id} is given by:

$$u_{ik} = k_{c1d}\left(\theta_{idk} - \theta_{ik}\right) - k_{c2d}\dot{\theta}_{ik} \quad (i = 1 \sim 4) \quad (7)$$

Manipulating velocities are obtained by integrating with sampling time.

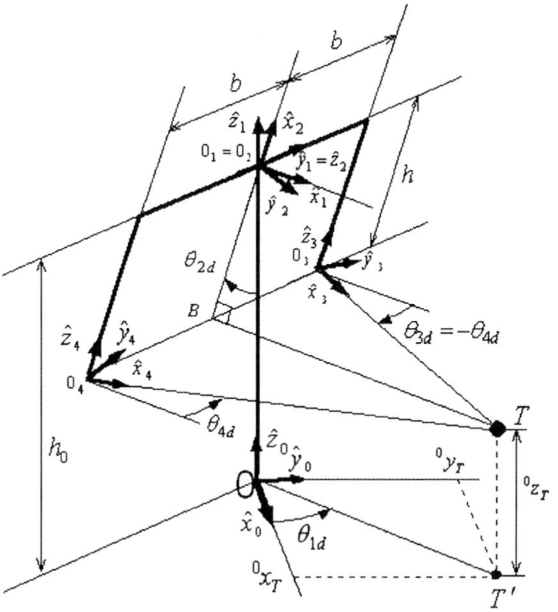

Fig. 4 Desired angles for gaze and tracking to control stereovision

2 Machine Vision for Boom Crane Control

Boom crane control using installed stereo vision is explained.

2.1 Boom Crane Control

Figure 5 shows a boom crane model. A crane-boom rotates θ_1, θ_2, to rotary and derricking directions. The crane has five-degree-of-freedom (boom θ_1, θ_2, rope length l, swing angle α, β to normal and tangential direction for rotary boom motion, respectively). A stereo vision with two cameras is installed on the rotary axis of the crane-boom. The stereo vision has three-degree-of-freedom (derricking for two cameras θ_1, rotation of left/right camera θ_2, θ_3). Then, this boom crane model has eight degree-of-freedom.

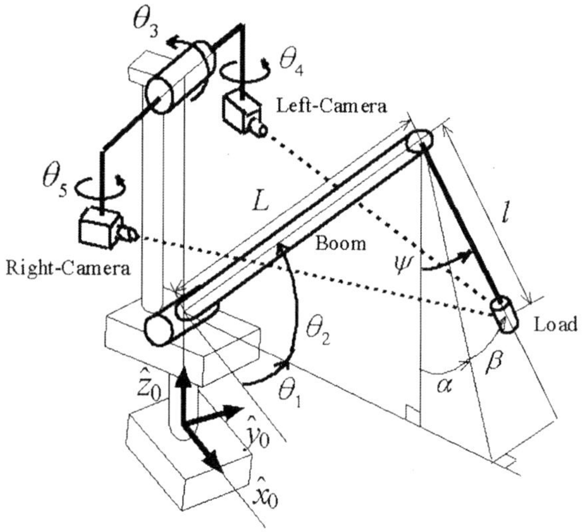

Fig. 5 Boom crane model

The equations of motion of the boom crane can be written as:

$$\begin{aligned} \ddot{\alpha} &= -\tfrac{g}{l}\alpha - 2\tfrac{\dot{l}}{l}\dot{\alpha} + \tfrac{L}{l}(\sin\theta_2)\ddot{\theta}_2, \\ \ddot{\beta} &= -\tfrac{g}{l}\beta - 2\tfrac{\dot{l}}{l}\dot{\beta} - \tfrac{L}{l}(\cos\theta_2)\ddot{\theta}_1. \end{aligned} \tag{8}$$

The first of the equation of motion eq. (8) expresses balance in α direction, and the second in β direction. These show dynamic coupling with boom transportation, rope hoisting and load swing, where, swing angles α, β of θ_2, θ_1 direction components are supposed to be small values. At first, control of rope length is thought about. The digital controller of hoisting velocity v_l by which the controlled length l follows the desired length l_d is given with gain k_{ld}:

$$v_{lk} = \dot{l}_{dk} + k_{ld}\left(l_{dk} - l_k\right) \tag{9}$$

where, $k = 1, 2, \ldots$ indicates time series with sampling time. Next, boom positioning and load swing suppression are thought about. The state equation associated with boom derricking/rotary angle and load swing is obtained from eq. (8), using control input $\ddot{\theta}_1 = u_{\theta 1}, \ddot{\theta}_2 = u_{\theta 2}$:

$$\frac{d}{dt}\tilde{\theta}_1 = A\tilde{\theta}_1 + b_1 u_{\theta 1}, \quad \frac{d}{dt}\tilde{\theta}_2 = A\tilde{\theta}_2 + b_2 u_{\theta 2} \tag{10}$$

where:

$$A = \begin{bmatrix} 0 & 0 & 1 & 0 \\ 0 & 0 & 0 & 1 \\ 0 & 0 & 0 & 0 \\ 0 & -\frac{g}{l} & 0 & -2\frac{\dot{l}}{l} \end{bmatrix}, \quad b_1 = \begin{bmatrix} 0 \\ 0 \\ 1 \\ -\frac{L}{l}\cos\theta_2 \end{bmatrix},$$

$$b_2 = \begin{bmatrix} 0 \\ 0 \\ 1 \\ \frac{L}{l}\sin\theta_2 \end{bmatrix}, \tilde{\theta}_1 = \begin{bmatrix} \theta_1 \\ \beta \\ \dot{\theta}_1 \\ \dot{\beta} \end{bmatrix}, \quad \tilde{\theta}_2 = \begin{bmatrix} \theta_2 \\ \alpha \\ \dot{\theta}_2 \\ \dot{\alpha} \end{bmatrix}$$

Digital regulator where boom angles follow desired angles and swing angles decrease to zero value can be designed as:

$$u_{\theta 1k} = G_d \tilde{e}_{\theta 1k}, \quad u_{\theta 2k} = G_d \tilde{e}_{\theta 2k} \tag{11}$$

where:

$$\tilde{e}_{\theta 1k} = \left[\theta_{1d} - \theta_{1k} \; -\beta_k \; -\dot{\theta}_{1k} \; -\dot{\beta}_k\right]^T, \quad \tilde{e}_{\theta 2k} = \left[\theta_{2d} - \theta_{2k} \; -\alpha_k \; -\dot{\theta}_{2k} \; -\dot{\alpha}_k\right]^T$$

From the closed loop relation between analog state and digital state that is redesigned to digitize the analog state using sample time T, digital gain $G_d = [k_{1d} \;\; k_{2d} \;\; k_{3d} \;\; k_{4d}]$ can be obtained from analog gain G_a determined by pole assignment.

$$G_d = \frac{1}{2}\left(1 + \frac{1}{2}G_a q\right)^{-1} G_a\left(I + p\right) \tag{12}$$

where:

$$p = \left(I - \frac{T}{2}A\right)^{-1}\left(I + \frac{T}{2}A\right), \quad q = T\left(I - \frac{T}{2}A\right)^{-1} b,$$

and vector b is b_1 or b_2 in eq. (10). Angular velocity of the crane boom control can be obtained from eq. (11) by integrating with sample time.

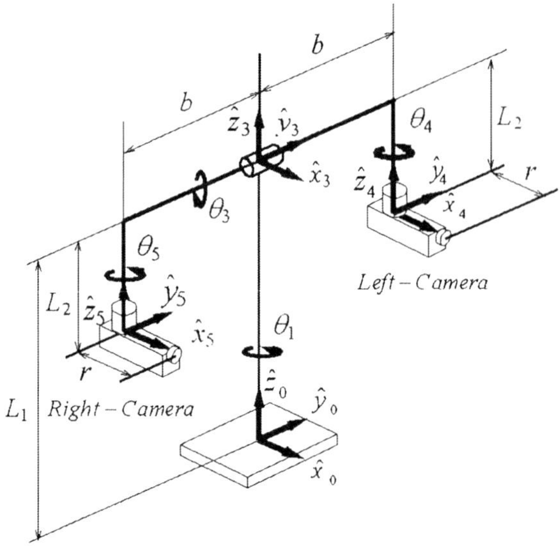

Fig. 6 Stereo vision installed on the boom crane

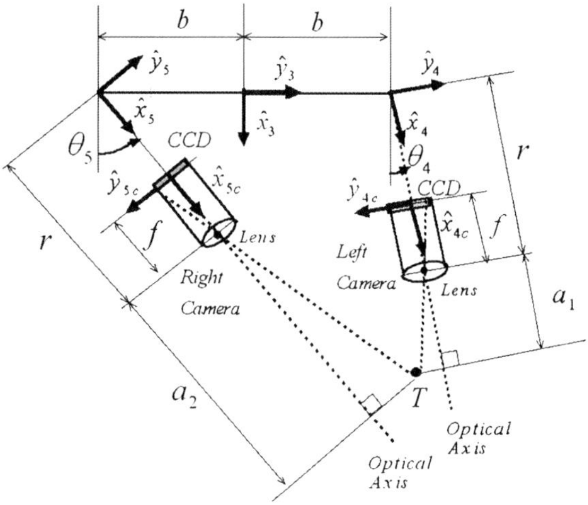

Fig. 7 Camera coordinates and optical axes

2.2 *Stereo Vision Control*

Figures 6 and 7 show the stereo vision installed on the boom crane and camera coordinates. Image positions of the crane load point T by camera coordinate 4c, 5c are indicated in the next expression:

$$ {}^{4c}P_T = [0 \quad y_{4c} \quad z_{4c}]^T, \quad {}^{5c}P_T = [0 \quad y_{5c} \quad z_{5c}]^T. \tag{13} $$

Positions by coordinate 4, 5 can be expressed using above camera image positions:

$$ {}^{4}P_T \equiv \begin{bmatrix} {}^4x_T \\ {}^4y_T \\ {}^4z_T \end{bmatrix} = \begin{bmatrix} r + a_1 \\ (a_1/f)\, y_{4c} \\ (a_1/f)\, z_{4c} \end{bmatrix}, \quad {}^{5}P_T \equiv \begin{bmatrix} {}^5x_T \\ {}^5y_T \\ {}^5z_T \end{bmatrix} = \begin{bmatrix} r + a_2 \\ (a_2/f)\, y_{5c} \\ (a_2/f)\, z_{5c} \end{bmatrix}. \tag{14} $$

where f is focus distance and a_1, a_2 are distances from the lens to the crane load in camera optical axis directions:

$$ \begin{aligned} a_1 &= \frac{2b\left(\frac{y_{5c}}{f} s_5 - c_5\right) - r\left\{\frac{y_{5c}}{f}(1-c_{45}) + s_{45}\right\}}{\frac{y_{4c}-y_{5c}}{f} c_{45} + \left(1 + \frac{y_{4c}y_{5c}}{f^2}\right) s_{45}}, \\ a_2 &= \frac{2b\left(\frac{y_{4c}}{f} s_4 - c_4\right) + r\left\{\frac{y_{4c}}{f}(1-c_{45}) - s_{45}\right\}}{\frac{y_{4c}-y_{5c}}{f} c_{45} + \left(1 + \frac{y_{4c}y_{5c}}{f^2}\right) s_{45}} \end{aligned} \tag{15} $$

where $c_i, s_i = \cos\theta_i, \sin\theta_i \; for \; i = 1 \sim 5$ and $c_{45}, s_{45} = \cos(\theta_4 - \theta_5), \sin(\theta_4 - \theta_5)$ are used above and hereafter.

Using ${}^4P_T$, the position expressed with base coordinate 0 is given by:

$$ \begin{aligned} {}^0P_T &= \begin{bmatrix} c_1 s_3 \left({}^4z_T - L_2\right) + c_1 c_3 \left(c_4 {}^4x_T - s_4 {}^4y_T\right) - s_1 \left(s_4 {}^4x_T + c_4 {}^4y_T + b\right) \\ s_1 s_3 \left({}^4z_T - L_2\right) + s_1 c_3 \left(c_4 {}^4x_T - s_4 {}^4y_T\right) + c_1 \left(s_4 {}^4x_T + c_4 {}^4y_T + b\right) \\ c_3 \left({}^4z_T - L_2\right) - s_3 \left(c_4 {}^4x_T - s_4 {}^4y_T\right) + L_1 \end{bmatrix} \\ &\equiv [x_0 \quad y_0 \quad z_0]^T \end{aligned} \tag{16} $$

Similar another expression of ${}^0P_T$ can be obtained using ${}^5P_T$. The three-dimensional position of the crane load is the function of image data, camera angles, focus distance and stereovision dimensions. It is gaze control that there is a target moving object such as the crane load always in the center of camera image. Then the gazing state is that the target is on a point of intersection of optical axes of both cameras. When an intersection point of two camera optical axes accords with the target point, stereo vision angles become desired angles for gazing and tracking control.

Desired angles as shown in Fig. 8 using measured position of eq. (16) are provided by inverse kinematics:

$$ \begin{aligned} \theta_{3d} &= A\tan 2\left(L_1 - z_0, x_0 c_1 + y_0 s_1\right) - \sin^{-1}\left(\frac{L_2}{\sqrt{(L_1 - z_0)^2 + (x_0 c_1 + y_0 s_1)^2}}\right) \\ \theta_{4d} &= -A\tan 2\left(b + S_1 x_0 - C_1 y_0, \sqrt{(L_1 - z_0)^2 + (C_1 x_0 + S_1 y_0)^2 - L_2^2}\right) \end{aligned} \tag{17} $$

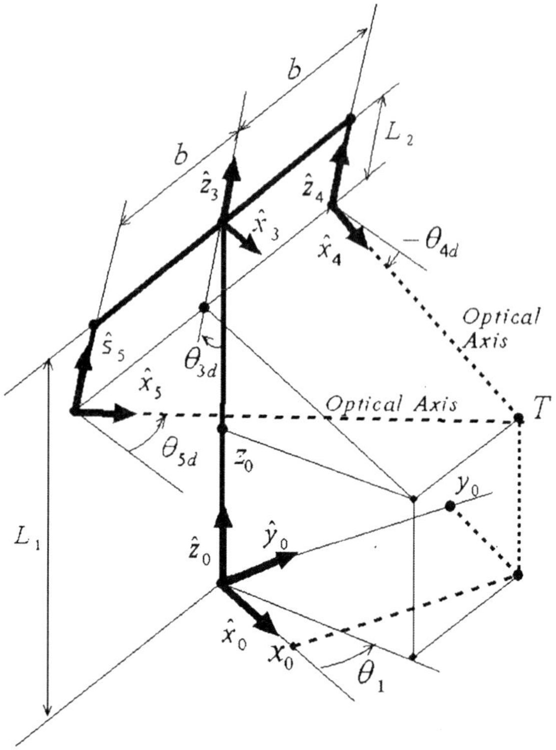

Fig. 8 Desired angles to control stereo vision for gazing and tracking

$$\theta_{5d} = A\tan 2\left(b - S_1x_0 + C_1y_0, \sqrt{(L_1 - z_0)^2 + (C_1x_0 + S_1y_0)^2 - L_2^2}\right)$$

Digital controller by which the controlled angle $\theta_3, \theta_4, \theta_5$ follows the desired angle $\theta_{3d}, \theta_{4d}, \theta_{5d}$ is given by:

$$u_{ik} = k_{c1d}(\theta_{idk} - \theta_{ik}) - k_{c2d}\dot{\theta}_{ik} \quad (i = 1 \sim 3) \tag{18}$$

Angular velocity of the stereo vision control can be obtained from eq. (18) by integrating with sample time. Figure 9 shows block diagram of boom crane control and stereo vision control. Half of the upper part shows crane control of the boom positioning and the load swing suppression, and half of the bottom shows stereo vision control of gazing and tracking to measure three dimensional position of the crane load.

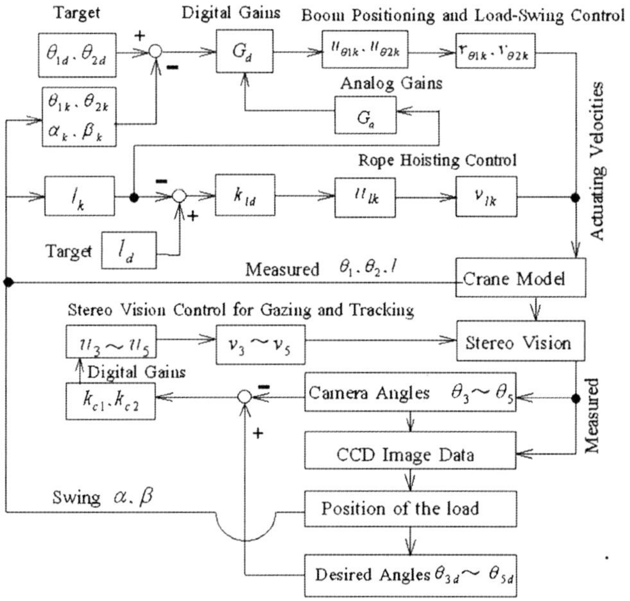

Fig. 9 Block diagram

3 Experimental Results

Figure 10 shows the photograph of an experimental boom crane model which installs a stereo vision. In experiments, 0.1 second was used as a sample time. Experiments are reported gaze and tracking control of stereo vision, derricking and rotary control of crane boom and swing suppression control of crane load.

3.1 Gaze Control of Crane Stereo Vision

Figure 11 shows controlled stereo vision angles to gaze the standstill boom crane load. Three dimensional position of the crane load is always calculated from the camera image data and thereafter, camera angles are controlled to gaze it. Time to gaze is about 0.6 seconds. Stereo vision's derrick angle θ_3 moves to downward and camera rotary angles of right and left move to each other's insides. Figure 12 shows controlled CCD image data. Both data of right and left images move to the center of CCD according to gaze control.

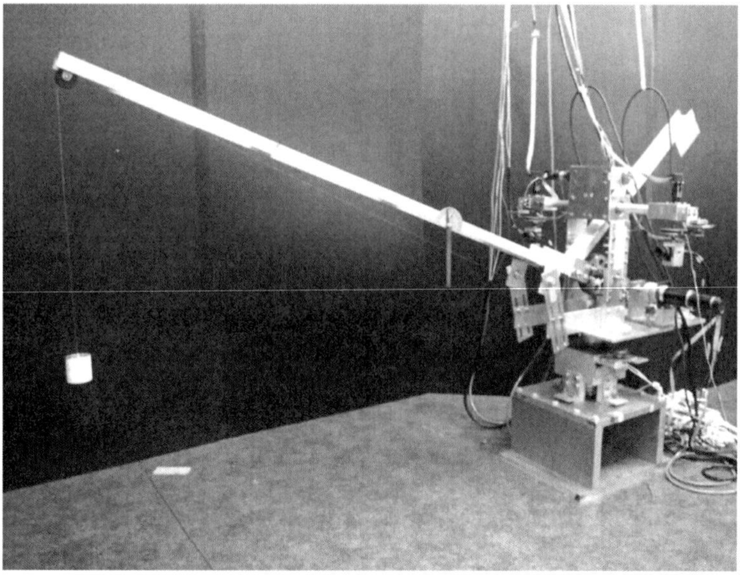

Fig. 10 Experimental model of boom crane and installed stereo vision

Fig. 11 Response of stereo vision angles (gaze control)

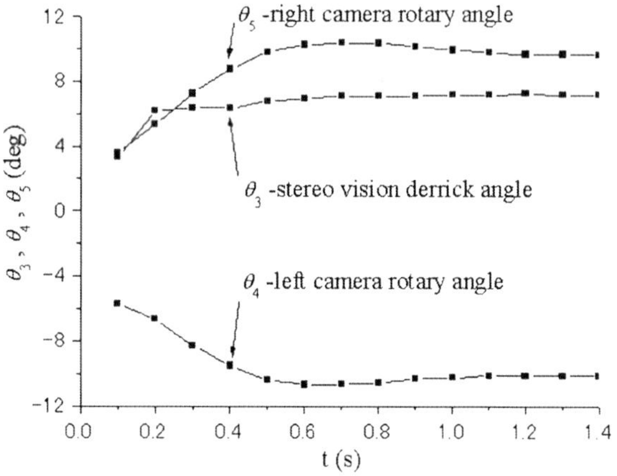

3.2 Derricking Control

Figures 13 and 14, experiments of derricking control of crane boom are shown, where rope length is 0.63 m constant. Figure 13 is the response of derrick angle θ_2 for target angle θ_d changing from 20 to 40 degrees. Transient response without swing control changes more than that with swing control. Figure 14 is the response of swing angle α and shows swing control is effective.

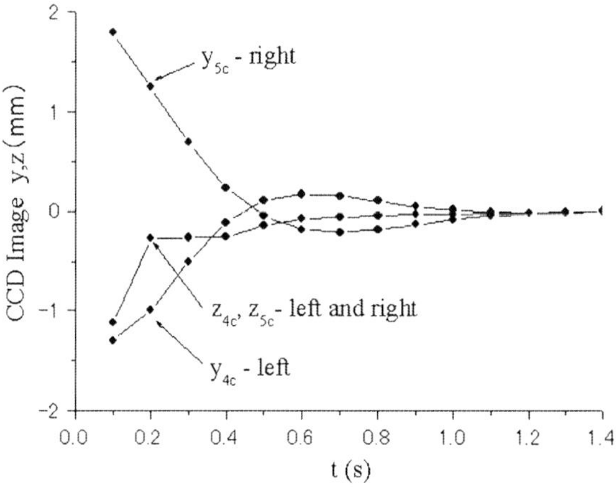

Fig. 12 Response of CCD image data (gaze control)

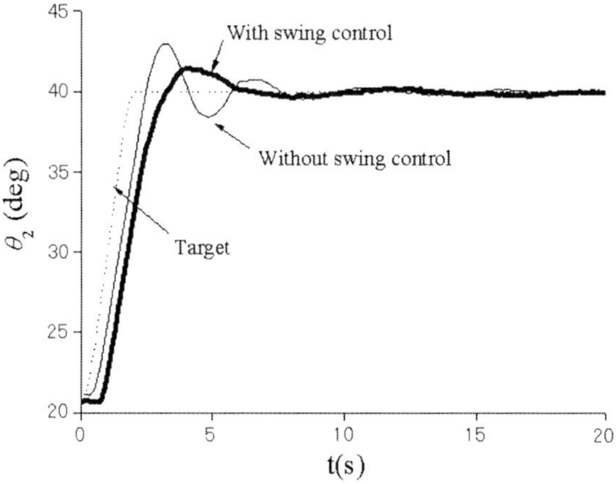

Fig. 13 Response of derrick angle (derricking control)

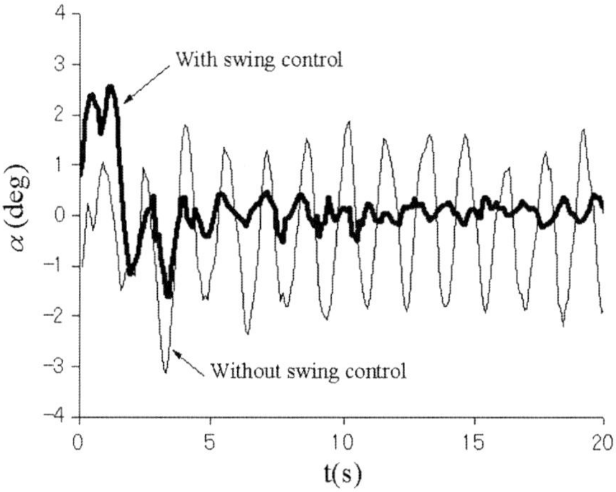

Fig. 14 Response of swing angle (derricking control)

Fig. 15 Response of rotary and derrick angles (simultaneous control of rotary and derricking, without swing control)

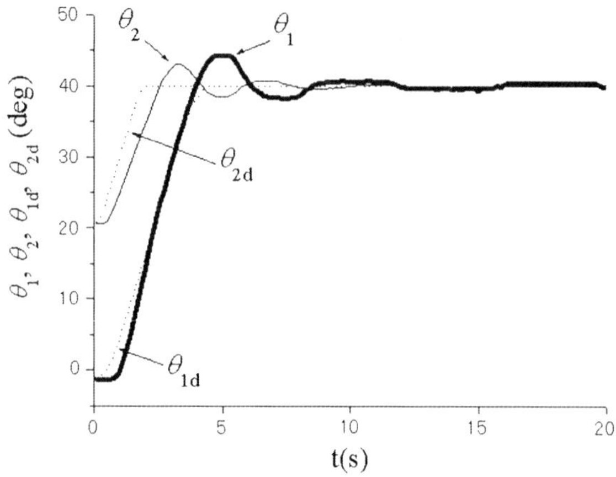

Fig. 16 Response of swing angles (simultaneous control of rotary and derricking, without swing control)

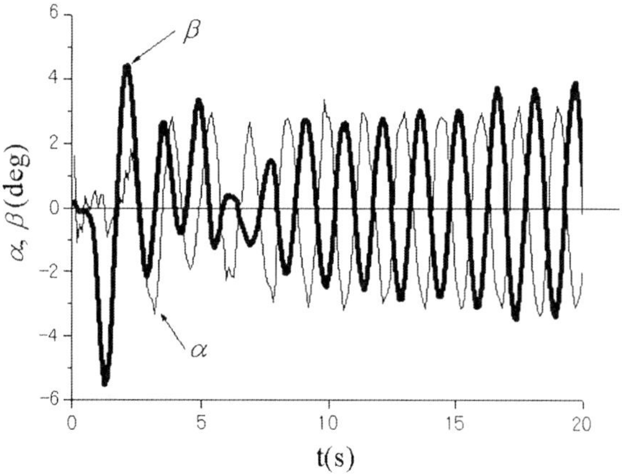

Fig. 17 Response of stereo vision angles (simultaneous control of rotary and derricking, without swing control)

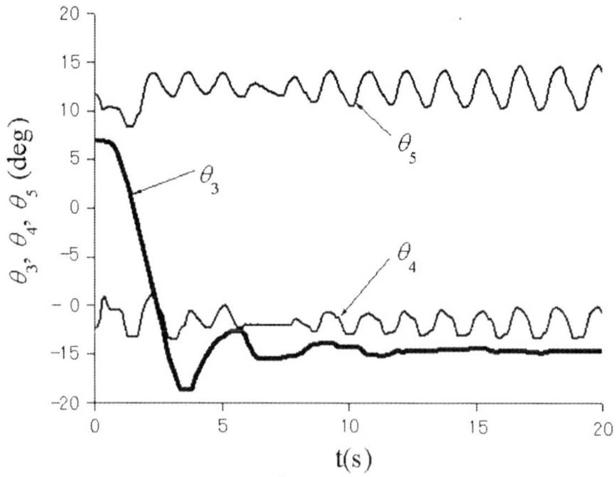

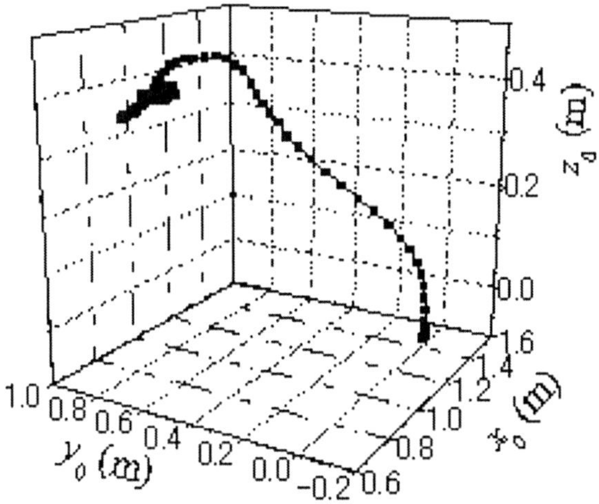

Fig. 18 Measured 3D position of a crane load by stereo vision (Simultaneous control of rotary and derricking, without swing control)

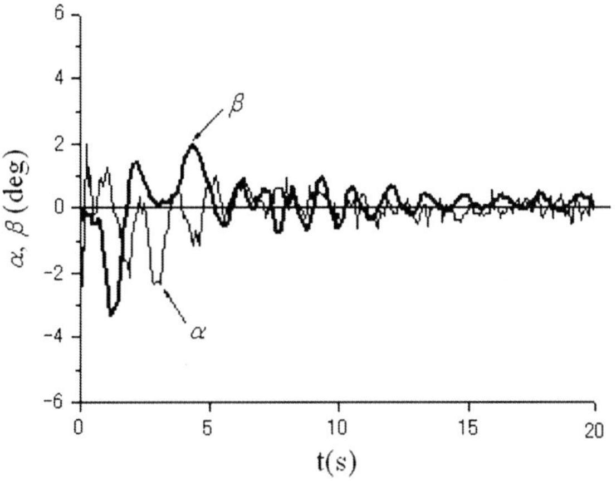

Fig. 19 Response of swing angles (simultaneous control of rotary and derricking, with swing control)

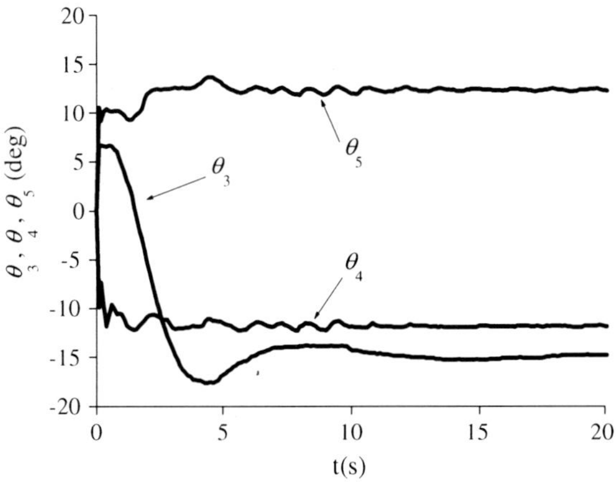

Fig. 20 Response of stereo vision angles (simultaneous control of rotary and derricking, with swing control)

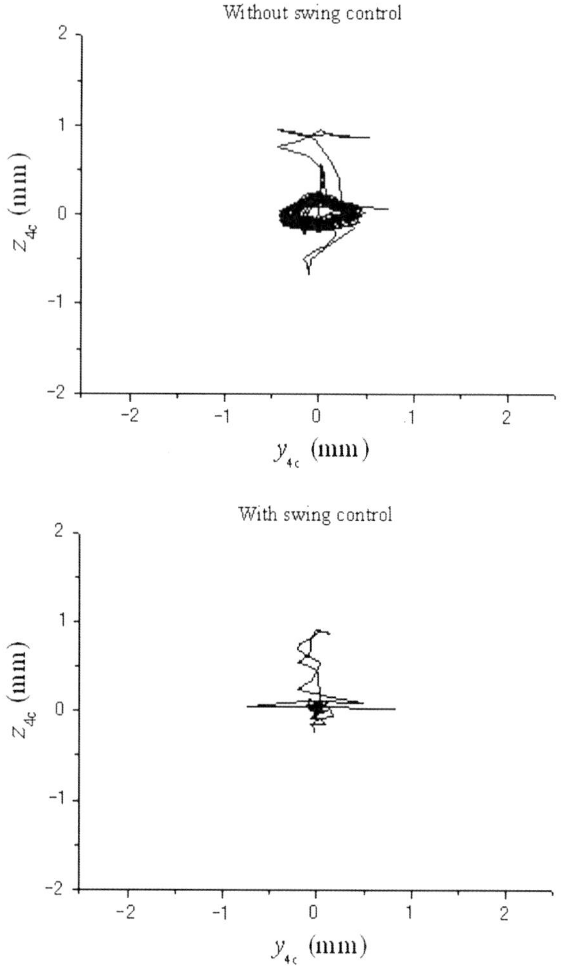

Fig. 21 CCD images of the left camera of stereo vision (simultaneous control of rotary and derricking, with and without swing control) **a** without swing control **b** with swing control

3.3 Simultaneous Control of Rotary and Derricking

From Fig. 15 to 18, experiments of simultaneous control of rotary and derricking are shown without swing control. Figure 15 is the response of rotary and derrick angles θ_1, θ_2 for target angles that are θ_{1d} from 0 to 40 degrees and θ_{2d} from 20 to 40 degrees. Figure 16 shows swing angles α, β without swing control and both swing amplitudes occur about 4 degrees. Figure 17 is the response of stereo vision angles. Stereo vision derrick angle θ_3 move to upward according to boom derrick angle θ_2 upward. Rotary angles θ_4, θ_5 move according to the crane load swing as in Fig. 16. Therefore, we understand stereo vision tracks the crane load using gaze. Figure 18

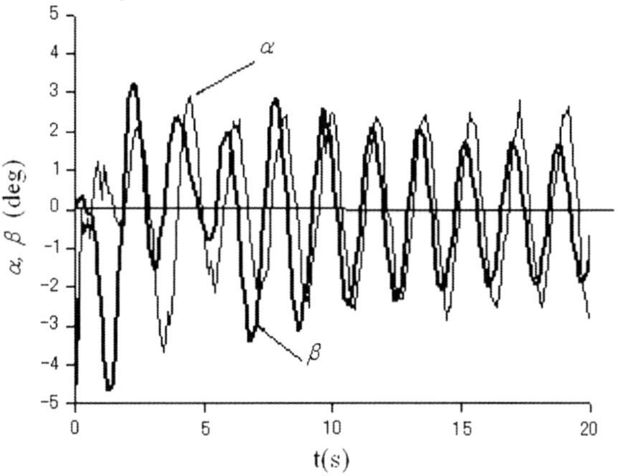

Fig. 22 Response of swing angles (simultaneous control of rotary and derricking, without swing control and rope length control)

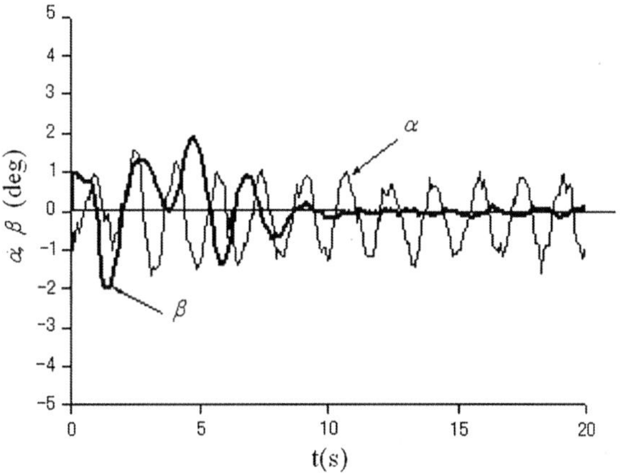

Fig. 23 Response of swing angles (simultaneous control of rotary and derricking, with swing control and rope length control)

shows the measured three dimensional position of the crane load by stereo vision. The position of the load is moved with rotating, lifting and swinging.

Figure 19 shows swing angles α, β with swing control and both swing amplitudes decrease to less than 1 degree. Figure 20 shows angles of stereo vision with swing control. As shown in Fig. 14, swing amplitude of the load is small; therefore rotary angles of stereo vision which tracks the swing of the load move smoothly.

Figure 21 shows CCD images of the left camera of stereo vision. Upper image of without swing control shows small circles which are caused by the delay of gaze control that tracks load swing and lower image with swing control shows no circle that means small swing of the load. Figures 22 and 23 show responses of swing angles with rope length control that is changing from 0.63 to 0.80 m and other experimental conditions are same as simultaneous control of rotary and derricking. Figure 22 is response of swing angles without swing control and Fig. 23 with swing control shows swing amplitudes decrease to small. The swing control is also effective in the case where rope length changes.

4 Conclusions

Gaze-controlled stereo vision and control of a boom crane using stereo vision were studied. The experimental results are as follows; A stereo vision installed on the crane's rotary axis that has gazing motion to track the suspended load was developed. The three dimensional position and the swing angle of the load are measured using this tracking stereo vision. The experimental results are presented concerning with gazing/tracking controls of the stereo vision, derricking/rotary positioning controls of the boom crane, and swing suppression control of the load. The control system of a boom crane using installed stereo vision was effective to boom positioning and load swing suppression.

References

1. L.E. Weiss, A.C. Sanderson, C.P. Neuman, Dynamic sensor-based control of robots with visual feedback. IEEE J. Robot. Autom. **RA-3**(5), 404–417 (1987)
2. N.P. Papanikopoulos, P.K. Khosla, K. Kaneda, Visual tracking of a moving target by a camera mounted on a robot: a combination of control and vision. IEEE Trans. Robot. Autom. **9**(1), 14–35 (1993)
3. P.K. Allen, A. Timcenko, B. Yoshimi, P. Michelman, Automated tracking and grasping of a moving object with a robot hand-eye system. IEEE Trans. Robot. Autom. **9**(2), 152–165 (1993)
4. G.D. Hager, A modular system for robust positioning using feedback from stereo vision. IEEE Trans. Robot. Autom. **13**(4), 582–595 (1997)
5. E. Samson, D. Laurendeau, M. Parizeau, s. Comtois, J. F. Allan, C. Gosselin, The agile stereo pair for active vision. Mach. Vision Appli. 17, 32–50 (2006)
6. Y. Ducrocq, S. Bahrami, L. Duvieubourg, F. Cabestaing, An effective active vision system for gaze control. $4th International Symposium on Advances in Visual Computing$, Las Vegas, USA, (2008)
7. E.M. Abdel-Rahman, A.H. Nayfeh, Z.N. Masoud, Dynamics and control of cranes: a review. J. Vibr. Control **9**, 863–908 (2003)
8. P.I. Corke, *Visual control of robot manipulators-a review, of the book Visual servoing: real-time control of robot manipulators based on visual sensory feedback* (World Scientific, Singapore, 1993)
9. S. Hutchinson, G.D. Hager, P.I. Corke, A tutorial on visual servo control. IEEE Trans. Robot. Automat. **12**(5), 651–670 (1996)
10. K. Kuhnlenz, M. Bachmayer, M. Buss, *A multi-focal high-performance vision system* (IEEE International Conference on Robotics and Automation, Orlando, Florida, USA, 2006), pp. 150–155
11. X. Clady, F. Collange, F. Jurie, P. Martinet, Object tracking with a pan tilt zoom camera: application to car driving assistance. IEEE International Conference on Robotics and Automation, Seoul, Korea **2**, 1653–1658 (2001)
12. L. Duvieubourg, S. Ambellouis, F. Cabestaing, Single-camera stereovision setup with orientable optical axes. Comput. Imaging Vision. pp 173–178, (2005)

13. J. Morat, F. Devernay, S. Cornou, *Tracking with stereo-vision system for low speed following applications* (IEEE Intelligent Vehicles Symposium, Iatanbul, Turkey, 2007)
14. Y. Yoshida, Visual feedback control of a three-dimensional overhead crane. Proceeding ASME 7th Biennial Conference on Eng. Sys. Design and Analysis, Manchester, UK, 2004, pp 1–7
15. Y. Yoshida, M. Hirano, T. Tomida, H. Teshima, Visual feedback control of traveling crane (in Japanese with English abstracts). Trans. Soc. Inst. Cont. Eng., Japan, **41**(6), 527–532, (2005)
16. Y. Yoshida, K. Tsuzuki, Visual tracking and control of a moving overhead crane load, in Proceeding IEEE 9th International Workshop on Advanced Motion Control, Istanbul, Turkey, 2006, pp 630–635
17. Y. Yoshida, H. Tabata, Visual feedback control of an overhead crane and its combination with time-optimal control. in Proceeding IEEE/ASME International Conference on Advanced Mechatronics, Xi'an, China, 2008, pp 630–635
18. T. Inukai, Y. Yoshida, Control of a boom crane using installed stereo vision. in Proceeding IEEE 6th International Conference on Sensing Technology, Kolkata, India, 2012, pp 189–194

Chapter 5
Integrated Determination of Tea Quality Based on Taster's Evaluation, Biochemical Characterization and Use of Electronics

P. Biswas, S. Chatterjee, N. Kumar, M. Singh, A. Basu Majumder and B. Bera

Abstract The summation of the desirable/positive attributes comprising aroma/flavor, strength, colour, briskness and character of infused tea leaves represent the quality of tea. Human experts have been traditionally assessing the tea quality by eye, nose and tongue approximations, but it suffers from biasness, inaccuracy and variability. Based upon the correlation of quality with chemical composition, chemists have developed chemical methods for the determination of tea quality. This needs highly sophisticated instruments like HPLC, GC, GC-MS, LC-MS etc, costly chemicals and expert man power. Moreover, these methods are very much time consuming. Thus, to find an accurate, cheep and real time method for tea quality assessment, recently electronic sensors mimicking the human vision, nose and tongue are being implemented for determining tea quality. All these methods have their own merits and demerits, when applied/developed independently. In this book chapter, a brief description of working procedure and usefulness of E-Vision, E-Nose and E-Tongue has been explained. In addition, certain possibilities have also been cited to highlight the productiveness of bio-mimicking sensors in quality assessment methodology and in the integration of perception for various physical attributes of tea, use of biochemical analysis and electronics.

Keywords Tea quality · E-nose · E-vision · E-tongue · Algorithms · Integration · Biochemical analysis · Tea Tasters scores

---

P. Biswas (✉) · N. Kumar · M. Singh
Darjeeling Tea Research and Development Centre, Tea Board of India, Kurseong 734203, India
e-mail: prjt.bsws@gmail.com

A. Basu Majumder · B. Bera
Tea Board of India, 14 BTM Sarani, Kolkata 700001, India

S. Chatterjee
Department of Physics, University of North Bengal, Siliguri 734013, India

A. Mason et al. (eds.), *Sensing Technology: Current Status and Future Trends II*, Smart Sensors, Measurement and Instrumentation 8, DOI: 10.1007/978-3-319-02315-1_5,

1 Introduction

Tea is one of the most popular beverages in the world for over 4,000 years. More than 3 million hectares of land in many parts of the world such as Japan, Taiwan, India, Nepal, Bangladesh, Srilanka, Kenya, Iran as well as Thailand have been used for tea plantation [1]. Due to various health beneficial attributes and medicinal properties as evident by a large number of published research works, its pharmaceutical use is also increasing day by day [2]. In spite of that ever increasing market demand, the tea industry has been facing difficulties in producing as well as marketing good quality tea because of various reasons. One of the most important reasons is the lack of an advanced quality assessment and monitoring method, as the quality consciousness among the consumers has also been increased during the recent years [3]. 'Tea quality' is a versatile term, refers to consumers' overall acceptability for a type of tea brew along with its food safety values. The consumers' choice or acceptability of the drink depends on the type of tea and its physical attributes [4]. It is an accepted fact that the market prices of teas depend primarily on their flavour and liquor characters [3]. Traditionally tea is evaluated by professional tea tasters considering mainly (a) the physical appearance of the made tea and the infused leaf, (b) the aroma, (c) liquor characters viz. colour, brightness, strength and briskness of tea brew [5]. This method of tea quality assesment by tea tasters suffers from few drawbacks like inconsistency and variability and human expertise. In order to overcome these deficiencies, many biochemical aspects of tea quality determination have been given due emphasis applying modern analytical tools like high-performance liquid chromatography, gas chromatography and spectrophotometry though these are costly and time-consuming. Quality depends on the chemical composition of the green leaves and the method of manufacture of tea from green leaves to finished product by following the proper method of manufacturing process with cautious and accurate monitoring [6]. Production of black tea involves certain well-defined, sensitive processing stages governed by many parameters. As followed in the other food industries, the tea industry also needs modern and intelligent production techniques to satisfy present market demands driven by the quality conscious cosumers [7, 8]. Recently, application of electronic sensor based devices with advanced data processing methodologies have also been used for tea quality determination and given due importance. These devices have been found successful and applied to different fields of food and beverage industries, such as tomato, and coffee [9]. E-Noses have been used for the prediction of tea quality, and monitoring of black tea fermentation process [10]. Computer vision systems are also gaining huge popularity in tea industry for identification of different tea, colour characterization and for monitoring fermentation process because of their cost effectiveness, consistency, superior speed and accuracy. However, the methods used so far for tea quality determination have their own merits and demerits as they have been developed as well as applied independently [11].

2 Factors Determining Tea Quality

Quality of tea starts getting defined right from the crop level with different types of plucking and its interval of plucking, ideally, only the bud with next two leaves are to be plucked [6]. Different processing stages, during manufacturing, affect the quality of the product. The black tea manufacturing consists of the following steps—withering, rolling, cutting-tearing-curling (for only CTC teas), fermentation and drying. Withering is mainly targeted to make the leaves permeable by removing moisture partially at ambient conditions; a strong airflow under shaded condition is usually used. During withering moisture content reduces to 30–40 % from initial value of 80 %. Some chemical changes also occur during withering which affects quality. In the next stage withered leaves are rolled to rupture the cell walls making the cell sap a homogenous mixture of all chemical components and bringing them in contact of enzymes. During fermentation step, enzymatic oxidation leads to chemical changes in tea leaves like polyphenols get oxidized; proteins get degraded; chlorophyll is transformed into pheophytins and pheophforbides; formation of some volatile organic compounds from lipids, amino acids, carotenoids and terpenoids also takes place. This process results in changing the colour of leaves from green to deep coppery red due to the formation of two groups of coloury compounds viz. Theaflavins (TF) and Thearubigins (TR) and manifestation of aroma starts [12].

The TF are golden yellow and TR are reddish-brown in colour. These two compounds together impart the characteristic colour to the tea liquor. Out of these quality descriptors, the brightness, briskness and strength are determined by the TF contents whilst body and colour are associated with TR contents. With the advance of process time, both TF and TR increase and colour of the tea also gradually changes. If the fermentation is prolonged, TF gets converted to TR. Concentration of TF reaches a peak value corresponding to which the ratio between TF and TR becomes almost 1:10 or 1:9 and the colour of processed tea becomes coppery red. This is the optimum fermentation point. Beyond this point, the concentration of TR goes on increasing with the increase of fermentation time and the body of liquor becomes thick. Over fermented tea have 'body' and lack other desirable characteristics of good cup of tea. The optimum ratio of TF and TR contents is desired in a quality cup of tea [13, 14].

Depending upon the extent of fermentation given, teas are categorized into three major groups viz. green, oolong and black tea. During green tea manufacturing, fresh green tea leaves are steamed or panned to deactivate the enzymes, so that no fermentation can take place. Black teas are produced by giving the required degree of fermentation. Oolong tea is intermediate of the two types described earlier; this is called semi fermented tea. This variation in magnitude of fermentation imparts a large difference in physical attributes of the three types of teas. Drying is the final step of tea manufacturing; which stops the ongoing chemical reactions by deactivating the enzymes. It has been found that an effective drying increases shelf life of the teas. Several researchers have also found that generation and retention of desired aroma is dependent on the efficiency of drying [15].

For the effective determination of tea quality, it has been observed that the tone of colour, brightness and briskness are influenced by the theaflavin content, while the thearubigin content influences the judgement of the depth of colour. A fairly good correlation has been established between tasters' evaluation of these liquor characters and the quantities of theaflavins and thearubigins of black tea. The briskness in a tea liquor is essentially caused by the association of theaflavins with caffeine. A relative measure of this character can be obtained from the estimation of caffeine and the ratio of thearubigins and theaflavins. The amount of un-oxidised polyphenols and amino acids also play important role on the taste of few types of tea. Where flavour is not a factor of importance, good and bad quality teas can easily be classified by studying their TF and TR values. Flavour is another factor which becomes points of reference during assessment of tea quality. However, it is the volatile flavoury constituents (VFC) which play a crucial role in the flavour intensity/type of tea [16].

3 Present Methods of Tea Quality Determination

Teas are mostly judged for quality in commercial purposes by organoleptic study. In this method human expert judges the quality parameters by assessing the physical attributes of liquor and infused leaf of the tea. As discussed earlier, a large number of biochemical compounds of tea have been found to correlate satisfactorily with quality terms, like liquor colour, brightness, strength, flavour and price. Using these correlations various methods employing sophisticated instruments have been developed to determine tea quality through study of chemical composition. Theaflavins and Thearubigins are usually analysed to determine the liquor colour and brightness using spectrophotometer and HPLC. Caffeine and amino acids having contribution to strength of liquor are measured by both of spectrophotometric and HPLC methods. Chlorophylls and carotenes are also quantified by both of spectrophotometric and HPLC methods. For assessment of flavour character/aroma profile volatile terpenoids and non-terpenoids are analyzed by GC/GC-MS systems. Water soluble solid content, moisture content, fibre content and crude lipid contents are usually analyzed by gravimetric methods.

4 Real Time Analysis of Tea Quality

The idea of artificial reproduction of human responses to external stimuli was first published in 1943 [17]. Later on, this concept was extended to build an "electronic brain" based on neural computing. The first analytical device on these concepts was an electronic nose capable of analyzing gases [18]. Electronic tongue was built few years later and very soon it proved itself as a promising device in both quantitative and qualitative analysis of multi-component matrices [19, 20]. Thereafter numerous types of sensors, devices and data processing methodologies have been developed and applied successfully in different fields. Several new ideas and theories have been implemented to make a really intelligent artificial brain, and the process is going on. Devices have been designed for studying the chemical compositions and reactions, diagnosis of diseases, controlling environment pollution, monitoring food quality and processing and few of them can be used in monitoring crimes also.

Thus viewing the success and speed of enrichment of the technology it may be felt that it is capable to provide more precise real time quality assessment methods including intelligent processing techniques/machineries for tea industry. In this book chapter the present status of tea quality determination by both of the conventional and sensor based methods, developed so far, have been critically reviewed and a possible route has been tried to make the electronic sensor based methods more precise by successful marriage of recent developments in this technology with physical attributes of tea through understanding the chemistry behind it. The mostly used devices in that field are Electronic Vision, Electronic Nose and Electronic Tongue

4.1 *Electronic Vision Systems*

The colour and texture are the important quality parameters of processed food and its raw materials. These two parameters also play important roles in monitoring the processing techniques of food stuffs [3]. Presently in that field of quality determination of food stuffs based on colour and texture analysis, electronic vision systems are gaining day by day more importance due to its low cost, fast operation time and ease of use [21, 22].

In general an electronic vision system works through the following steps:

Image capturing → preprocessing → colour/texture analysis → classification → template matching → output of results

The set up of a typical electronic vision system generally consists of digital camera, signal conditioning unit and image capturing add on card, interfaced with a PC/laptop. A specially designed illumination arrangement is used to maintain fixed illumination at the sampling area. Usually, incandescent, fluorescent, lasers, X-ray tubes and infrared lamps are used as light sources. In modern E-Vision systems solid state charge coupled devices (CCD) are usually used as image sensors and several advanced image processing algorithms are implemented to convert the images into digit format and further processing to bring out results [3, 23, 24].

A number of new ideas and their successes regarding the colour and texture analysis by E-Vision systems have been reported till the date. Histogram Intersection technique by colour indexing for object identification through image processing on the basis of their colour, proposed by Swain and Ballard in 1991, may be considered among the most important developments in this field [25]. This method involves matching colour-space histograms of images. The method is largely different from traditional object recognition methods based on geometric properties [26]. This algorithm is remarkably robust, and accuracy of this is only slightly effected by the variations such as change in orientation, partial occlusion, change in shape, shift in viewing position, change in the scene background etc. However, it is very sensitive to the lighting conditions. This technique has many drawbacks viz. it does not preserve the spatial information in the image and also does not incorporate spatial adjacency of pixels in the image which may lead to inaccuracies. Although different images may have similar colour histograms, this technique is quite reliable for applications where colours are compared and where the prime important characteristic is the colour content only. Mojsilovic and Hu [27] defined a metric called optimal colour composition distance (OCCD), in a major extension beyond colour histogram analysis which measures the difference between two images in terms of colour composition based on the optimal mapping between the two corresponding sets of colour components called quantized colour component (QCC) set. It was claimed that the algorithm has a potential to overcome limitations of Swain and Ballard's colour indexing which has been highly sensitive to quantization boundaries in addition to having its discrimination ability dependent on the selection of colour quantization method [27]. On the other hand, texture of an image represents the spatial information contained in object surfaces and identifies the class of segments. Texture properties represent the set of informations that permits the human eye to differentiate between image regions.The features, which define the texture, include uniformity, density, coarseness, roughness, regularity, linearity, directionality, frequency and phase. There are many other defintitions of textures defined for different applications but no universally acceptable texture discrimination model is yet available. The methods of extracting textural features can be broadly categorised as statistical methods, geometrical, model-based and transform-based methods [23].

For tea particularly, both of colour and texture, which greatly influence the cash valuation, are analyzable by the E-vision systems. It has been found that tea images contain some definite patterns or textures, which computer vision technique uses to study its quality. Image processing applications such as classification, detection and segmentation are applied optimally to determine the parameters [28].

4.1.1 Liquor Character Determination

The liquor colour is one of the most important quality parameters of tea. During cash valuation, where flavour is not important, the liquor colour is considered to be the qualifying factor for quality, as in those cases the liquor colour alone can represent the other physical attributes of tea. An interesting fact regarding the liquor colour is that in few parts of the world the consumers are biased only to some particular liquor colour of tea and it varies from country to country. Furthermore like other commercial sectors some trade disputes also exist in tea. Measurements on liquor colour help a lot to monitor/rectify such problems also. But uunfortunately there is no standard method other than sensory evaluation available to study the tea liquor colour.

As stated earlier during the process of fermentation in black and oolong tea manufacture, enzymatic oxidation of tea catechins takes place leading to formation of a series of coloured chemical compounds viz. TFs and TRs, which impart characteristic colour to the black tea infusion. The variation of liquor colour from greenish yellow to deep brown or blackish brown is mainly due to the extent of TF and TR formation during manufacturing of the teas, and the formation of these two compounds are influenced by so many factors like difference in manufacturing techniques, regional agro climatic condition, seasonal effects, leaf character etc. Very recently, based upon that basic principle, few interesting findings regarding the use of E-Vision systems for standard colour measurement of tea liquor and its use for discrimination of the teas have been reported [29, 30].

A study on Darjeeling tea liquor colour by electronic vision system has revealed a wide range of color variation from greenish yellow to deep brown. The liquor images taken through E-Vision system were correlated with the tea tasters scores on infusion colour of the samples. It was found that the liquor colour of teas with highest colour scores are of greenish yellow and that of the lowest scored teas are of deep brown. The trend of gradual change of liquor colour from greenish yellow → Yellow → Yellowish brown → Deep brown was found to be linearly related with the decrease of tasters colour scores on liquor, and is represented in the Fig. 1 [11].

The study also shows that this analytical methodology may be used to characterize different grades of teas. The liquor colours of different grades of Darjeeling teas were found to vary systematically. When the images of liquor colour of four major grades were analyzed (Fig. 2) in R G B colour space, it was found that the R G B values of the liquor images decrease in a gradual manner with decreasing grade size [11].

It has also been reported that characterization of green, oolong and black teas is also possible by this technique. The basic difference between green, oolong and black tea is acquired during fermentation stage of manufacturing. During green tea

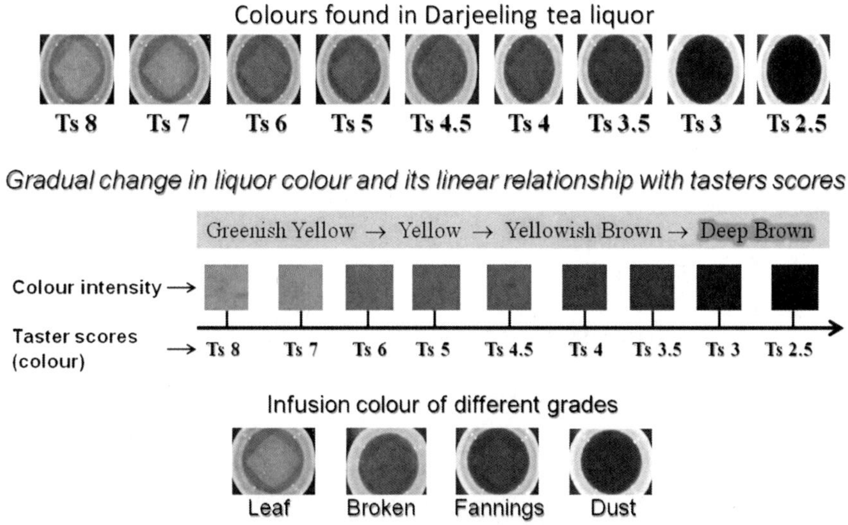

Fig. 1 Observations on liquors of Darjeeling tea [11]

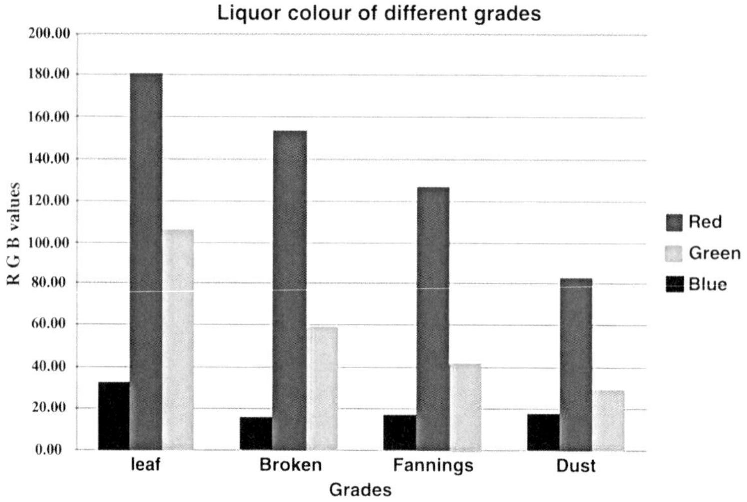

Fig. 2 Liquor colour of different grades of Darjeeling tea [11]

manufacturing fermentation is arrested by deactivating the enzymes to preserve the green colour of tea leaves. Tea shoots are steamed or roasted to deactivate various enzymes of the tea shoots to inhibit the conversion of tea polyphenols and chlorophyll. On the other hand black tea is fully fermented and oolong tea is given partial fermentation. As the coloured compounds are generated during fermentation due to oxidation of tea catechins and the measure of those compounds produced are dependent on the extent of fermentation, so the liquor colour of the three types of teas are usually different.

Liang et al. [29] applied the $L\ a\ b$ concept of colour analysis of images to characterize these three kinds of teas by E-Vision system based upon their liquor colour. The ΔL results showed that infusions of black teas were darkest and those of green teas lightest, while that of oolong teas is intermediate. According to the Δa, all the black tea infusions were red while the green tea infusions were green. The Δa could easily differentiate between green and black teas. However, oolong tea infusions were different with 38 % of samples showing red and 62 % green. The Δb showed that the infusions of the three kinds of tea were yellow, although the deepness of the yellow colour changed with various kinds of tea. The total colour difference, ΔE, was the largest for the fully fermented black tea and the smallest for the non-fermented green tea, with oolong tea intermediate. The results showed that the ΔL value decreased but the Δa, Δb and ΔE increased with the degree of fermentation of the three kinds of tea and the Δa of tea infusions is a reliable descriptor to differentiate among green and black teas [24, 29].

In the Sect. 4 we have discussed that the major difference in colour is found in green, oolong and black tea, but within the same group the difference in liquor colour also plays an important role to determine quality. In case of Darjeeling tea the quality

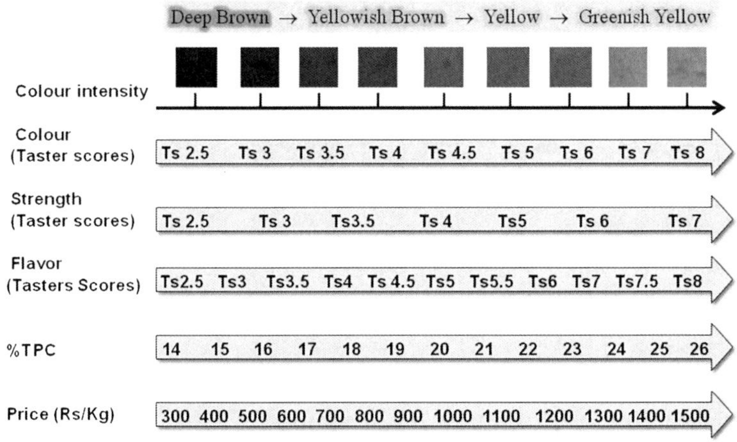

Fig. 3 Correlation of liquor color of Darjeeling tea with different quality parameters [11]

desceictors like total polyphenol content and organoleptic scores on liquor colour, strength, flavor and price have been found to corelate positively with the colour of liquors (Fig. 3). Darjeeling orthodox black teas with greenish yellow liquor colour have been found to be superior in quality and that with deep brown liquor colour was found to be inferior [11].

In the work of Liang et al. [29] it was reported that the lightness of colour represented by ΔL correlates negatively to various quality attributes of black tea but correlates positively with quality attributes of green tea and oolong tea. According to the study that quality is improved for deeper infusion colour for black tea and lighter for green or oolong tea. On the other hand, Δa, Δb and ΔE correlate negatively to quality of green and oolong tea but positively for black tea [29].

4.1.2 Grade Percentage Determination

At the final stage of tea processing the bulk material is sorted into different grades according to the size of the particles because the quality as well as price has been found to depend largely on the size. An application of E-Vision system for discrimination of different grades of CTC (cutting, tearing, and curling) tea applying Wavelet Texture Analysis (WTA) was proposed by Borah et al. [31]. This technique conjugates the feature information of one group of images along with the information of rest of the groups. This was executed by considering range of different groups of images of the same granule size. The ranges were estimated using the existing statistical texture features. Daubechies' wavelets transform (WT) based sub-band images were utilised for calculating these statistical features. For calculating the final feature set, a simplified version of Mahalanobis distance calculation was adopted. This provided the difference between two images in terms of texture variations.

For data visualisation, principal component analysis (PCA) was used to visualise the existing classes of textures and distinguishable characteristics were observed among the new feature sets. Authors claimed that the unsupervised clustering algorithm viz. self organising map (SOM) successfully classified the images efficiently into appropriate clusters. Two neural networks namely multi-layer perceptron (MLP) network and learning vector quantization (LVQ) used for texture classifications exhibited substantial classification accuracy. The results obtained by MLP and LQV outperformed the results obtained by using existing statistical texture features. This algorithm mainly focused on analysis using tailor made uniform sized samples and did not consider other attributes such as presence of stalk and bold, etc., in the tea sample [31].

4.1.3 Determination of Optimum Fermentation

Fermentation is the most important step in black and oolong tea manufacturing. Incompletely fermented teas are considered to be dull in taste and over fermented teas loose strength thereby affecting the quality. The colour of the tea shoots changes from green to coppery brown during fermentation. Traditionally human experts monitor the extent of fermentation by eye estimation of the developed colour. But sometimes it suffers from inconsistency and inaccuracy. During the past decade several research works have been done on application of electronic vision to develop a standard method for monitoring the fermentation process. A number of colour matching algorithms have been reported for image retrieval purposes where colour of the images are treated as a prominent parameter and these techniques may prove effective in image colour matching during tea fermentation and classification [27, 32, 33]. A histogram comparison method for colour comparison was developed by Borah (2003) [34] and applied to fermenting tea images. Histogram comparison method for finding the dissimilarity pixel value (DPV) for R, G, and B colour spaces of images of fermenting tea leaves using the standard colour image of well fermented tea leaves (deep coppery red in colour) has given promising results. Prominent variation of DPV for the R colour space shows a good correlation between the sensory panel judgment and the computer-aided approach and a poorer correlation have been found for other colour spaces (G and B). The images having R values bellow dissimilarity threshold pixel level indicates the completion of the fermentation process [34].

Borah et al. (2002) [35] applied a two layer Artificial Neural Network model with Perceptron learning algorithm for training phase for studying tea fermentation process by image matching technique. The Hue and Saturation data set of the images were used as the inputs to the feature vectors and outputs were the decision—Similar and Dissimilar. Based upon sensory panel decision, two sets of image data were used, one for the standard colour images (Similar) and the other is for the different (Dissimilar) consisting under fermented and over fermented samples. A satisfied level of accuracy was obtained for several images, when tested with this ANN-based computer aided approach [35].

Table 1 Physical changes in the sensor active films and the sensor devices used to transduce them into electrical signals [36]

Physical changes	Sensor devices
Conductivity	Conductivity sensors
Mass	Piezoelectric sensors
Optical	Optical sensors
Work function	MOSFET sensors

4.2 Electronic Nose systems

The E-nose system is a fast and cost effective solution to the problems associated with both of the conventional methods of tea quality determination based on organoleptic study and chemical analysis. The system consists of an array of gas sensors which record the characteristic response for a particular odour and the pattern recognition methodologies, in form of software to identify the odour using the sensor response data set [36].

The E-Nose systems are usually designed to mimic the functioning of mammalian olfactory systems. A typical E-Nose device consists of an array of sensors, signal conditioning unit and pattern recognition software. Using different sampling techniques like head space sampling, diffusion methods, bubblers or pre-connectors the odour molecules are carried on to the sensor array. The sensing materials upon contact with the odour molecules suffer some physico-chemical changes, leading into the change in their electrical properties such as conductivity. Each sensor in the array responds to different odour molecule in a varying degree and results into difference in the responses. The total set of responses from all of the sensors are then transduced into electrical signals and used for pattern recognition after pre-processing and preconditioning [37, 38].

In modern e-nose systems a variety of sensors like Conductivity sensors, Conducting polymer composite sensors, Intrinsically conducting polymers, Metal oxide sensors, Piezoelectric sensors, Surface acoustic wave sensor, Quartz crystal microbalance sensor, Optical sensors, Metal-oxide-semiconductor field-effect transistor sensor etc. are usually used in the array depending upon the chemical composition of the odour/odour mixtures. Gas molecules interact with sensors and forms a thin or thick films on the sensor materials by adsorption or chemical reactions. These changes in the sensing materials are measured as electrical signals and considered as responses. The most common types of changes utilised in e-nose sensor systems are shown in Table 1 along with the classes of sensor devices used to detect these changes.

In the first stage of odour analysis by E-Nose system a reference gas, usually inert gases like nitrogen are flushed through the array of sensors, and then the odour is allowed to reach the sensors. After data recording is completed the reference gas is again flushed through the sensors. The reference gas is flushed before and after the analysis to prepare the base line of responses. The period of time when sensors are exposed to odour is called response time, and the time when reference gas is passed through it, is called the recovery time. Once the responses are recorded, it is then processed for pattern recognition after the base line correction [37].

The basic data analysis techniques employed for pattern recognition in E-Nose system fall into three major types:

1. **Graphical analysis:** Bar chart, profile, polar and offset polar plots.
2. **Multivariate data analysis (MDA):** Principal component analysis (PCA), canonical discriminate analysis (CDA), featured within (FW) and cluster analysis (CA).
3. **Network analysis:** Artificial neural network (ANN) and Radial basis function (RBF).

The choice of method depends on the type of responses obtained and purpose of its use [39].

4.2.1 Quality Determination of Tea by E-Nose Systems

In 2008 Borah et al. [38] proposed that Neural Network based E-Nose, comprising of an array of four tin-oxide gas sensors, can assist tea quality monitoring during quality grading. principal component analysis (PCA) was used to visualise the different aroma profiles. In addition, K-means and Kohonen's self organising map (SOM) cluster analysis was done. multi layer Perceptron (MLP) network, radial basis function (RBF) network, and constructive probabilistic neural network (CPNN) were used for aroma classification [38].

Kashwan et al. (2003) [40] have used a metal oxide sensor (MOS) based electronic nose (EN) to analyze five tea samples with different qualities, namely, drier month, over fired drier month, well fermented normal fired in oven, well fermented over fired in oven, and under-fermented normal fired in oven using an array of 4 MOSs, each of which has an electrical resistance that has partial sensitivity to the headspace of tea, and a suitable interface circuitry for signal conditioning. The data were processed using principal components analysis (PCA), Fuzzy C Means algorithm (FCM). The use of a self-organizing map (SOM) method along with a radial basis function network (RBF) and a Probabilistic Neural Network classifier was also explored. Using FCM and SOM feature extraction techniques along with RBF neural network 100 % correct classification was achieved for the five different tea samples with different qualities viz. over-fermented, overfired, under-fermented, etc. [40].

Chen et al. [41] proposed support vector machine algorithm to be superior to KNN, and ANN model for green tea quality gradation by E-nose with MOS sensors, and it could be successfully used in discrimination of green tea's quality. The study was under taken for classification of four different grades of green tea classified by human sensor panel. The principal components (PCs), as the input of the discrimination model, were extracted by principal component analysis (PCA) and three different linear or nonlinear classification tools viz., K-nearest neighbors (KNN), artificial neural network (ANN) and support vector machine (SVM) were compared in developing the discrimination model [41].

Table 2 Gradient of each sensor calculated from individual response while placed in VOCs coming out from heated tea samples

Sensors	S1	S2	S3	S4	S5	S6	S7	S8
Response	R1	R2	R3	R4	R5	R6	R7	R8
Gradient	19.29	6.81	26.23	−16.66	−4.10	−0.22	−12.59	−8.04

Yu et al. [42] employed the electronic nose (E-nose) for identification of quality grades of green tea applying BPNN, PNN and cluster analysis (CA). Principal components obtained by principal component analysis (PCA) were extracted as the inputs to the BPNN and PNN. Results of CA showed that the classification of the different green tea samples is possible using the response signals of the E nose. BP neural network was reported to give better experimental results. For the two neural networks BPNN and PNN, the classification success rates were reported 100 and 98.7 % for the training set, and 88 and 85.3 % for the testing sets respectively. As the overall results of the experiment, it was reported that the two neural networks are usable for identification of the different green tea samples [42].

Bipan Tudu et al. [43] proposed, the incremental-learning fuzzy model to classify black tea using E-Nose instrument, since incremental learning is not possible in commonly used classification algorithms. The proposed incremental-learning fuzzy model promises to be a versatile pattern classification algorithm for black tea grade discrimination using electronic nose. The algorithm has been tested in some tea gardens of northeast India, and encouraging results have been obtained [43].

Biswas et al. [11] shows the calibration of E-nose instrument, originally calibrated for using in CTC tea flavour indexing. When Tasters scores of Darjeeling Orthodox black tea samples were plotted with machine given "Normalized Aroma Index", it shows an inverse relationship with tea tasters flavour score and a substantially high scatter of data. So, a function is constructed for the total sensor response (S) which is a linear sum of individual response multiplied by the individual gradient value of each sensor response.

$$S = \sum_{i=1}^{8} R_i \times M_{Si} \tag{1}$$

where M_{Si} ($i = 1–8$) is the individual response of each attributes factor, and R_i is the weight coefficient corresponding to each attribute. The gradients of individual responses, obtained from plotting each sensor responses with tasters flavour score, are taken as the weight coefficient of that sensor element (Table 2).

The resulting function is plotted with the Taster's scores obtained for Darjeeling orthodox black tea and a very good linear relationship is observed with very small scattered data exhibiting a significantly high value of the gradient.

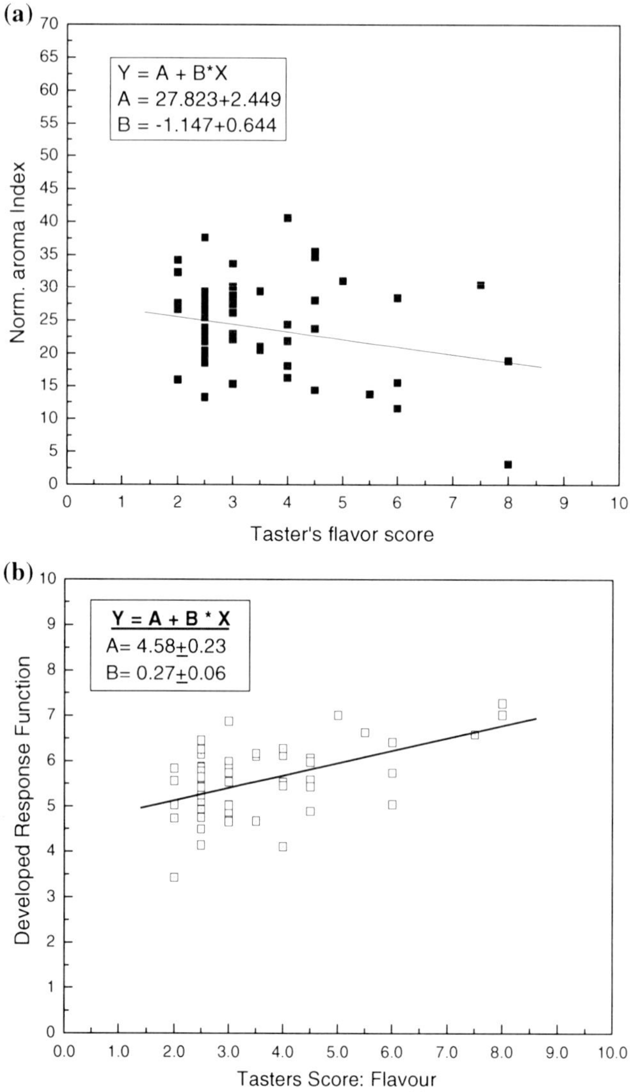

Fig. 4 **a** Relationship of the sensor responses with tasters scores on flavor for CTC black tea; **b** Calibration of the sensors using a function built from individual sensor responses for Orthodox black tea

4.3 Optimum Fermentation Time Detection Based on Aroma Generation

Black tea fermentation is essentially an oxidation process. During this process, the tea leaves change colour from green to coppery brown and the grassy smell gets transformed to floral smell. It is critical that the leaves be allowed to ferment only up to the desired limit in pursuit of maximum aroma development as both under and over fermentation result in deteriorated quality of black tea affecting the flavour index [44]. Out of the two detectable parameters (colour and aroma), aroma is very important since a strong and unique fragrance emanates from the leaves once leaves are optimally fermented.

Bhattacharya et al. (2007a) [45] carried out a study on electronic nose-based monitoring of volatile emission pattern during black tea fermentation process and detection of the optimum fermentation time on the basis of peaks recorded in the sensor outputs. A good correlation was found between biochemical test of fermentation and E-Nose study. During colorimetric biochemical test to determine the optimum fermentation time two peaks in absorbance versus time plot are obtained, among which the second one corresponds to the optimum fermentation time. Further it was reported that the plot of E-Nose responses against time also shows two peaks and the second peak corresponds to the optimum fermentation period of tea [45]. In another experiment, Bhattacharya et al. (2007b) [46] applied two methods namely the 2-Norm method (2NM) and the Mahalanobis distance method (MDM). These results were correlated with the results of colorimetric tests and human expert evaluation and a good correlation was found [46]. In the year 2008 these workers applied five different time-delay neural networks (TDNNs), named as multiple-time-delay neural networks (m-TDNN) to find out the optimum fermentation time of tea using E-Nose. They also investigated the possibility of existence of different aroma stages during the fermentation runs of black tea processing using self-organizing map (SOM), and then used three TDNNs for different aroma stages. The results were promising and the E-Nose setup used for the study, were recommended for use in the tea industry [47].

The experiments discussed above were employed in various seasons at various locations of India. Most of the Electronic Nose readings were matching accurately with colorimetric as well as human panel data regarding optimum fermentation time. The success of the experiments, mentioned above, promises that the electronic nose can be used for determining the completion of black tea fermentation by monitoring volatile emission pattern with a very high degree of accuracy, reliability and repeatability. Electronic nose-based new method as discussed above may be highly beneficial for black tea processing dispensing with human expert dependence and cumbersome chemical or indirect tests which are invasive and offline.

4.4 Electronic Tongue

An electronic tongue is defined as a bio-mimicking sensor based analytical device consisting of an array of nonspecific/low selective, chemical sensors with considerable stability and cross-sensitivity to different compounds in solution, and an appropriate method for pattern recognition and/or multivariate calibration of data processing [48].

The main elements of an electronic taste-sensing system are sensor array, signal conditioning/processing unit and pattern recognition. Sensors, when interact with the molecules in the test solution, a potential change is initiated at the sensor surfaces and signals are generated. The signals obtained are recorded by the computer and compared with the taste patterns, already existing in the computer memory.

Electronic tongue generally uses nonspecific or low specific sensors mimicking the non specific taste systems in mammals. Till now mainly three types of devices have been developed, based on potential, impedance spectroscopy, and voltametry [20]. Several new techniques have been applied for betterment of sensor arrays such as screen printing for thick-film and electron beam evaporation, thermal vacuum deposition and pulsed laser deposition for thin-film technique [49], micro-fabrication techniques to prepare a sensor array for use in a voltametric e-tongue by depositing gold (Au), platinum (Pt), iridium (Ir) and rhodium (Rh) on a silicon wafer [50] and colorimetric sensor array for non-volatile molecules that are still under development. These are made by immobilizing nano-porous pigment onto porous modified silicate and printing on a hydrophilic membrane [51]. A lipid membrane is often used as a recognition element for changing taste relevant substances into electric potential charge across the membranes [52].

4.4.1 Tea Quality Determination by E-Tongue

An application of electronic tongue coupled with multivariate data analysis in the analysis of liquid phase foodstuffs has been increasing quickly in the last decade [53, 54]. The Electronic tongue has been attempted to classify tea categories. The influence of different applied waveforms on discrimination ability between green and black teas by means of pulse volumetric electronic tongue has been investigated [55]. An e-tongue with multivariate calibration techniques (PCA-ANN) was found able to quantify catechins and caffeine in green tea in contrast to reverse phase HPLC [56].

He et al. [58] reported that a potentiometric sensor array in combination with principal component analysis (PCA) can classify tea samples from different geographical origins and quality types based upon the study of different sensory attributes in tea taste. A good correlation among instrumental results and tea tasters assessments was also found, which signifies that the instrument can be used to study the taste properties of teas [57].

Astringency and umami tastes are imparted in tea mainly due to tea-tannis and amino acids respectively. Chang et al. (2010) attempted to measure these two tastes by e-tongue technique employing the two reactions viz. quenching the fluorescence of 3-aminophthalate by tannin and the fluorogenic reaction of o-phthalaldehyde (OPA) with amino acids. A fluorescence sensing system with same excitation and emission wavelengths (340/425 nm) was used as the device to study both of the reactions. Partially fermented oolong teas with different flavour categories were successfully distinguished using ratio of umami to astringency as a taste indicator [59].

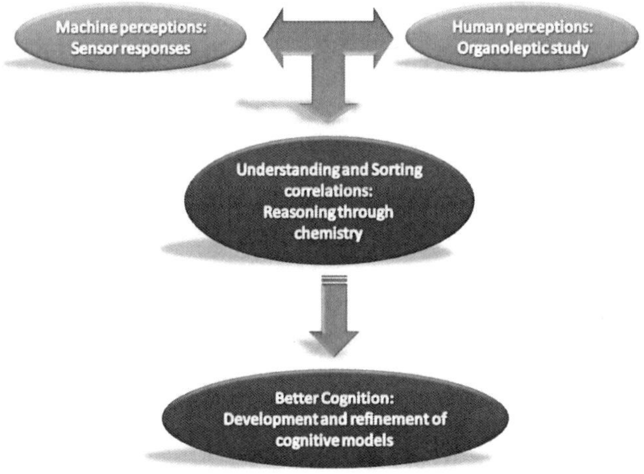

Fig. 5 Integration of methodologies

An application of disposable all-solid-state potentiometric electronic tongue micro-system for the discrimination of Korean green tea was reported. An amperometric electronic tongue was also employed as a sensors array system for the detection of flavor related phenols [53].

Silicon transistor sensors with organic coating and an Ag/AgCl reference electrode have also been found to be successful in predicting sensory properties and flavour qualities of tea [60].

5 Further Challenges and Possibilities for Real Time Tea Quality Monitoring Systems

Though the real-time methods have so many benefits over the traditional methods but these are not as accurate as the lab based methods. So development in the area of new sensing systems and allied techniques are still needed. Works are underway to develop learning techniques to standardize obtained results with reference to human perceptions of tea. Advanced data processing methodologies like Principal Component Analysis, Multiple Linear Regressions, Artificial Neural Networks, Fuzzy systems etc. are worth mentioning in this regard. New approaches like fusion of sensors and integration of chemical studies of the object under study, through understanding and sorting the correlations during selection of sensors and data processing methodologies may be suggested in this regard (Fig. 5).

As an example, the results obtained by E-Vision studies on tea liquor colour coupled with chemical analysis may assist remarkably the E-Tongue and E-Nose studies/developments. As discussed earlier, golden yellow coloured Theaflavins and rusty brown coloured Thearubigins formed during fermentation impart characteristic colour to black tea liquor. It has also been reported that these Theaflavins and Thearubigins play an important role in determining cash valuation of the CTC black teas. Theaflavins in combination with Thearubigins determine the colour and the strength, and in combination with Caffeine determine the briskness of the black tea liquor. Moreover, liquor characters are also derived upon the relative proportion of un-oxidized polyphenols to that of the oxidized products i.e. Theaflavins and Thearubigins. The black teas with high proportion of the former have been reported to be very brisk, whereas in the opposite case, the teas are said to lack briskness and usually termed as dull [5, 6]. Thus, it may be concluded that the relative proportion of these compounds greatly affects the quality of the black teas.

Now, if we combine the three facts—(i) Theaflavins and Tearubigins are mainly responsible for varying intensity of black tea liquor (Fig. 6), (ii) Quantity of un-oxidised polyphenols in black tea is inversely proportional to the Theaflavins and thearubigins content as polyphenols only generate Theaflavins and Thearubigins after getting oxidized and (iii) Caffeine is a colourless compound, then a hypothesis easily can be made that a gross estimation of colour intensity of black tea liquor can give the ideas about its strength, in addition to appearance. When the pertinency of the hypothesis was tested for Darjeelig tea, it was found that not only strength but flavor also increases substantially with increase in the ratio of Theaflavins (TF) to Thearubigins (TR). The results are plotted in Fig.7a, b.

A study on Kangra orthodox black tea by Akuli et al. [24, 61] represented that E-Vision system can definitely be used for studying tea quality parameters like TC, TB, TR, TF/TR of orthodox black tea samples. Standard colour models like HSI

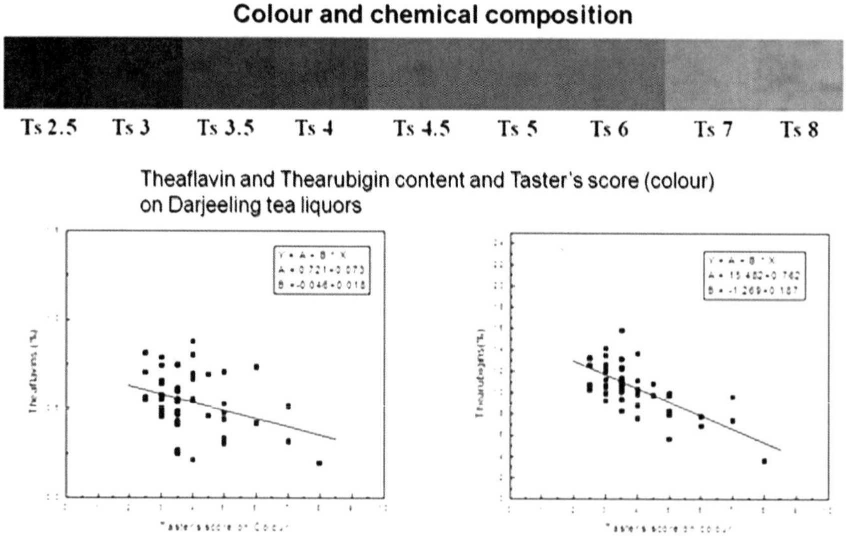

Fig. 6 Relation of Theaflavin and Thearubigin with intensity of liquor colour

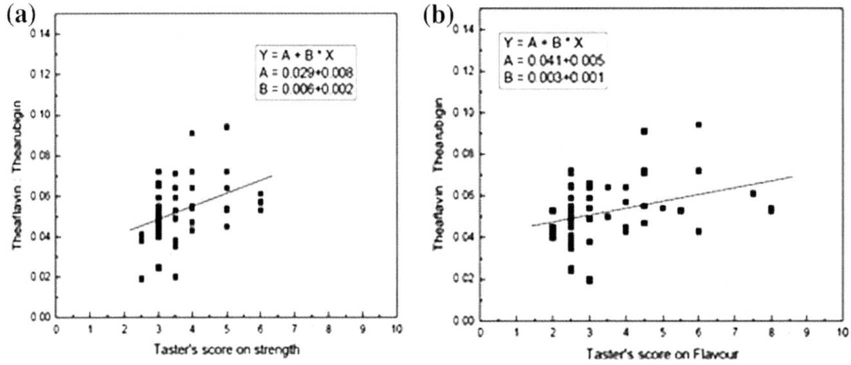

Fig. 7 **a** Relation of TF to TR ratio with strength and **b** relation of TF to TR ratio with flavor

Table 3 Comparison of different accuracy level for biochemical parameters

Modelling technique	Parameter	Accuracy	(Average absolute error)
MLR	Tea colour	75	0.100
MLR	Brightness	82	5.000
MLR	Thearubigins	75	0.400
MLR	TF/TR	770.022	

(Hue, Saturation, Intensity), CIE L*a*b etc were used to analyze the images for input data to the PCA. The data was analyzed by MLR data analysis technique for different quality attributes of tea like TC, TB, TR, TF/TR and matched with biochemical analysis. PCA analysis was done on the sample image data to find out the hidden clusters with TB, TC, TR and TF/TR ratio. The input images analyzed using HSI colour model showed formation of three distinct clusters with TC data and four distinct clusters with TR data after PCA analysis. Applying L*a*b colour model the input image analysis data showed four distinct clusters with parameter TB and two distinct clusters for TF/TR data after PCA analysis. From the PCA analysis it was found that the data set is clearly classifiable. For classification MLR technique has been applied. Table 3 shows the comparison of different accuracy level for TC, TB, TR, TF/TR analysis [24, 61].

6 Conclusions

Though the electronic sensor based techniques for tea quality determination have benefits over the traditional methods, but the use of the traditional methods are still predominant because of the lower accuracy of the electronic systems. So many facts are left till now to develop to make an electronic instrument which will satisfy every concern of R&D and industry needs for tea quality determination. The selection of sensors, application of data processing algorithms and instrument training strategies still needs a careful research and developments for electronic systems, so that it can give precise results with high resolution, capable of categorizing the teas from different region, planting material, season, manufacturing practices etc.

Since tea quality is a summation of various attributes derived mainly from appearance, liquor color, taste, mellowness, astringency, flavour and aroma, the electronic devices viz; E-Nose, E-Vision, E-Tongue once developed may be integrated into a single system. Such a system developed and trained by an integrated approach of human and electronic perception analysis guided by chemical investigations may provide an accurate, cost effective and simple tool for objective assessment of tea quality.

Acknowledgments Authors are thankful to CDAC, Kolkata for providing the instrument E-Nose and E-Vision to Darjeeling Tea Research and Development Center.

References

1. R. Ravichandran, R. Parthiban, The impact of processing techniques on tea volatiles. Food Chem. **62**, 347–353 (1998)
2. N. Khan, H. Mukhtar, Tea polyphenols for health promotion. Life Sci. **26**(81(7)), 519–533 (2007)
3. S. Borah, M. Bhuyan, A computer based system for matching colours during the monitoring of tea fermentation. Int. J. Food Sci. Technol. **40**, 675–682 (2005)
4. K.I. Tomlins, A. Mashingaidze, Influence of withering, including leaf handling, on the manufacturing and quality of black teas—a review. Food Chem. **60**(4), 573–580 (1997)
5. D.N. Barua, D.N. Borbora, S.B. Deb, R. Choudhury, B. Banerjee, What is quality? Two and a Bud **13**(4), 150 (1966)
6. S. Bhatia, Composition of leaf in relation to liquor characters of made tea. Two and a Bud **32**(1&2), 1–4 (1985)
7. P.K. Mahanta, Biochemical basis of colour and flavour of black tea, in *30th Tocklai Conference*, pp. 124–134 (1988)
8. P.K. Mahanta, M. Hazarika, Chlorophyll and degradation products in orthodox and CTC black teas and their influence on shade of colour and sensory quality in relation to thearubigins. J. Sci. Food Agric. **36**, 1133–1139 (1985)
9. T. Takeo, P.K. Mahanta, Why C.T.C. tea is less fragrant? Two and a Bud **30**, 76–77 (1983)
10. B.G. Kermani, S.S. Schiffman, H.T. Nagle, Performance of the Levenberg-Marquardt neural network training method in electronic nose applications. Sens. Actuators B Chem. **110**, 13–22 (2005)
11. P. Biswas, N. Kumar, M. Singh, A. Basu Majumder, B. Bera, I.B. Karki, S. Chatterjee, Correlation of tasters scores with biochemical and electronic sensor data for Darjeeling orthodox black tea, in *2012 Sixth International Conference on Sensing Technology (ICST)* pp. 769–774 (2012)
12. I.S. Bhatia, The role of chemistry in tea manufacturer. Two and a Bud **11**(4), 109–117 (1964)
13. I.S. Bhatia, Application of chemical test in manufacturing experiment. Two and a Bud **7**(4), 18 (1960)
14. I.S. Bhatia, Biochemical investigations in relation to tea manufacture. Two and a Bud **11**(1), 8–15 (1964)
15. D.J. Wood, Manufacture in relation to regional environment. Two and a Bud **5**(1), 14 (1958)

16. P.K. Mahanta, M. Hazarika, Improve flavour quality assured. Two and a Bud **32**(1 and 2), 25–29 (1985)
17. Y. Vlasov, A. Legin, A. Rudnitsaya, C.D. Natale, A.D. Amico, Nonspecific Sensor Arrays (electronic Tongue) for chemical analysis of liquids. Pure Appl. Chem. **77**, 1965–1983 (2005)
18. K. Persaud, G. Dodd, Analysis of discrimination mechanisms in the mammalian olfactory system using a model nose. Nature **299**, 352–355 (1982)
19. A. Legin, A. Rudnitskaya, C.D. Natale, E. Mazzone, A.D. Amico, Application of electronic tongue for qualitative and quantitative analysis of complex liquid media. Sens. Actuators B Chem. **65**, 232–234 (2000)
20. P.K. Lakshmi, Electronic tongue: an analytical gustatory tool. J. Adv. Pharm. Technol. Res. **3**(1), 3–8 (2012)
21. M.S.M. Alfatni, A.R.M. Shariff, H.Z.M. Shafri, O.M.B. Saeed, O.M. Eshanta, Oil palm fruit bunch grading system using red, green and blue digital numbers. J. Appl. Sci. **8**, 1444–1452 (2008)
22. B.S. Anami, V.C. Burkpalli, Colour based identification and classification of boiled food grain images. Int. J. Food Eng. **5**(5), 1556–3758 (2009)
23. G.S. Gill, A. Kumar, R. Agarwal, Monitoring and grading of tea by computer vision—a review. J. Food Eng. **106**, 13–19 (2011)
24. R. Akuli, T. Joshi, A. Dey, A. Pal, N. Gulati, Bhattacharyya, A new method for rapid detection of total colour (TC), theaflavins (TF), thearubigins (TR) and brightness (TB) In orthodox teas, in *2012 Sixth International Conference onSensing Technology (ICST)*, pp. 23–28 (2012)
25. M.J. Swain, D.H. Ballard, Colour indexing. Int. J. Comput. Vision **7**, 11–32 (1991)
26. B.V. Funt, G.D. Finlayson, Colour constant colour indexing. IEEE Transaction on PAMI **17**, 522–529 (1995)
27. I. Mojsilovic, J. Hu, A method for colour content matching of images, in *2000 International Conference on Multimedia and Expo (ICME)*, pp. 649–652 (2000)
28. M. Pietikainen, T. Ojala, Z. Xu, Rotation invariant texture classification using feature distributions. Pattern Recogn. **33**, 43–52 (2000)
29. Y. Liang, J. Lu, L. Zhang, S. Wu, Y. Wu, Estimation of tea quality by infusion colour difference analysis. J. Sci. Food Agric. **85**, 286–292 (2005)
30. L.F. Wang, J.Y. Lee, J.O. Chung, J.H. Baik, S. So, S.K. Park, Discrimination of teas with different degrees of fermentation by SPME-GC analysis of the characteristic volatile flavour compounds. Food Chem. **109**, 196–206 (2008)
31. S. Borah, E.L. Hines, M. Bhuyan, Wavelet transform based image texture analysis for size estimation applied to the sorting of tea granules. J. Food Eng. **79**, 629–639 (2007)
32. C. Colomboa, A. Bimbo, Colour-induced image representation and retrieval. Pattern Recogn. **32**, 1685–1695 (1999)
33. S.L. Michae, S. Nicu, S.H. Thomas, Improving visual matching, in *International IEEE Conference on Computer Vision and Pattern Recognition (2000)*, pp. 2058–2065 (2000)
34. S. Borah, M. Bhuyan, Non-destructive testing of tea fermentation using image processing. INSIGHT- non-destructive testing and condition monitoring. J. British inst. Nondestr. Test. **45**, 55–58 (2003)
35. S. Borah, M. Bhuyan, H. Saikia, ANN based colour detection in tea fermentation, in *presented at ICVGIP 2002: 3rd Indian Conference on Computer Vision, Graphics and Image Processing*, Ahmadabad, India, December 16–18 (2002)
36. K. Arshak, E. Moore, G.M. Lyons, J. Harris, S. Clifford, A review of gas sensors employed in electronic nose applications. Sens. Rev. **24**(2), 181–198 (2004)
37. T.C. Pearce, S.S. Schiffman, H.T. Nagle, J.W. Gardner, *Handbook of Machine Olfaction* (Wiley-VCH, Weinheim, 2003)
38. S. Borah, E. L. Hines, M. S. Leeson, D. D. Iliescu, M. Shuyan, J. W. Gardner, Neural network based electronic nose for classification of tea aroma. Sens. Instrumen. Food Qual. **2**(1), 7–14 (2008)
39. E. Schaller, J. O. Bosset, F. Esher Electronic noses and their application to food. Lebensm.-Wiss.Ul.-Technol. **31**, 305–316 (1998)

40. K.R. Kashwan, M. Bhuyan, E.L. Hines, J.W. Gardner, R. Dutta, Electronic nose based tea quality standardization. Neural Networks **16**, 847–853 (2003)
41. Q. Chen, J. Zhao, Z. Chen, H. Lin, Z.D.A. Zhao, Discrimination of green tea quality using the electronic nose technique and the human panel test, comparison of linear and nonlinear classification tools. Sens. Actuators B Chem. **159**(1), 294–300 (2011)
42. H. Yu, J. Wang, C. Yao, H. Zhang, Y. Yu, Quality grade identification of green tea using E-nose by CA and ANN. LWT Food Sci. Technol. **41**(7), 1268–1273 (2008)
43. B. Tudu, A. Jana, A. Metla, D. Ghosh, N. Bhattacharyya, R. Bandyopadhyay, Electronic nose for black tea quality evaluation by an incremental RBF network. Sens. Actuators B Chem. **138**(1), 90–95 (2009)
44. N. Kumar, P. Biswas, R. Rai, M. Sing, B. Bera, Fermentibility of tea clones popularly grown in Darjeeling hills in relation to flavour during manufacture of orthodox black tea, in *Recent Trends in Plant and Microbial Research (2013)*, p. 19, (2013)
45. N. Bhattacharyya, S. Seth, B. Tudu, P. Tamuly, A. Jana, D. Ghosh, R. Bandyopadhyay, M. Bhuyan, S. Sabhapandit, Detection of optimum fermentation time for black tea manufacturing using electronic nose. Sens. Actuators B Chem. **122**(2), 627–634 (2007a)
46. N. Bhattacharyya, S. Seth, B. Tudu, P. Tamuly, A. Jana, D. Ghosh, R. Bandyopadhyay, M. Bhuyan, Monitoring of black tea fermentation process using electronic nose. J. Food Eng. **80**(4), 1146–1156 (2007b)
47. N. Bhattacharya, B. Tudu, A. Jana, D. Ghosh, R. Bandhopadhyaya, M. Bhuyan, Preemptive identification of optimum fermentation time for black tea using electronic nose. Sens. Actuators B Chem. **131**(1), 110–116 (2008)
48. I. Bratov, N. Abramova, A. Ipatov, Recent trends in potentiometric sensor arrays—a review. Anal. Chim. Acta **678**(2), 149–159 (2010)
49. K. Twomey, E. Alvarez de Eulate, J. Alderman, D.W.M. Arrigan, Fabrication and characterization of a miniaturized planar voltammetric sensor array for use in an electronic tongue. Sens. Actuators B Chem. **140**(2), 532–541 (2009)
50. S. Iiyama, S. Ezaki, K. Toko, Sensitivity-improvement of taste sensor by change of lipid concentration in membrane. Sens. Actuators B Chem. **141**(2), 343–348 (2009)
51. V. Parraa, Á.A. Arrietaa, J.A. Fernández-Escuderob, H. Garcíab, C. Apetreia, M.L. Rodríguez-Méndeza, J.A. de Sajac, E-tongue based on a hybrid array of voltammetric sensors based on phthalocyanines, perylene derivatives and conducting polymers: discrimination capability towards red wines elaborated with different varieties of grapes. Sens. Actuators B Chem. **115**(1), 54–61 (2006)
52. C. Zhang, D.P. Bailey, K.S. Suslick, Colorimetric sensor arrays for the analysis of beers: a feasibility study. J. Agric. Food Chem. **54**, 4925–4931 (2006)
53. C.J. Musto, S.H. Lim, K.S. Suslick, Colorimetric detection and identification of natural and artificial sweeteners. Anal. Chem. **81**, 6526–6533 (2009)
54. L. Lvova, S.S. Kim, A. Legin, Y. Vlasov, J.S. Yang, G.S. Cha, All-solidstate electronic tongue and its application for beverage analysis. Anal. Chim. Acta **468**, 303–314 (2002)
55. L. Lvova, A. Legin, Y. Vlasov, G.S. Cha, H. Nam, Multicomponent analysis of Korean green tea by means of disposable all-solid-state potentiometric electronic tongue microsystem. Sens. Actuators B Chem. **95**, 391–399 (2003)
56. P. Ivarsson, S. Holmin, N.E. Hojer, C. Krantz-Rulcker, F. Winquist, Discrimination of tea by means of a voltammetric electronic tongue and different applied waveforms. Sens. Actuators B Chem. **76**, 449–454 (2000)
57. Q. Chen, J. Zhao, Z. Guo, X. Wang, Determination of caffeine content and main catechinscontents in green tea (Camellia sinensis L.), using taste sensor technique and multivariatecalibration. J. Food Compos. Anal. **23**, 353–358 (2010)
58. W. He, X. Hu, L. Zhao, X. Liao, Y. Zhang, M. Zhang, J. Wu, Evaluation of Chinese tea by the electronic tongue: correlation with sensory properties and classification according to geographical origin and grade level. Food Res. Int. **42**(10), 1462–1467 (2009)
59. K.-H. Chang, R.L.C. Chen, B.-C. Hsieh, P.-C. Chen, H.-Y. Hsiao, C.-H. Nieh, T.-J. Cheng, A hand-held electronic tongue based on fluorometry for taste assessment of tea. Biosens. Bioelectron. **26**, 1507–1513 (2010)

60. E.A. Baldwin, J. Bai, A. Plotto, S. Dea, Electronic noses and tongues: applications for the food and pharmaceutical industries. Sensors **11**, 4744–4766 (2011)
61. I. Akuli, A. Pal, A. Ghosh, N. Bhattacharyya, Estimation of theaflavins (TF) and thearubigin (TR) ration in black tea liquor using E-vision system, in *Olfaction and Electronic Nose* (2011), vol. 1362, pp. 253–254 (2011)

Chapter 6
Electronic Nose and Its Application to Microbiological Food Spoilage Screening

M. Falasconi, E. Comini, I. Concina, V. Sberveglieri and E. Gobbi

Abstract Electronic Nose (EN) is a machine designed for detecting and discriminating complex odours using an array of broadly specific chemical sensors by mimicking the working mechanism and the main building blocks of biological olfaction. ENs are valuable candidates to be applied in various areas of food quality control, including microbial contamination diagnosis. In this chapter the EN technology is presented and its exploitation for microbiological screening of food products is reviewed. Two paradigmatic examples are presented. Both advantages and drawbacks of sensor technology in food quality control are discussed. Despite of many successful results, the high intrinsic variability of food samples together with persisting limits of the sensor technology still impair ENs trustful applications at the industrial scale thus further research efforts and technology improvements are required.

Keywords Olfaction · Electronic nose · Metal oxide sensors · Food quality control · Microbial spoilage

M. Falasconi (✉) · E. Comini
Dip. di Ingegneria dell'Informazione, Università di Brescia, Via Branze 38, 25133 Brescia, Italy
e-mail: matteo.falasconi@ing.unibs.it

V. Sberveglieri
University of Modena and Reggio Emilia, Deptartment of Life Sciences, Via Amendola 2–Padiglione Besta, 42122 Reggio Emilia, Italy

E. Gobbi
Dip. di Medicina Molecolare e Traslazionale, Università di Brescia, Viale Europa, 11, 25123 Brescia, Italy

M. Falasconi · E. Comini · V. Sberveglieri · E. Gobbi · I. Concina
CNR-IDASC, SENSOR Lab, ViaBranze, 45, 25123 Brescia, Italy

A. Mason et al. (eds.), *Sensing Technology: Current Status and Future Trends II*, Smart Sensors, Measurement and Instrumentation 8, DOI: 10.1007/978-3-319-02315-1_6,

1 Introduction

The sense of smell long remained the most enigmatic of our senses. The genetic bases of biological olfaction were not understood until Richard Axel and Linda Buck discovered a family of about one thousand genes that encode olfactory receptor neurons [1]. For their pioneering studies Axel and Buck were awarded in 2004 with the Nobel Prize in Physiology or Medicine.

Olfaction is very important for every individual's quality of life and plays a central role for most living species. For example, smell is absolutely essential for a newborn mammalian pup to find the teats of its mother and obtain milk—without olfaction the pup does not survive unaided. Olfaction is also of paramount importance for many adult animals, since they observe and interpret their environment largely by sensing smell. All living organisms can detect and identify chemical substances in their environment. It is obviously of great survival value to be able to identify suitable food and to avoid putrid or unfit foodstuff. To lose the sense of smell is a serious handicap—we no longer perceive the different qualities of food and we cannot detect warning signals, for example smoke from a fire or edible food and drinks.

Electronic Nose (EN), or Electronic Olfactory Systems (EOS), is a machine that is designed for detecting and discriminating among complex odors using an array of broadly-tuned (non-specific) chemical sensors [2]. EN takes its inspiration from the working mechanism of biological olfaction. Therefore, the EN instruments comprise: (1) an odor sampling unit (*nose*) to deliver the gaseous mixture to (2) the sensor array (*olfactory receptor cells*), (3) an electronic circuitry to collect the sensor responses (*axons, olfactory bulb, mitral cells and olfactory nerve*) and (4) software tools for statistical data analysis (*brain*). An odor stimulus generates a characteristic fingerprint from the sensor array. Known odor samples are initially measured and then used to build a database for training the EN to recognize unknown samples that can be subsequently identified. Many scientific reviews have been published on this subject; a comprehensive overview on EN technology is provided for instance in the classical "Handbook of machine olfaction" [3].

Attempts to measure odors with electronic instruments were made in the early 1960s [4, 5], but the modern era of artificial olfaction began in 1982 with the work of Persaud and Dodd [6], who used a small array of gas-sensitive metal-oxide devices to classify odors. The expression "electronic nose" however appeared for the first time in 1987 [7] while its current definition was given by Gardner in 1988 [8]. Commercial instruments became available in early 1990s with original machines developed by Alpha Mos (France) and Aromascan (UK).

EN systems were designed (in principle) to be used in a broad range of fields; however, food analysis has been, and still seems to be, one of the most promising applications of EN because of its simplicity of use, low cost, rapidity and good correlation with sensory panels [9, 10]. ENs have been applied in various food contexts, such as: process monitoring, freshness evaluation, shelf-life investigation, authenticity determination, product traceability [11–17]. In fact, aroma is one of the key parameters of food and the characteristic bouquet of volatile organic compounds, the so called fingerprint, may provide information and act as an indicator of food safety and quality. Unpleasant smells (off-flavors) can also include substances originating from the metabolism of spoilage microorganisms, bacteria or fungi, which naturally or accidentally contaminate the food product [18].

Microbial contamination affects most of foodstuffs consumed in the world, often as a mandatory step of the food production chain. For instance, a residual bacterial charge is commonly accepted and even wished in some foods (for instance fermented milk and derivatives) and the fungal presence is a characteristic pursued in some cases (as for some cheese varieties and salami). Yet, in many cases the presence of unwanted microbial contaminants can be a serious problem depending on the nature and level of contamination.

Apart from health problem, microorganisms can cause unacceptable organoleptic alterations of taste and flavor of the final products, resulting in economic damages for the food producers. The availability of reliable, fast, easy to use tools for early screening of food microbiological contamination is therefore a target both for customer's safeguard and production improvement. Traditional quality control tests include microbiological and physical/chemical techniques (microbiological cell counting, gas and liquid chromatography, mass spectrometry, optical spectroscopic techniques). Although effective and accurate, these have some usual drawbacks such as: high costs of implementation, long time of analysis, low samples throughput, need of a highly qualified manpower, and cannot be used for on-line production monitoring. On the other side, trained human sensory panels, often employed for food quality assessment, are also not suitable for routine industrial controls because they suffer from lack of objectivity and reliability due to human fatigue or stress, requiring long training time and high implementation costs.

Many recent works have evidenced that ENs can be exploited to screen microbial contamination of food by analyzing the pattern of volatile compounds produced by microbial metabolism. The fingerprint variation can be due to either the appearance of new chemical compounds (primary or secondary metabolites) or to changes in the relative amount of the original volatile compounds without changes in the qualitative composition. Detection of food contamination by using standard microbial plate count methods involves time and extensive sample preparation. Also, improper sampling of the food product may give misleading results since the culture-based methods rely on the site of sampling. The use of EN can provide rapid and accurate means of sensing the incidence of food contaminant bacteria with little or no sample preparation.

The following of the Chapter aims to present an overview of EN applications in the field of food microbiology and will focus in particular on recent achievements obtained with the commercial electronic nose EOS (SACMI IMOLA scarl, Imola, Italy) that was developed in collaboration with SENSOR research group in Brescia [2].

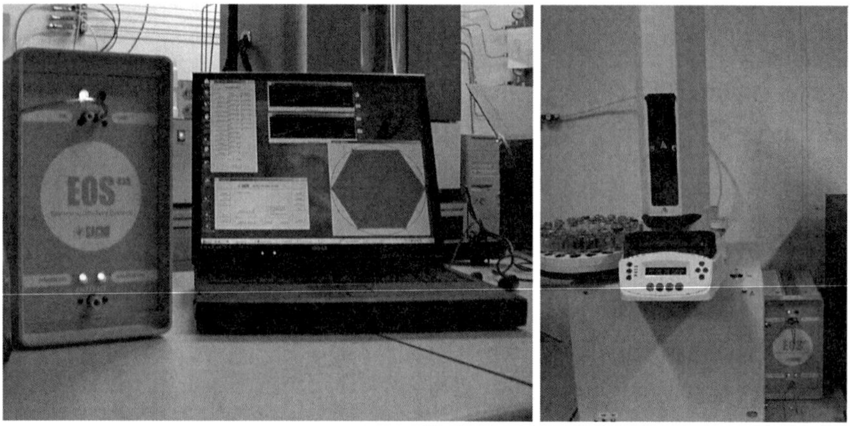

Fig. 1 EOS835 electronic nose (*left*) equipped with static headspace sampling unit HT200H (*right*)

2 EOS Electronic Nose

The EOS (Fig. 1) consists of a pneumatic assembly for dynamic sampling (pump, electro-valve, electronic flow meter), a thermally controlled sensor chamber of 20 ml internal volume, an electronic board for controlling the sensor heaters and measuring the sensing layers, and software for data acquisition and signal processing. The instrument remote control and the data acquisition can be performed by an external lap-top through standard communication port RS232. Two models are currently available: EOS835 and EOS507; the latter being a recent upgrade of the system with a humidity control device of the baseline air and a more accurate sensor read-out electronics especially developed for stand-alone applications.

2.1 Odor Sampling Systems

The EOS supports dynamic or static headspace sampling unit (optionally with an auto sampler HT200H, HTA srl, Italy) (Fig. 1). Static headspace has clear advantages in terms of reproducibility and repeatability. The HS generation parameters (incubation temperature, time and so on) can be fully and accurately controlled. Besides, the HS analysis is carried out without perturbing the equilibrium conditions—this ensures there are no artifacts in the sensor response due to changes of HS concentration during the measurement. Finally, static headspace may be used to perform long runs of measurements, thus improving the training set collection and the device calibration. Nevertheless, the use of static headspace sampling strongly limits the EN sensitivity due to the small amount of available headspace (about 5 ml) and consequently low carrier flow rate values (10 ml/min). Therefore, in some applications, dynamic headspace is to be preferred; it basically consists of a pump and a flow controller that conveys the odor sample under investigation from a vessel (typically 100 ml in volume) into the sensor array chamber.

2.2 MOX Sensors: Thin Films and Nano-Wires

Metal oxides are among the most used active materials for conductometric chemical sensors. They have a wide variety of electrical properties spanning from insulator to quasi metallic behavior. The discovery of their sensing properties was made more than five decades ago, thereafter the interest of researchers was focused on nanostructured materials. These materials may give a greater modulation of the electrical properties for the interaction with the surrounding atmosphere thanks to the higher surface to volume ratio.

Oxides are normally stable at the operating temperatures necessary to enhance the interaction between their surface and the gas phase, much more stable compared to organic materials. They are normally operated between 500 and 800 K where the conduction is electronic and oxygen vacancies are doubly ionized. Different oxides have been proposed for conductometric chemical sensors, the most studied is by far tin dioxide that has also been commercialized in form of thick film sensors. Other oxides studied are titanium oxide, tungsten oxide, zinc oxide, indium oxide and iron oxide, first in form of thick and then in form of thin films. Furthermore, the use of mixed oxides, as well as the addition of noble metals, has been studied to improve not only selectivity but also stability.

One of the most important articles in metal oxide chemical sensing was published in 1991 by Yamazoe [19]. The sensing performances are enhanced with the decrease in crystal dimensions. The greater challenge remains the achievement of metal oxide with small crystallite, but stable over long-term operation at the high temperatures required for chemical sensing. In 2001 a new preparation methodology and morphology have been proposed by Prof. Z. L. Wang [20]. Single crystal tin dioxide nanobelts were prepared by the simple evaporation of the commercial oxide powder at high temperatures and its following condensation at lower temperatures on the substrates. These materials have great potential thanks to their reduced lateral dimensions, high degree of crystallinity, stoichiometry, both for fundamental study and for potential nanodevice applications. After the first publication in 2002 presenting the possible integration of nanowires in conductometric chemical sensors [21], plenty of literature has been published regarding this topic.

For the preparation of conductometric chemical sensors the nanowires electrical properties have to be monitored during the exposure to gases, therefore electrical contacts have to be deposited over the nanowires bundles or the single nanowire. Moreover the nanowires have to be kept at the desired working temperature for the target chemical species. The most used substrates are alumina or silicon. For the latter since the substrate is not insulating an additional layer is needed to measure the electrical properties of the oxide layer. The simplest configuration is achieved with two electrical contacts deposited over the nanowire layers to measure resistance changes and a heater on the back side of the substrate to regulate the operating temperature of the metal oxide layer. Alumina is stable in a wide range of deposition condition such as high temperature, pressure and aggressive environments; therefore it may be used in almost all the preparation procedures proposed in literature. On the other hand flexible substrates may not be compatible with high temperature deposition techniques.

Concerning the preparation techniques, there are different approaches from vapor or liquid phase. The critical aspects, that have to be taken into account, are the reliability of the preparation process, the quality of the nanostructures prepared and the integration into final devices. Among the most promising techniques there are catalyst assisted *vapor phase transport* and *thermal oxidation*. Vapor phase technique consists in the evaporation of the oxide powder in a furnace with controlled atmosphere. In general the pressure is lower than 100 mbar to ease the vaporization of the oxide powder and an inert gas carrier is used to facilitate the mass transport from the source to the substrates, where the vapors condense in form of nanowires.

Catalyst assisted growth incorporates the use of metal clusters in order to ease the formation of the seeds for the unidirectional growth of the oxide. Nobel metals clusters or the same metal of the metal oxide, that has to be prepared, may be used. The dimension of the cluster has a strong influence on the final distribution in morphology, furthermore the catalyst may be used for an easy patterning of the growth region on the substrate.

Shadow masking can be used to pattern the nanowires network and leave a clean surface for the contact deposition. Figure 2a report the different steps involved in the preparation of nanowires based chemical sensor with the catalyst assisted vapor phase transport deposition.

Concerning the thermal oxidation process (Fig. 2b), the formation of metal oxide nanowires is achieved with the oxidation of a metal film in controlled environment. Pressure, temperature and atmosphere composition are crucial for the unidirectional

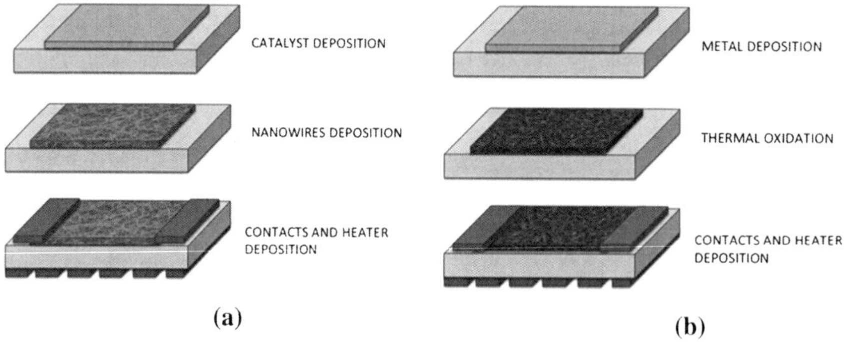

Fig. 2 Different steps for the preparation of a chemical sensor device based on metal oxide nanowires by catalyst assisted vapor phase transport (**a**) and thermal oxidation (**b**)

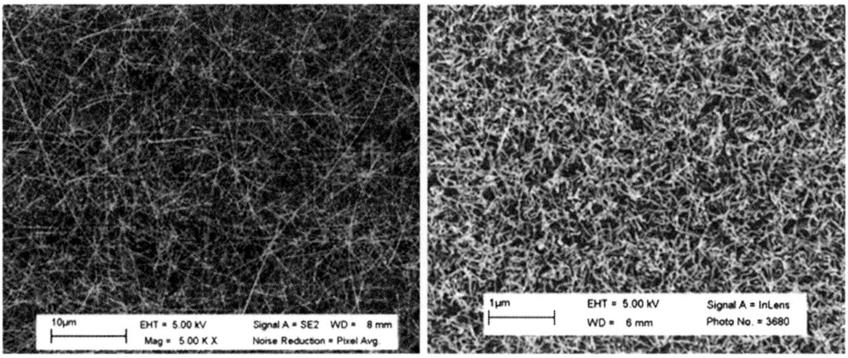

(a) (b)

Fig. 3 Tin dioxide (**a**) and zinc oxide (**b**) nanowires prepared by catalyst assisted vapor phase transport

growth and depend on the specific oxide selected. It is a reliable process and it guarantees an easy patterning of the growth region since the nanowires form where the metal source layer is present. The only drawback is the present of a thin oxide films at the bottom of the nanowires.

The morphology of these nanostructures strongly depends on the technique selected for the preparation and on the operating conditions. Figure 3 reports, as an example, some of the morphologies that may be achieved for tin and zinc oxide. The most important features of these nanowires for chemical sensors are their high surface to volume ratio and their single crystalline nature confirmed by transmission microscopy.

2.3 Data Analysis

Data generated with the EOS are elaborated by Exploratory Data Analysis (EDA) software, a written-in-house software package based on MATLAB® [22]. The EDA software includes the usual (univariate or multivariate) descriptive statistics functions among which Principal Component Analysis (PCA) [23], with the additional utilities for easy data manipulation (e.g. data sub sampling, data set fusion) and plots customization.

Supervised classification is typically carried out by different pattern recognition algorithms. In the referred case studies Support Vector Machines (SVM) with linear kernel [24] and k-Nearest Neighbors (1NN) classifier were applied. Five-fold cross-validation (CV) is usually implemented to get more robust classification results. Supervised regression is performed by Partial Least Squares (PLS) [25]; this technique can be used for training the EN to predict the colony forming unit concentration (expressed as cfu/ml), taking as covariates the results of microbiological counts.

3 Applications of EN to Microbial Food Spoilage Screening

3.1 Literature Review

In the food safety framework, the use of EN devices for rapid and reliable testing pathogenic bacteria contamination in foods is widespread. ENs have been successfully used to detect spoilage of a large variety of food categories and food products as shown in Table 1. The table reports the application of various chemical sensor systems, based on different sensor technologies, catalogued by food category (down to food products) and nature of the screened contaminants (either microbial or toxins).

Several applications are reported for quality control of grains by odor mapping techniques. These controls are routinely performed by a human olfactory panel constituted by inspectors smelling the grain odor. EN could then represent a valid method of choice for its rapidity, simplicity and low cost. More importantly, this would overcome the potential health hazard to the human tasters caused by repeated exposures to mold spores and mycotoxins and the low predictive ability of odor classification system for certain mycotoxins contamination. Hybrid sensor technology has shown to be able determining the mycological quality of barley grains [33] and wheat [30, 31], as well as to detect some mycotoxins classes such as Fumonisins and Aflatoxins [29, 30]. Much emphasis has been given to Maize grains, whose contamination by mycotoxigenic fungi such as *Fusaria* and by Fumonisins, the mycotoxins they produce, has been thoroughly investigated [27, 28].

Similar to cereal grain contamination is that related with coffee grains or beans. EN has been widely applied in the past along the entire coffee production chain. For instance, various EN sensing technologies have been used to distinguish different types of coffee beans, or to identify various brands and mixtures, or again to classify commercial coffee blends and samples with different roasting levels [24]. In spite, the investigation of microbial contamination of coffee grains has received little attention [36]; this case study is presented afterwards in this Chapter.

Fungal spoilage is an important issue in bakery products. Some companies use the measurement of the water activity of the final products as an index for fungal spoilage prediction and batches rejection. However, the need for reliable alternative methods has been evidenced. Marín et al. [37] used a MS-based electronic nose to detect fungal spoilage in samples of bakery products. These were inoculated with different *Eurotium*, *Aspergillus* and *Penicillium* species. Once the headspace was sampled, ergosterol content was determined in each sample in order to have a reference technique for EN training and testing. Both the EN signals and ergosterol levels were used to build models for prediction of ergosterol content. This model has shown excellent regression performance (between 87 and 96 % in some cases) confirming the EN as a reliable method.

Needham and co-workers [38] applied a commercial EN (Bloodhound BH-114) for early detection and differentiation of both bacteria (*Bacillus subtilis*) and fungi (*Penicillium verrucosum* and *Pichia anomala*) spoilage of bread analogues.

Table 1 Applications of chemical sensors in the food microbiology sector (data from [26] updated with recent publications)

Food category	Specific product	Instrumental system and/or sensor technology	Nature of the contamination	Reference
Grains	Maize	EOS835, Sacmi (thin film MOX)	Fungi and fumonisins	[27, 28]
		MOX	Aflatoxins	[29]
	Wheat	Libranose (QuartzMicrobalance)	Fungi	[30]
		Polymers	Fungi	[31]
	Barley	Cyranose-320^{TM} (Carbon-black polymer sensors)	Fungi	[32]
		VCM 422 S-SENCE (MOSFET, tin oxide Taguchi sensors)	Fungi	[33]
		VCM 422 S-SENCE (MOSFET, tin oxide Taguchi sensors)	Fungi, Ochratoxin A	[34]
	Oats rye and barley	MOSFET and Taguchi	Fungi and bacteria	[35]
Coffee	Green coffee grains	EOS835, Sacmi (thin film MOX)	Fungi (*Aspergillus*)	[36]
Bakery products	Bakery analogous	MS-enose (mass spectroscopy)	Fungi	[37]
	Bread analogous	Bloodhound BH-114	Bacteria, yeasts and fungi	[38]
Meat	Beef strip loins	Cyranose-320^{TM} (Carbon-black polymer sensors)	Bacteria	[39]
	Beef strip loins	MOX Taguchi sensors	Bacteria (*Salmonella typhimurium*)	[40]
	Beef and sheep meat	MOX Taguchi sensors	Bacteria	[41]
	Beef fillet	LibraNose, Technobiochip, Napoli, Italy) –quartz crystal microbalance (QMB)	*Pseudomonas* spp., *Brochothrix-thermosphacta*, lactic acid bacteria, *Enterobacteria*	[42]
	Packaged beef steaks	MOX, Bio-imaging and Sensing Center (NDSU)	Bacteria (*Salmonella typhimurium*)	[43]
	Pork	MOX	Bacteria	[44]

(continued)

Table 1 (continued)

Food category	Specific product	Instrumental system and/or sensor technology	Nature of the contamination	Reference
Fish	Sardines	Doped tin oxide	Bacteria	[45]
	Alaska pink salmon (*Oncorhynchus-gorbuscha*)	Cyranose 320 (Carbon-black polymer sensors)	Bacteria	[46]
	Fresh Atlantic salmon (*Salmosalar*)	AromaScanTM (conductive polymers)	Bacteria	[47]
	Cold smoked Atlantic salmon (*Salmosalar*)	FishNose (GEMINI - 6 MOS sensor)	Bacteria	[48]
Milk and dairy product	Milk	MOX, MOSFET	Bacteria	[49]
	Ewe milk	MOX, MOSFET	Aflatoxin B1	[50]
	Milk	Polymer sensors	Bacteria, yeasts	[51]
	Milk	MOX	Bacteria	[52]
Processed vegetables/fruit	Onion	Smith Detection Inc., Pasadena, CA (Polymer sensors)	Bacteria, fungi	[53]
	Tomatoes	EOS835, Sacmi (thin film MOX)	Fungi, bacteria, yeasts	[54]
	Fruit juices	EOS835, Sacmi (thin film MOX) NST3320 type electronic nose (Applied Sensor, A.G., Sweden)	Bacteria (*Alyciclobacillus*)	[55, 56] [57]
	Orange	Thin film conductive polymers	Fungi (*Penicillium digitatum*)	[58]
Drinks	Soft drinks	EOS835, Sacmi (thin film MOX)	Bacteria (*Alyciclobacillus*)	[59]
	Red wine	FOX 3000 Alpha MOs (MOX sensors)	Yeast (*Brettanomyces*)	[60]

Cluster analysis led in this case a differentiation between microbial and physiological (lipoxygenase) spoilage after 48 h.

Much work has been done on meat products. Meat is an ideal growth medium for several groups of pathogenic bacteria (such as *Salmonella* spp., *Escherichia coli* or *Listeria monocytogenes*). Estimation of meat safety and quality is usually based on microbial cultures. Bacterial strain identification requires a number of different growth conditions and biochemical tests with overnight or large incubation periods and skilled personnel, which means that testing may not be frequently performed.

Panigrahi and co-workers [39], analyzed the headspace from fresh beef strip loins kept at 4 °C for 10 days by a commercially available Cyranose-320^{TM} with conducting polymer sensors. They developed various classification models using radial basis function neural networks that enable to identify (with accuracies of 100 %) spoiled and unspoiled meat samples. In this case the type of bacteria was not identified but results were correlated with total viable counts (TVC). The same research group [40] also investigated the ability of home-made EN (Taguchi sensors based) for screening the contamination of beef samples by *Salmonella typhimurium*. In this case the results obtained suggested that the use of higher-order statistical techniques, like Independent Component Analysis (ICA), could help in improving the performance of the sensor system.

Other groups, like El Barbri et al. [41], analyzed both beef and sheep meat (stored at 4 °C for up to 15 days) by a laboratory EN based on Taguchi sensors aiming to develop a protocol for the quality control of red meat. The EN, coupled to SVM, could discriminate between unspoiled/ spoiled beef or sheep meats with a success rate above 96 %. Good correlation between the EN signals and the bacteriological data were also obtained.

More recently [42], the performance of one commercial QMB sensors based EN (LibraNose, Technobiochip, Napoli, Italy) has been evaluated in monitoring aerobically packaged beef fillet spoilage at different storage temperatures. The obtained results demonstrated good performance in discriminating meat samples; overall classification accuracies of prediction obtained for the three sensory classes regardless of storage temperature was above 89 %. Experiments were conducted with MOX based EN of Bio-imaging and Sensing Center (NDSU) to evaluate the performance of integrated sensor system towards identification of *Salmonella* contaminated beef packages [43]; high classification rates (around 90 %) were obtained using wavelet packet transform for feature extraction from sensor array responses and radial basis function network (RBFN) pattern recognition algorithms.

Wang et al. [44] used an EN equipped with MOX sensors together with support vector machine (SVM) and partial least squares (PLS) to predict the total viable counts in chilled pork samples. The achieved correlation coefficients for training and validation were close to 90 %, which suggested that the EN system could be used as a simple and rapid technique for absolving the task.

Most freshness and spoilage investigations with ENs have involved studies with fish or fish products.

Quality changes of cold smoked salmon from four different smokehouses in Europe were monitored by a prototype MOS sensors array system, the so called FishNose [48]. The responses of the gas-sensors correlated well with sensory analysis of spoilage odor and microbial counts suggesting that they can detect volatile microbially produced compounds causing spoilage odors in cold-smoked salmon during storage. In this case, gas-sensor selection was optimized for the detecting of changes in the highly volatile compounds mainly representing microbial metabolism during spoilage. The system was therefore ideal for fast quality control related to freshness evaluation of smoked salmon products.

Regarding fresh salmon fillets, the feasibility of using an AromaScan^{TM} EN to assess seafood quality and microbial safety was assessed by Du et al. [47]. AromaScan mappings of these fillets were compared to their time related changes in microbial counts, histamine contents, and sensory panel evaluations. Promising results were obtained and authors concluded that the EN can be used as an assisting instrument to a sensory panel in evaluating the seafood microbial quality and safety.

The ability of a portable hand-held EN (Cyranose 320^{TM}, composed of 32 individual thin-film carbon-black polymer sensors) in detecting spoilage of salmon under different storage conditions (at 14°C and in slush ice) was also investigated by Chantarachoti et al. [46]. As a result of these experiments, a predictive model may be developed for spoilage of whole Alaska pink salmon by analyzing belly cavity odors using the e-nose. This could be easily extended to other types of fish. In fact, an EN system based on a 4-element, integrated, micro-machined, MOX gas sensor array was used in [45] to assess the evolutionary stages of freshness in sardine samples stored up to one week at 4 °C.

Research with ENs in the area of milk and other dairy products have ranged from detecting adulteration/contamination of milk to determining the geographical origins of cheese. An important aspect related to milk quality and safety is the detection of contaminants, including aflatoxins, in milk. For instance, Benedetti et al. [50] studied the feasibility of using a commercial sensor array system, comprising 12 MOS and 12 MOSFET sensors, to detect the presence of aflatoxin M1 (AFM1). In this study, twenty-four raw milk samples collected from two different groups of ewes fed with a formulated feed containing increasing amounts of aflatoxin B1, and six non-contaminated ewe milk samples were analyzed. The results obtained by using the head space sensor array, processed by statistical methods, made it possible to group the samples according to the presence or the absence of aflatoxin M1.

Finally, various promising results have been also achieved with EN for microbial screening of fresh and processed vegetables.

Canned or packed processed tomatoes (tomato pulp or juice) are a food category extremely exposed to safety risks related with the presence of both chemical residuals, like pesticides and herbicides, but also microbial contaminants among which bacteria and fungi. Concina et al. [52] have investigated the ability of EOS835 to perform early diagnosis of microbial contamination of canned peeled tomatoes aiming to design an analytical protocol for an objective quality control at the end of the production chain. A main challenge for the EOS was early screening of contaminated samples subjected to multiple microbial contaminants: 3 bacteria (*Escherichia coli*, *Enterobacter cloache* and *Lactobacillus plantarum*), 1 yeast (*Saccharomyces cerevisiae*) and 2 molds (*Aspergillus carbonarius* and *Penicillium puberulum*) were used in this study to contaminate the product. Supervised classification tests were performed by implementing a 5-fold cross-validated k-NN classifier which provided the 83 % of correct classification rate for contaminated samples, upon careful feature (principal components) selection, just after 48 h of incubation.

The screening of fruit juices [55–57] or soft drinks [59] contaminated by *Alicyclobacilli* seems to be a very promising industrial application; this problem will be presented in more detail in what follows.

3.2 *Example 1: Screening of Fruit Juices Spoilage by Alicyclobacillus spp.*

In 1982, spoilage of aseptically filled apple juice from Germany was attributed to a new type of thermophilic acidophilic bacteria, later classified into a new genus, named *Alicyclobacillus* (ACB) [61]. *Alicyclobacillus* spp. are aerobic, Gram-positive, endospore-forming, non-pathogenic, thermoacidophilic bacteria isolated mainly from soil and hot springs. The spores are resistant to high temperature, thus they can survive ordinary pasteurization regimes used in the juice industry.

Contamination by *Alicyclobacillus* spp. was firstly detected in apple juice but, since then, a larger variety of fruit juices, soft drink, fortified with minerals products have been found to be contaminated. *A. acidoterrestris* and *A. acidocaldarius* are the most common specie able to cause typical off-flavors (medicine-like taints) in fruit juices; this was related to the production of 2-methoxyphenol (guaiacol), 2-6-dibromophenol, 2-6 diclorophenol [62, 63] which are retained to be the markers of contamination.

Spoilage of fruit juices by *Alicyclobacilli* is extremely difficult to be revealed at early stages; therefore, this task is regarded as a relevant industrial issue that requires effective control measures to be developed. Culture-dependent conventional microbiological methods present some drawbacks, mainly related to the high detection limit and to the underestimation of the true microbial community. Traditionally, gas-chromatography (GC) and mass-spectrometry (MS) provide accurate measurements of the volatile fraction and are useful for specific identification of off-flavors compounds; nevertheless these methods remain still rather complex and expensive, being more suitable for laboratory quality control than for routine industrial analyses which often require faster, simpler and massive screening of large product batches.

Different types of commercial fruit juices (orange, peach, pear and apple) artificially contaminated by *Alicyclobacillus* spp. were tested with the EOS in an experimental work that last for almost two years.

Preliminary results reported in Gobbi et al. [55] were very promising, showing that EOS has good detection capabilities, being able to early reveal the presence of *Alicyclobacillus* spp. just after a growth time of 24 h. The sensors showed some specificity related with the juice matrix, indeed contamination was easily identified in orange and peach juices (where detection threshold was around 100 cfu/ml) whereas it was impossible to correctly classify the contaminated apple juice samples.

The juice samples were contaminated both with *A. acidoterrestris* and *A. acidocaldarius* to test whether intragenus specificity could affect the EOS results. In fact the peach juice samples contaminated by the two species were clustered closely together but it was still possible to perfectly discriminate the contaminated samples from the not contaminated ones. Thus, the intragenus specificity of EOS was much lower than the genus specificity. This can be an advantage because from a practical point of view it is possible to perform the EN training to diagnose bacterial contamination over a limited numbers of species without having the classification capability substantially affected by other species.

Table 2 Classification results of contaminated samples and detection thresholds ([56])

Type of juice	Classification rate		Detection thresholds	
	SVM %	1NN %	Growth time (hours)	cfu/ml
Orange	86	78	24	10^3
Pear	90	84	24	10^2–10^3
Apple	60	63	72	10^5

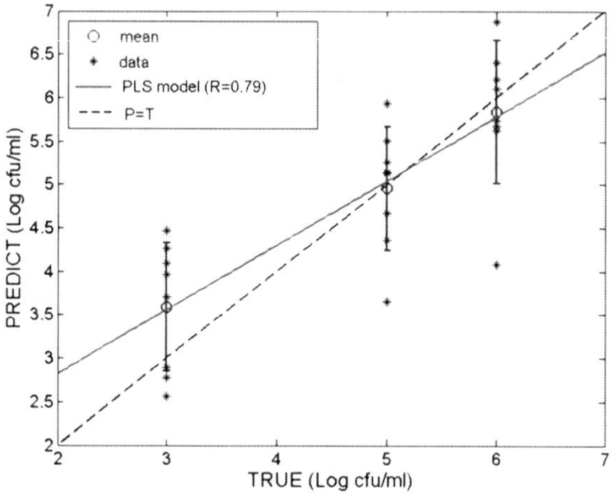

Fig. 4 PLS regression of EOS data for contaminated orange juice samples owing to predict the amount of contamination (reproduced from [26] with permission)

These results were subsequently confirmed and ameliorated by Cagnasso et al. [56]. The EOS showed good classification performance of contaminated samples (Table 2), up to 90 % for pear juices classified by SVM. The system detection limits and the required growth times were consistent with the results for orange and apple juices formerly obtained by Gobbi et al. and still appeared to be specific of the juice matrix with apple juice being the most challenging case. Following these results, the authors argued that the detection of *Alicyclobacillus* spp. was favored by the strong change of specific volatile compounds present in the juice matrix, for instance limonene on orange juice, although this hypothesis was not assesses through dedicated analytical studies.

Very fast though not very accurate capability to predict the amount of contamination was also observed (Fig. 4). The number of cfu/ml predicted by the EOS correlated quite well with the true value as measured by the microbiological essays (the correlation coefficient scored about 0.80). The discrepancies can be associated with three facts: first, the EOS overestimated the cfu/ml at low values, the predicted mean value are about 3 times larger; second, the PLS model has been built with the mean Log (cfu/ml) value whilst the EOS measurements refer to the actual concentration of individual samples which is not known; third, the model suffers from the accuracy with which the microbiological counts are determined that can be realistically of one order of magnitude.

EOS results were also correlated with GC-MS quantification of the claimed chemical markers associated with *A. acidoterrestris* presence, i.e. guaiacol, but outcomes were inconsistent. The authors then argued that the gas sensors are sensitive to the change of the global olfactory fingerprint induced by *A. acidoterrestris* presence more than to the guaiacol content of the samples. Yet, this lack of correlation can be regarded as a limitation of the technology since target sensors could certainly facilitate the detection of bacteria in apple juice while enhancing the specificity of the technique.

3.3 Example 2: Fungal Contamination of Green Coffee Beans

Fungal growth on green coffee beans can occur along the entire distribution chain. This may cause organileptic defects on roasted coffee thus impairing its consumption, the deterioration of sensorial properties is often due to the production of exoenzymes during growth, or can also bring about health hazards to consumers because of production of toxic metabolites like mycotoxins [64].

The main toxigenic fungal genera (*Aspergillus*, *Penicillium* and *Fusarium*) are natural coffee contaminants and are present from the field to the warehouse [65]. OTA is produced by species of two genera of fungi, *Aspergillus* and *Penicillium* that grown naturally on cereals, grapes, coffee and cocoa. In coffee, the most important OTA-producers are *Aspergillusochraceus*, *A.carbonarius* and strain of *A. niger*.OTA is so dangerous that FAO/WHO experts have set a maximum tolerable limit for humans of 100 billionths of a gram per kilogram of body weight per week. In 2004, the EU set maximum permissible limits for OTA of 5 ppb in roasted and ground coffee and 10 ppb in instant coffee.

Therefore, it is necessary to develop strategies to distinguish and quantify fungal infection, and possibly toxins production, at early stages. One of the most promising techniques is the analysis of volatile compounds which are released by the coffee in the headspace gas surrounding the samples. For this reason, the ability of the EOS to early detect microbial contamination of Arabica green coffee was evaluated [36].

Two species of the genus *Aspergillus* (*A. niger* type strain A733 and *A. ochraceus* type strain DSM 824) were selected. Coffee beans were preliminarily sterilized with UV light in order to remove any undesired contamination and to be sure of having sterile control samples. The green coffee beans were first contaminated and then incubated in a moist chamber at 27 °C for 11 days (analyzed at 0, 2, 5,6,8, 9,13 and 15 days after inoculations) in order to promote the growth of fungi inoculated and to standardized at 0,6 the activity water (Aw). The samples were maintained at a constant temperature and humidity for the entire duration of the analysis in order to be certain that the difference between the types of samples (contaminated and not contaminated) was not due to physical parameter, independent of the growth of molds.

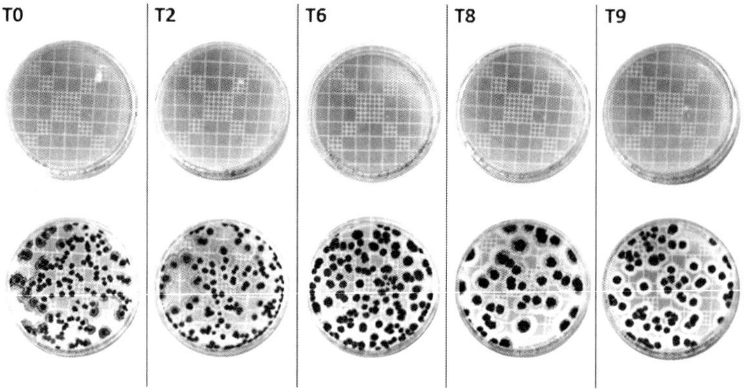

Fig. 5 Petri plates of coffee samples collected over different days of incubation after fungal inoculation. *Upper row* control plates (uncontaminated coffee), *lower row* plates of contaminated samples

Coffee samples were analyzed with classical microbiological isolation techniques to monitor the fungal growth and quantify the actual contamination in the days after inoculation (Fig. 5) and verify that nothing was growing on uncontaminated coffee beans. Analytical chemical techniques, like Gas chromatography coupled with Mass spectroscopy (GC-MS) with Solid phase micro-extraction (SPME), were also used for detecting the formation of volatile secondary metabolites. GCMS analyses evidenced quite different chemical volatile profiles for uncontaminated and contaminated coffee beans. In particular, a relevant presence of carbon dioxide, 5-methyl-2-phenylindole, which is considered to be typical microbial metabolites, emerged in the headspace of contaminated samples. Samples contamination was also confirmed by the appearance of anisole, recognized as one of the major volatile markers of fungal contaminated coffee [66].

For the EOS analyses 3 g of green coffee were placed in 20 ml chromatographic sterile vials sealed Teflon septa. Measurements were performed in a static HS fashion using the HT200H unit. The sensors baseline was obtained using synthetic chromatography air with a continuous flow of 10 ml/min. Each vial was incubated at 40 °C for 10 min in the HT200 oven with continuous stirring. The sample headspace (4 ml) was then extracted from the vial and injected into the flow (speed 4 ml/min) through the HT injector. The EOS was able to successfully discriminated contaminated samples of green coffee from non-contaminated ones, as shown in the PCA score plot (Fig. 6) and classification tests by kNN reported about 96 % score with only few false negative—patterns of contaminated samples that are also visibly overlapped on the PCA with the uncontaminated class.

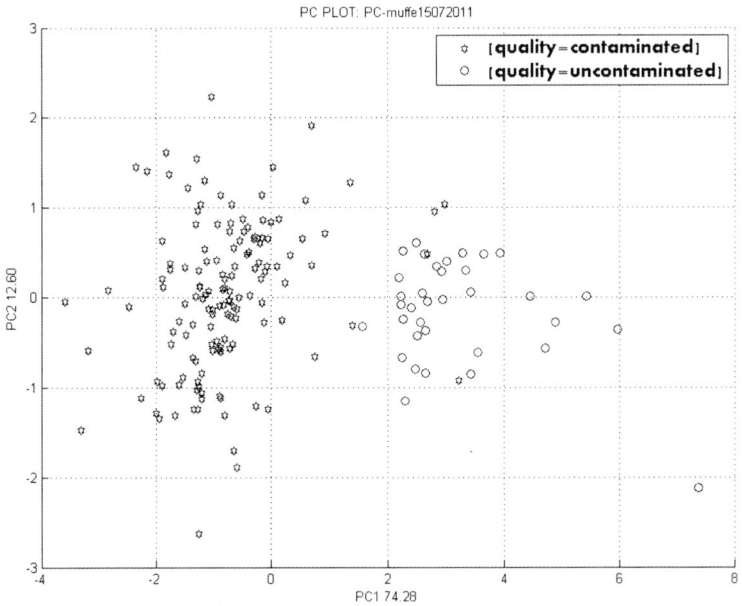

Fig. 6 PCA score plot of olfactory patterns of contaminated green coffees as measured by the EOS

4 Discussion and Future Potential Improvements

Electronic nose could be a very valuable tool to evaluate microbiological food quality and safety. Strengths of the electronic nose include good sensitivity and correlation with data from microbiological tests. It might have other advantages regarding portability, price, and ease of use. Therefore, it has the potential to move from well-equipped chemical laboratories to industrial routine at-line controls. At present ENs present some downsides that must be overcome, however.

The major limitation of currently available chemical sensor based ENs remains the independence (cross-correlation) and selectivity of the sensing devices. Sensors with poor selectivity affect adversely the discriminating power of the array. Moreover, the EN use of semi-selective sensors prevents any real identification or quantification of the individual compounds present in the headspace of a food sample, as it can be done classical analytical techniques, with which sometimes can be crucial for the end-user. In fact, whenever specific contamination markers are present (as shown in our first case study) their detection allows more reliable, faster and more replicable results than training an EN. To tackle this issue, in recent years, classical chemosensor technologies were complemented by new emerging technologies [67]. In particular, machine olfaction has benefited from developments in several fields ranging from optical technologies developed by the telecommunications industry to the improvements in analytical chemistry such as: gas chromatography, mass spectroscopy and ion mobility spectrometry. This trend has also narrowed the gap between traditional

ENs—used as a black box—and classical analytical techniques which aim to quantify individual volatile components. Nonetheless, it is unrealistic to envisage a universal electronic nose that is able to cope with every odor type, conversely data processing and instrumentation must be specifically designed for each application.

The EN training procedure still remains lengthy and laborious, and finally, the lack of sensors stability and reproducibility over time can put at risk on the use of previously collected databases, which are compulsory for data comparison purposes and for the classification of new unknown samples. The problem of chemical sensor stability over time is known as "sensor drift". It consists of (more or less) small and non-deterministic temporal variations of the sensor response when it is exposed to the same analytes under identical conditions. This is generally attributed to sensors aging or thermo-mechanical degradation, but it can also be influenced by a variety of sources including environmental factors. The main result is that sensors selectivity and sensitivity slowly decrease with time. Drift correction is perhaps one the most relevant issues in the field of chemical sensors. Indeed, in spite of constant improvements in micro/nano fabrication techniques that allowed the production of sensing devices with superior stability, it is still impossible to fabricate chemical sensors without drift. This issue is currently approached on one hand by improving the sensors performances with novel sensing materials [68], on the other hand by adopting various approaches for compensating sensors drift to increase pattern recognition accuracy [69, 70].

Another major drawback, partially connected to the previous issue, is that in case of sensor breakage it would be very difficult to replace the sensor by another one having exactly the same behavior. This dramatically jeopardizes the use of previously collected databases, which are compulsory for data comparison purposes and for the classification of new unknown samples. This problem has been addressed from different points of view: while researchers are trying to understand the mechanisms related to materials and device processing to improve the sensors reproducibility; at industrial level this problem is fixed by producing a large number of sensors and then by selecting the most similar on the basis of an application specific test protocol. The problem of database recovering after sensors replacement in an electronic nose has been also approached by means of multivariate calibration methods [71]. Such methods permit to alleviate the problems due to lack of sensor-to-sensor reproducibility and allow recovering good classification rates.

5 Conclusions

In this work, some relevant applications of electronic nose technology to microbiological food quality control have been reviewed. The literature review has been accompanied by some significant case studies previously performed in this field by the same authors, in order to provide the reader a better insight of the EN application. All the reported case studies showed promising results, thus confirming that EN could represent a rapid mean for controlling and improving the microbiological quality of food. Further investigations to induce more sensitive specific response of the gas sensors may help both in the qualitative and quantitative analysis of microbial contamination, however.

Before the EN can be treated as a completely reliable, industrial instrument, to be used in the food field much improvement is still needed on the technology side, for instance: improve sensor selectivity, reduce interferences (e.g. to humidity), compensate drift effects and handle with sensor replacement.

Progresses could also be made on the application side by better investigation of available technologies. At present, the major part of EN applications are represented by limited feasibility studies, often present as preliminary results, with poor validation especially in terms of reproducibility and predictive ability. Few replication or confirmation studies—like the two cases (a) and (c) presented here—are reported in the literature; conversely for people working in the field would be very beneficial to see long term studies results with extensive investigation of data reproducibility and system stability.

Keeping in mind advantages and limitations ENs do not allow to replace human panels or analytical techniques, as long as their ability to smelling odors rather than detecting and quantifying specific volatiles is still far from required standards. In spite, they can be used in parallel to those techniques, or even considered as valuable alternatives, to perform quick 'go-no go' product tests or occasionally replace human panels when non-odorous, irritant or potentially toxic volatile substances need to be detected.

Acknowledgments Authors acknowledge CAFIS project POR-FERS 2007/2013 and Consorzio Casalasco del Pomodoro Soc.Agr.Coop (Cremona, Italy) for financial support.

References

1. L. Buck, R. Axel, A novel multigene family may encode odorant receptors: a molecular basis for odor recognition. Cell **65**(1), 175–187 (1991)
2. M. Pardo, G. Sberveglieri, Electronic olfactory systems based on metal oxide semiconductor sensor arrays. Mrs Bull. **29**(10), 703–708 (2004)
3. T.C. Pearce, S.S. Schiffman, H.T. Nagle, J.W. Gardner, *Handbook of Machine Olfaction* (Wiley, NY, 2002)
4. W.F. Wilkens, J.D. Hartman, An electronic analogue for the olfactory process. Ann. NY Acad. Sci. **116**, 608–612 (1964)
5. R.W. Moncrieff, An instrument for measuring and classifying odours. J. Appl. Physiol **16**, 742–749 (1961)
6. K.C. Persaud, G.H. Dodd, Analysis of discrimination mechanisms of the mammalian olfactory system using a model nose. Nature **299**, 352–355 (1982)
7. H. Shurmer, Development of an electronic nose. Phys. Technol. **18**(4), 170–176 (1987)
8. J.W. Gardner, Pattern recognition in the Warwick electronic nose, 8th International Congress of the European Chemoreception Research Organisation, Coventry (UK), (1988)
9. Linda M. Reid, Colm P. O'Donnell, Gerard Downey, Recent technological advances for the determination of food authenticity. Trends Food Sci. Technol. **17**(7), 344–353 (2006)
10. K. Arora, S. Chand, B.D. Malhotra, Recent developments in bio-molecular electronics techniques for food pathogens. Anal. Chim. Acta **568**(1–2), 259–274 (2006)

11. E. Schaller, J.O. Bosset, F. Escher, Electronic noses and their application to food. Food Sci.Technol. Lab. **31**(4), 305–316 (1998)
12. M. Ghasemi-Varnamkhasti, S. SaeidMohtasebi, M. Siadat, Biomimetic-based odor and taste sensing systems to food quality and safety characterization: an overview on basic principles and recent achievements. J. Food Eng. **100**, 377–387 (2010)
13. A. Berna, Metal oxide sensors for electronic noses and their application to food analysis. Sensors **10**, 3882–3910 (2010). doi:10.3390/s100403882
14. A.D. Wilson, M. Baietto, Applications and advances in electronic-nose technologies. Sensors **9**, 5099–5148 (2009). doi:10.3390/s90705099
15. M. Peris, L. Escuder-Gilabert, A 21st century technique for food control: electronic noses. Anal. Chim. Acta **638**(1), 1–15 (2009)
16. N. Sahgal, R. Needham, F.J. Cabanes, N. Magan, Potential for detection and discrimination between mycotoxigenic and non-toxigenic spoilage moulds using volatile production patterns: a review. Food Addit. Contam. **24**(10), 1161–1168 (2007)
17. N. Magan, P. Evans, Volatiles as an indicator of fungal activity and differentiation between species, and the potential use of electronic nose technology for early detection of grain spoilage. J. Stored Prod. Res. **36**, 319–340 (2000)
18. M.J. Saxby, *Food Taints and Off-flavours*, 1st edn. (Blackie Academic and Professional, Cambridge, 1993)
19. N. Yamazoe, New approaches for improving semiconductor gas sensors. Sens. Actuators B **5**, 7–19 (1991)
20. Z.W. Pan, Z.R. Dai, Z.L. Wang, Nanobelts of semiconducting oxides. Science **291**, 1947–1949 (2001)
21. E. Comini, G. Faglia, G. Sberveglieri, Z.R. Pan, Z.L. Wang, Stable and highly sensitive gas sensors based on semiconducting oxide nanobelts. Appl. Phys. Lett. **81**, 1869–71 (2002)
22. M. Vezzoli, A. Ponzoni, M. Pardo, M. Falasconi, G. Faglia, G. Sberveglieri, Exploratory data analysis for industrial safety application. Sens. Actuators B Chem. **131**(1), 100–109 (2008)
23. D.L. Massart, B.G.M. Vandeginste, L.M.C. Buydens, S. De Jong, P.J. Lewi, J. Smeyers-Verbeke, in *Handbook of Chemometrics and Qualimetrics* Part A, Chap. 17, (Elsevier, Amsterdam, 1997)
24. M. Pardo, G. Sberveglieri, Classification of electronic nose data with support vector machines. Sens. Actuators B Chem. **107**(2), 730–737 (2005)
25. D.L. Massart, B.G.M. Vandeginste, L.M.C. Buydens, S. De Jong, P.J. Lewi, J. Smeyers-Verbeke, in *Handbook of Chemometrics and Qualimetrics* Part B, Chap. 37, (Elsevier, Amsterdam, 1997)
26. M. Falasconi, I. Concina, E. Gobbi, V. Sberveglieri, A. Pulvirenti, and G. Sberveglieri, Electronic nose for microbiological quality control of food products. Int. J. Electrochem. **2012**, Article ID 715763, p. 12, (2012). doi:10.1155/2012/715763
27. M. Falasconi, E. Gobbi, M. Pardo, A. Bresciani, G. Sberveglieri, Detection of toxigenic strains of Fusariumverticillioides in corn by electronic olfactory system. Sens. Actuators B Chem. **108**, 250–257 (2005)
28. E. Gobbi, M. Falasconi, E. Torelli, G. Sberveglieri, Electronic nose predicts high and low fumonisin contamination in maize cultures. Food Res. Int. **44**, 992–999 (2011)
29. F. Cheli, A. Campagnoli, L. Pinotti, G. Savoini, V. Dell'Orto, Electronic nose for determination of aflatoxins in maize. Biotechnol. Agron. Soc. Environ. **13**, 39–43 (2009)
30. R. Paolesse, A. Alimelli, E. Martinelli, C. Di Natale, A. D'Amico, M.G. D'Egidio, G. Aureli, A. Ricelli, C. Fanelli, Detection of fungal contamination of cereal grain samples by an electronic nose. Sens. Actuators B Chem. **119**(2), 425–430 (2006)
31. P. Evans, K.C. Persaud, A.S. McNeish, R.W. Sneath, N. Hobson, N. Magan, Evaluation of a radial basis function neural network for the determination of wheat quality from electronic nose data. Sens. Actuators B Chem. **69**, 348–358 (2000)
32. S. Balasubramanian, S. Panigrahi, B. Kottapalli, C.E. Wolf-Hall, Evaluation of an artificial olfactory system for grain quality discrimination. Lwt-Food Sci. Technol. **40**(10), 1815–1825 (2007)

33. J. Olsson, T. Börjesson, T. Lundstedt, J. Schnürer, Volatiles for mycological quality grading of barley grains: determinations using gas chromatography-mass spectrometry and electronic nose. Int. J. Food Microbiol. **59**, 167–178 (2000)
34. J. Olsson, T. Borjesson, T. Lundstedt, J. Schnurer, Detection and quantification of ochratoxinA and deoxynivalenol in barley grains by GC-MS and electronic nose. Int. J. Food Microbiol. **72**, 203–214 (2002)
35. A. Jonsson, F. Winquist, J. Schnurer, H. Sundgren, I. Lundstrom, Electronic nose for microbial quality classification of grains. Int. J. Food Microbiol. **35**, 187–193 (1997)
36. V. Sberveglieri, I. Concina, M. Falasconi, E. Gobbi, A. Pulvirenti, P. Fava, Early detection of fungal contamination on green coffee by a MOX sensors based electronic nose, in AIP Conference Proceedings, *14th International Symposium on Olfaction and Electronic Nose, ISOEN 2011*, vol. 1362, (New York City, 2011) pp. 19–120
37. S. Marin, M. Vinaixa, J. Brezmes, E. Llobet, X. Vilanova, X. Correig, A.J. Ramos, V. Sanchis, Use of a MS-electronic nose for prediction of early fungal spoilage of bakery products. Int. J. Food Microbiol. **114**(1), 10–16 (2007)
38. R. Needham, J. Williams, N. Beales, P. Voysey, N. Magan, Early detection and differentiation of spoilage of bakery products. Sens. Actuators B Chem. **106**(1), 20–23 (2005)
39. S. Panigrahi, S. Balasubramanian, H. Gu, C. Logue, M. Marchello, Neural-network-integrated electronic nose system for identification of spoiled beef. Lwt-Food Sci. Technol. **39**(2), 135–145 (2006)
40. S. Balasubramanian, S. Panigrahi, C.M. Logue, C. Doetkott, M. Marchello, J.S. Sherwood, Independent component analysis-processed electronic nose data for predicting Salmonella typhimurium populations in contaminated beef. Food Control **19**(3), 236–246 (2008)
41. N. ElBarbri, E. Llobet, N. El Bari, X. Correig, B. Bouchikhi, Electronic nose based on metal oxide semiconductor sensors as an alternative technique for the spoilage classification of red meat. Sensors **8**, 142–156 (2008)
42. Olga S. Papadopoulou, Efstathios Z. Panagou, Fady R. Mohareb, George-John E. Nychas, Sensory and microbiological quality assessment of beef fillets using a portable electronic nose in tandem with support vector machine analysis. Food Res. Int. **50**(1), 241–249 (2013)
43. L.R. Khot, S. Panigrahi, C. Doetkott, Y. Chang, J. Glower, J. Amamcharla, C. Logue, J. Sherwood, Evaluation of technique to overcome small dataset problems during neural-network based contamination classification of packaged beef using integrated olfactory sensor system. LWT-Food Sci. Technol. **45**(2), 233–240 (2012)
44. D. Wang, X. Wang, T. Liu, Y. Liu, Prediction of total viable counts on chilled pork using an electronic nose combined with support vector machine. Meat Sci. **90**(2), 373–377 (2012)
45. N. ElBarbri, J. Mirhisse, R. Ionescu, N. El Bari, X. Correig, B. Bouchikhi, E. Llobet, An electronic nose system based on a micro-machined gas sensor array to assess the freshness of sardines. Sens. Actuators B Chem. **141**, 538–543 (2009)
46. J. Chantarachoti, A.C.M. Oliveira, B.H. Himelbloom, C.A. Crapo, D.G. McLachlan, Portable electronic nose for detection of spoiling Alaska pink salmon (Oncorhynchusgorbuscha). J. Food Sci. **71**, S414–S421 (2006)
47. W.X. Du, C.M. Lin, T. Huang, J. Kim, M. Marshall, C.I. Wei, Potential application of the electronic nose for quality assessment of Salmon fillets under various storage conditions. JFS: Food Microbiol Saf. **67**, 307 (2002)
48. G. Olafsdottir, E. Chanie, F. Westad, R. Jonsdottir, C.R. Thalmann, S. Bazzo, S. Labreche, P. Marcq, F. Lundby, J.E. Haugen, JFS S: Sens. Nutr. Qualities Food **70**, S563 (2005)
49. J.E. Haugen, K. Rudi, S. Langsrud, S. Bredholt, Application of gas-sensor array technology for detection and monitoring of growth of spoilage bacteria in milk: a model study. Anal. Chim. Acta **565**(1), 10–16 (2006)
50. S. Benedetti, S. Iametti, F. Bonomi, S. Mannino, Head space sensor array for the detection of Aflatoxin M1 in Raw Ewe's milk. J. Food Prot. **68**, 1089–1092 (2005)
51. N. Magan, A. Pavlou, I. Chrysanthakis, Milk-sense: a volatile sensing system recognises spoilage bacteria and yeasts in milk. Sens. Actuators B Chem. **72**, 28–34 (2001)

52. S. Labreche, S. Bazzo, S. Cade, E. Chanie, Shelf life determination by electronic nose: application to milk. Sens. Actuators B Chem. **106**, 199–206 (2005)
53. C. Li, N.E. Schmidt, R. Gitaitis, Detection of onion postharvest diseases by analyses of headspace volatiles using a gas sensor array and GC-MS. LWT-Food Sci. Technol. **44**, 1019–1025 (2011)
54. I. Concina, M. Falasconi, E. Gobbi, F. Bianchi, M. Musci, Mattarozzi, M. Pardo, A. Mangia, M. Careri, G. Sberveglieri, Early detection of microbial contamination in processed tomatoes by electronic nose. Food Control **20**, 873–880 (2009)
55. E. Gobbi, M. Falasconi, I. Concina, G. Mantero, F. Bianchi, M. Mattarozzi, M. Musci, G. Sberveglieri, Electronic nose and Alicyclobacillus spp. spoilage of fruit juices: an emerging diagnostic tool. Food Control **21**(10), 1374–1382 (2010)
56. S. Cagnasso, M. Falasconi, MP. Previdi, B. Franceschini, C. Cavalieri, V. Sberveglieri, P. Rovere, Rapid screening of *Alicyclobacillus acidoterrestris* spoilage of fruit juices by electronic nose: a confirmation study. J. Sens. (2010). Article ID 143173. doi:10.1155/2010/143173
57. Piroska Hartyáni, István Dalmadi, Dietrich Knorr, Electronic nose investigation of Alicyclobacillus acidoterrestris inoculated apple and orange juice treated by high hydrostatic pressure. Food Control **32**(1), 262–269 (2013)
58. J. Gruber, H.M. Nascimento, E.Y. Yamauchi, R.W.C. Li, C.H.A. Esteves, G.P. Rehder, C.C. Gaylarde, M.A. Shirakawa, A conductive polymer based electronic nose for early detection of Penicillium digitatum in post-harvest oranges. Mater. Sci. Eng. C **33**(5), 2766–2769 (2013)
59. I. Concina, M. Bornšek, S. Baccelliere, M. Falasconi, E. Gobbi, G. Sberveglieri, Alicyclobacillus spp.: detection in soft drinks by electronic nose. Food. Res. Int. **43**(8), 2108–2114 (2010)
60. A.Z. Berna, S. Trowell, W. Cynkar, D. Cozzolino, Comparison of metal oxide-based electronic nose and mass spectrometry-based electronic nose for the prediction of red wine spoilage. J. Agric. Food Chem. **56**, 3238–3244 (2008)
61. M. Walker, C.A. Phillips, *Alicyclobacillus acidoterrestris*: an increasing threat to the fruit juice industry? Int J. Food Sci. Technol. **43**(2), 250–260 (2008)
62. S.S. Chang, D.H. Kang, Alicyclobacillus spp. in the fruit juice industry: history, characteristics, and current isolation/detection procedures. Crit. Rev. Microbiol. **30**(2), 55–74 (2004)
63. N. Jensen, F.B. Whitfield, Role of Alicyclobacillusacidoterrestris in the development of a disinfectant taint in shelf-stable fruit juice. Lett. Appl. Microbiol. **36**(1), 9–14 (2003)
64. R. Etzel, Mycotoxins. JAMA **287**, 527–528 (2002)
65. M. Nakajima, H. Tsubouchi, M. Miyabe, Y. Ueno, Survey of aflatoxin B1 and ochratoxinA in commercial green coffee beans by high performance liquid chromatography linked with immunoaffinity chromatography. Food Agric. Immunol. **9**, 77–83 (1997)
66. K. Karloshøj, P.V. Nielsen, T.O. Larsen, Differentiation of closely related fungi by electronic nose analysis. J. Food Sci. **72**, 187–192 (2007)
67. F. Röck, N. Barsan, U. Weimar, Electronic nose: current status and future trends. Chem. Rev. **108**(2), 705–725 (2008)
68. E. Comini, G. Faglia, G. Sberveglieri, *Solid State Gas Sensing*, (Springer, Berlin, 2008). ISBN: 978-0-387-09664-3
69. S. Di Carlo, M. Falasconi, E. Sanchez, A. Scionti, G. Squillero, A. Tonda, Increasing pattern recognition accuracy for chemical sensing by evolutionary based drift compensation. Pattern Recogn. Lett. **32**(13), 1594–1603 (2011)
70. M. Padilla, A. Perera, I. Montoliu, A. Chaudry, Drift compensation of gas sensor array data by orthogonal signal correction. Chemometr. Intell. Lab. Syst. **100**(1), 28–35 (2010)
71. O. Tomic, T. Eklov, K. Kvaal, J.E. Haugen, Recalibration of a gas-sensor array system related to sensor replacement. Anal. Chim. Acta **512**(2), 199–206 (2004)

Chapter 7
Multiclass Kernel Classifiers for Quality Estimation of Black Tea Using Electronic Nose

P. Saha, S. Ghorai, B. Tudu, R. Bandyopadhyay and N. Bhattacharyya

Abstract Electronic nose (e-nose) is a machine olfaction system that has shown significant possibilities as an improved alternative of human taster as olfactory perceptions vary from person to person. In contrast, electronic noses also detect smells with their sensors, but in addition describe those using electronic signals. An efficient e-nose system should analyze and recognize these electronic signals accurately. For this it requires a robust pattern classifier that can perform well on unseen data. This research work shows the efficient prediction of black tea quality by means of modern kernel classifiers using the e-nose signatures. As kernel classifiers, this work investigates the potential of state of the art support vector machine (SVM) classifier and very recently developed nonparallel plane proximal classifier (NPPC) and vector-valued regularized kernel function approximation (VVRKFA) technique of multiclass data classification to build taster-specific computational models. Experimental results show that VVRKFA and one-versus-rest (OVR) SVM models offer high accuracies to predict the considerable variation in tea quality.

P. Saha (✉) · S. Ghorai
Department of Applied Electronics and Instrumentation Engineering, Heritage Institute of Technology, Kolkata 700107, India
e-mail: pradip.saha@heritageit.edu

S. Ghorai
e-mail: santanu.ghorai@heritageit.edu

B. Tudu · R. Bandyopadhyay
Department of Instrumentation Electronics Engineering, Jadavpur University, Kolkata 700107, India
e-mail: bt@iee.jusl.ac.in

R. Bandyopadhyay
e-mail: rb@iee.jusl.ac.in

N. Bhattacharyya
Centre for Development of Advanced Computing (CDAC), Kolkata 700107, India
e-mail: nabarun.bhattacharyya@cdac.in

A. Mason et al. (eds.), *Sensing Technology: Current Status and Future Trends II*, Smart Sensors, Measurement and Instrumentation 8, DOI: 10.1007/978-3-319-02315-1_7,

Keywords Black tea · Electronic nose · Feature selection · Support vector machine.

1 Introduction

Electronic-nose (e-nose) devices have gained substantial importance in the field of sensor technology during the past 20 years, mostly because of the innovation of several applications in a number of commercial industries like agricultural, food, environmental, biomedical, cosmetics, manufacturing, pharmaceutical and military etc. This new technology has many advantages. Human panels testing for quality evaluation in different food and agro products is highly subjective. Various human factors like individual variability decrease in sensitivity due to prolonged exposure and adverse mental state at times causes problems like inaccuracy, non-repeatability and high time consumption. Electronic-nose [1] is an electronic alternative of human tasters that has shown promising ability to encounter such problems. It has been successfully employed for the recognition and quality analysis of different food and agro products, viz., coffee [2], meat [3], wine [4], fish [5], etc. The potential of this instrument in classifying black tea aroma in different processing stages is also established by Dutta et al. [1].

Flavor and aroma are two most important quality attributes of black tea that are conventionally evaluated by the human panel, known as tea tasters, in the tea industry. The tea tasters rank the tea quality on a scale of 1–10. Different parameters like weather, plantation type, number of flushes, manufacturing methods, etc. causes quality variations from plant to plant. An e-nose system for black tea quality prediction has two phases, as shown in the Fig. 1, training phase and testing phase. Both the phases consist of three parts, of which the first two parts are same; an array of sensors or sensor unit to acquire the data and signal pre-conditioning unit or electronic unit to make the data suitable for recognition. The third part of the training phase, as shown in Fig. 1a, is to build a pattern classifier model according to the scores given by the human tester to the tea samples under test by e-nose sensors.

On the other hand, the third part of the testing phase, as shown in Fig. 1b, is the prediction of the scores or quality by the trained classifier model. It is desirable that an e-nose system should use regularly to evaluate the quality of black tea. But practically it is difficult due to the non-availability of some generalized computational model capable of predicting accurate tea taster like scores for the quality of tea samples. Developing such universal computational model requires huge experimentation over different seasons and at different tea producing plants. This make the data collection a tough but crucial job because if the pattern classifier is trained with the data set available in one garden during a particular flush then the same model may not give the correct result when subjected to data from a different garden or in a different flush. To encounter this problem the author reported development of computational model using artificial neural network [6] and incremental RBF [7] network, respectively.

Most of the real world data classification problems, such as speaker recognition, diagnosis of diseases, text classification, optical character recognition, etc., require

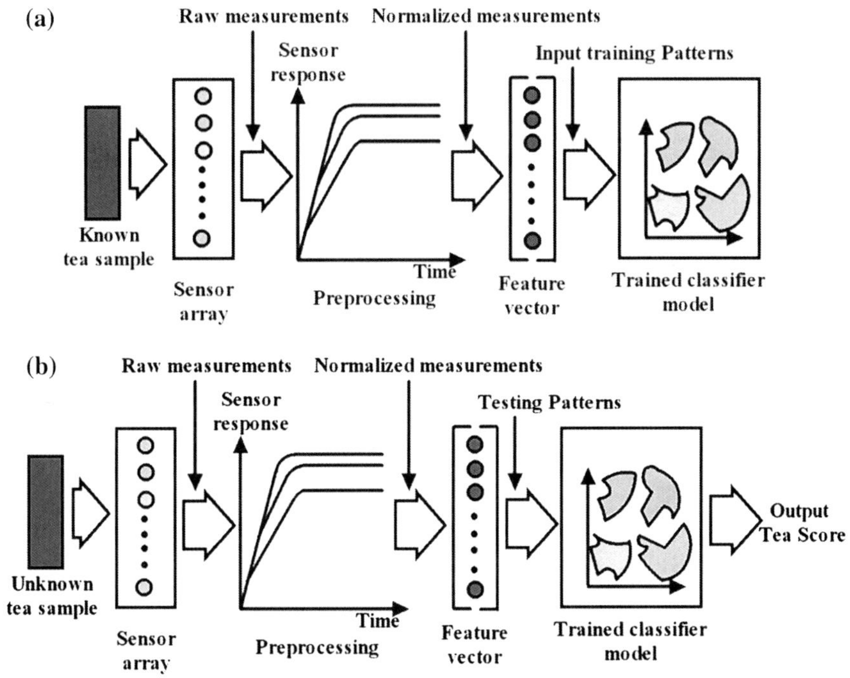

Fig. 1 Typical electronic nose system describing **a** the training phase and **b** testing phase

multiclass data classifier. The tea classification problem is also a multiclass data classification problem. The classification of multiclass data is tricky compared to binary classifier. We can roughly categorize the multiclass data classification approaches in to two groups. Pattern classification algorithms like nearest neighbourhoods [8, pp. 463–483], linear discriminant analysis (LDA) [8, pp. 106–119], neural networks [9], regression and decision trees including C4.5 [10] and CART [11], etc belong to first group of algorithms which gets natural extension from its binary predecessor. The second group of classifiers decomposes a multiclass problem into a number of binary classification problems [12]. A number of binary to multiclass extension approaches, such as, one versus rest (OVR) [13], one versus one (OVO) [14], directed acyclic graph (DAG) [15] and all at once [16], etc. are there in literature. Several binary classifiers can be used in this frame work. But among them support vector machine (SVM) [17, 18] is the most popular binary classifier used in this framework [19]. Machine learning community has observed the growth of this kernel classifiers in the last decade.

In this work, we have revealed the consequence of three types of multiclass kernel classifiers. Among these two classifiers use decomposition technique of multiclass data classification and remaining one classifier uses recently developed regression based multiclass data classification approach for the prediction of black tea quality using e-nose signature. For the first category we have used three different

decompositions of multiclass-support vector machine (SVM) classifiers [19] and Multiclass NPPC [20–22]. For the later type we have used vector-valued regularized kernel function approximation (VVRKFA) [23, 24]. SVM is a widely used machine learning algorithm that has used in almost all applications. Different applications of electronic nose reveal use of SVM methods [25–27] for successful classification of odors of ethanol, methanol, acetone and benzene [28]. Application of least-square SVM classifier also has been reported in order to recognize the gas category [29, 30]. In [31], sensor subset selection for electronic nose was done with one versus one multiclass SVM method. SVM algorithm was applied to electronic nose for predicting the sensor response to odor mixtures [32]. But performance of multiclass-SVM, multiclass NPPC and VVRKFA in tea quality prediction is still unexplored. In this paper we have shown that three decomposition techniques, namely one-versus-rest (OVR), one-versus-one (OVO) and DAG of both SVM and NPPC, and regression-based approach VVRKFA can be highly useful for evolution of universal correlation model between electronic nose signatures and tea tasters' scores.
The rest of the paper is organized as follows: the e-nose setup for data collection is described in Sect. 2. Section 3 describes basic SVM classifier and its three extensions to multiclass classifiers as well as NPPC and VVRKFA method of classification. Section 4 describes the results and associated discussions. Finally, Sect. 5 concludes the work.

2 Data Collection by E-nose

2.1 *Customized Electronic Nose Setup for Black Tea*

The sensor array of an e-nose is one of the most important parts. In this study five gas sensors from Figaro, Japan i.e., TGS-832, TGS-823, TGS-2600, TGS-2610 and TGS-2611 have been used for quality evaluation of black tea aroma. The experimental conditions for black tea quality evaluation have been optimized on the basis of repeated trials and sustained experimentation under the conditions as given in Table 1. During the experiments, dry tea samples have been used to avoid the effect of humidity.

2.2 *Sample Collection*

Tea samples have been used for the experiments from the tea gardens located across north and north-east India of the following industries:

- Glenburn tea estate
- Fulbari tea estate
- Khongea tea estate
- Mateli tea estate

Table 1 Experimental setup

Amount of black tea sample	50 gm
Temperature	60± 3 °C
Headspace generation time	30 s
Collection time	100 s
Purging time	**100 s**
Air-flow rate	**5 mL/s**

Samples of tea produced in their gardens are sent regularly to the tea-tasting centers for quality assessment by the tea tasters. For our experiments, one expert tea taster was deputed by the respective industries to provide taster's score to each of the samples and these scores considered for the correlation study with the computational model. A sample tea taster score sheet is given in Table 2. Here scores given against "aroma" have been considered only to find correlation with electronic nose. Aroma scores basically indicate the smell and flavor of the tea samples whereas liquor scores indicate the combined perception of taste, briskness and astringency of the samples. The scores assigned to leaf quality and infusions are based on visual inspection of the samples by the experts or testers. The aroma scores of Table 2 are used for the training of the computational models of multiclass SVM classifiers as described in the next section.

3 Multiclass Kernel Classifiers

3.1 Support Vector Machine Classifier

Support vector machine (SVM) is originally a binary supervised classification algorithm, introduced by Vapnik and his co-workers [13, 32], based on statistical learning theory. Instead of traditional empirical risk minimization (ERM), as performed by artificial neural network, SVM algorithm is based on the structural risk minimization (SRM) principle. In its simplest form, linear SVM for a two class problem finds an optimal hyperplane that maximizes the separation between the two classes. The optimal separating hyperplane can be obtained by solving the following quadratic optimization problem:

$$\begin{aligned} &\text{Min } Q(w) = \tfrac{1}{2}||w||^2 \\ &\text{s.t. } \; y_i\left(w^T x_i + b\right) \geq 1 \; for \; i = 1, 2, \ldots, m \end{aligned} \tag{1}$$

where, $x_i \in \Re^n$ is the ith training pattern in n-dimensions, $y_i \in \{-1, 1\}$ is the class label of the ith training pattern, $w \in \Re^n$ and $b \in \Re$, respectively, are the normal to the hyperplane and bias term and m is the number of training patterns. This formulation is known as hard SVM and it can be used only when the patterns are linearly separable. In general, most of the practical problems are not linearly

Table 2 Sample tea taster's score sheet

Sample code	Scores (1–10)			
	Leaf quality	Infusion	Liquor	Aroma
KON240804-01	7	5	3	5
KON240804-10	6	5	5	6
KON190704-07	5	4	4	7
KON280704-03	7	5	6	6
KON050604-01	7	5	3	4
MAT070504-01	8	8	8	8
FUL150604-01	7	6	5	7
GLN180604-01	8	7	7	8

separable. As a result modified version of this formulation, known as soft SVM, is used for practical problems. The objective function of the soft SVM is modified by introducing a slack variable as follows:

$$\begin{aligned} &\text{Min } \tfrac{1}{2}||w||^2 + C\sum_{i=1}^{m}\xi_i \\ &\text{s.t. } y_i\left(w^T x_i + b\right) \geq 1 - \xi_i \; for \; i = 1, 2, \ldots.., m \end{aligned} \quad (2)$$

where $\xi(\geq 0)$ is non-negative slack variables or soft margin errors of the ith training sample and $C(\geq 0)$ is a penalty or regularization parameter that determines the trade-off between the maximization of the margin i.e., small $||w||^2$ and minimization of the classification error i.e., small $\sum_{i=1}^{m}\xi_i$. The large value of C leads to a small number of misclassified training samples and the margin width decreases. On the other hand, a small value of C allows large training error and increases the margin width. In this way, C regulates the trade-off between the margin width and the generalization performance. The solution of the quadratic programming (QP) problem (2) is obtained by forming its Lagrangian dual as follows:

$$\begin{aligned} &L(\alpha) = \sum_{i=1}^{m}\alpha_i - \tfrac{1}{2}\sum_{i,j=1}^{m}\alpha_i\alpha_j y_i y_j (x_i^T x_j) \\ &\text{s.t. } \sum_{i=1}^{m}\alpha_i y_i = 0, c \geq \alpha_i \geq 0 \; for \; i = 1, 2, \ldots.., m \end{aligned} \quad (3)$$

where $\alpha_i(0 < \alpha_i < c)$ is the Lagrange multiplier. Finally, solving the above dual problem the expression of the decision hyperplane is obtained as:

$$f(x) = \text{sgn}(w^T x + b) = \text{sgn}(\sum_{i=1}^{m}\alpha_i y_i (x_i^T x) + b) \quad (4)$$

But if the training data are not linearly separable, the obtained classifier may not have high generalization ability although the hyperplanes are determined optimally. In this case, kernel trick is employed to learn nonlinear SVM classifier. In this technique the patterns are mapped from the input space to the high dimensional feature space by using a transformation function $\Phi(.,.)$ and thereby implementing a linear hyperplane that allows linear separation in the feature space. This approach again leads to solving a QP problem similar to (3) with all x_i replaced by $\Phi(x_i)$. Then of course the training algorithm would depend only on the dot product of the patterns in the feature space. This dot product can be computed by using some kernel function:

$$k(x_i, x_j) = \Phi(x_i)^T \Phi(x_j) \tag{5}$$

Thus, using the kernel trick, the Lagrangian dual problem in the feature space will maximize the following problem:

$$\begin{aligned} &\text{Max} \quad L(\alpha) = \sum_{i=1}^{m} \alpha_i - \tfrac{1}{2} \sum_{i,j=1}^{m} \alpha_i \alpha_j y_i y_j k(x_i, x_j) \\ &\text{s.t.} \quad \sum_{i=1}^{m} \alpha_i y_i = 0, \quad c \geq \alpha_i \geq 0 \; for \; i = 1, 2, \ldots., m \end{aligned} \tag{6}$$

The corresponding decision function becomes:

$$f(x) = \text{sgn}\left(\sum_{i=1}^{m} y_i \alpha_i (\Phi(x_i)^T \Phi(x)) + b \right) \tag{7}$$

The above decision function can be expressed in terms of kernel trick as follows:

$$f(x) = \text{sgn}\left(\sum_{i=1}^{m} y_i \alpha_i k(x_i, x) + b \right) \tag{8}$$

The expression (8) is used for testing a new pattern by the trained classifier. There are many possible kernels, such as linear, Gaussian, polynomial and multilayer perceptron etc. In this study, we have used polynomial and Gaussian (RBF) kernel functions, respectively, of the form as given in (9) and (10) below:

$$k(x_i, x_j) = (1 + x_i^T x_j)^d \tag{9}$$

$$k(x_i, x_j) = \exp(-\gamma \left\| x_i - x_j \right\|^2) \tag{10}$$

Originally SVM was developed for binary data classification. The excellent performance of it tempted the researchers to find out the efficient way to extend it as multiclass classifier [19]. Brief descriptions of the three commonly used extension methods are explained below.

3.2 One-Versus-Rest Method

In OVR method N numbers of binary classifier models are constructed for a N class problem [13, 19]. In doing so ith binary classifier model is trained with all the training data of ith class as positive samples and all other classes data as the negative samples. In the Fig. 2 OVR method of decomposition of a multiclass problem is illustrated containing four different classes. For a four class problem, four binary classifier models are trained with taking each class once as positive samples and the other as negative samples. To test a new pattern it is presented to all the binary models and assigned a class label which has the largest value of the decision function obtained by (4).

3.3 One-Versus-One (OVO) method

In contrast to OVR, OVO method trains ${}^{N}C_2 = N(N-1)/2$ binary classifier models considering each possible pair for a N class problem. Fig. 3 illustrates how six binary classifiers are trained in OVO method for a four class problem. The testing of a sample in OVO method is performed by max win strategy [33]. By this technique each of the trained binary classifier can deliver one vote for its favored class and the class with maximum votes specifies the class label of the sample. Thus as the number of classes increases the training and testing time also increase in this method.

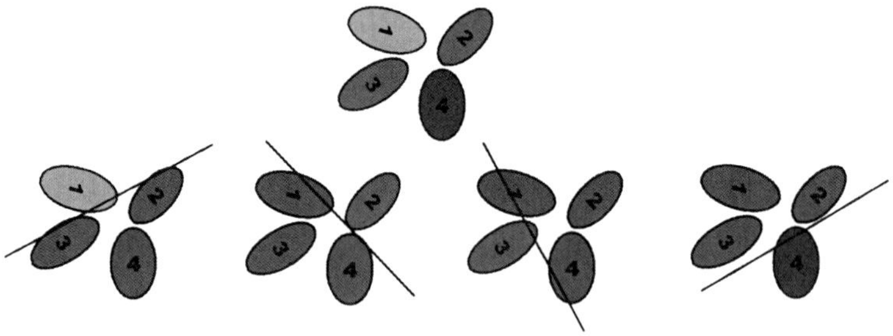

Fig. 2 One-versus-rest method of decomposition of a multiclass problem consisting of four different classes. For a four class problem, four binary classifiers are trained

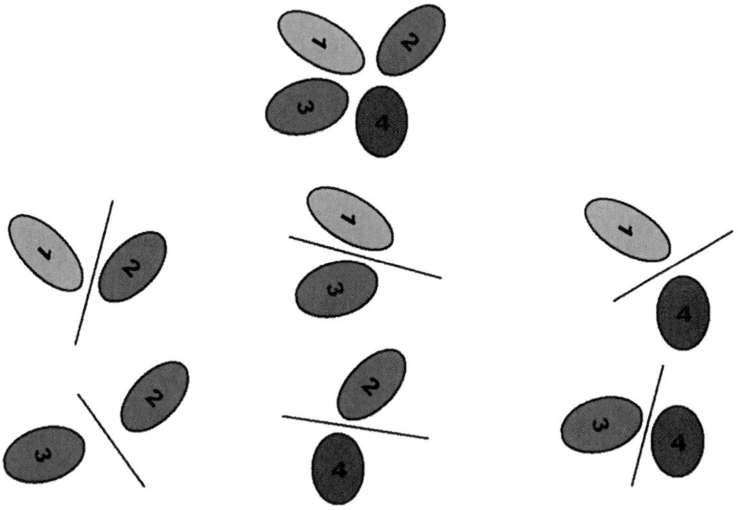

Fig. 3 OVO method of decomposition of a multiclass classification problem containing four different classes. For a four class problem it trains six binary classifiers with taking two different classes at a time

3.4 Directed Acyclic Graph Method

To reduce the testing time of OVO method Platt introduced DAG method [15]. The training part of the DAG method is same as that of the OVO method. But DAG method of testing follows a rooted binary directed acyclic graph to make a decision as shown in Fig. 4. The final decision is made when a test sample reaches to the leaf node. Thus, both OVO and DAG methods have the same training phase but during testing DAG uses only (N-1) evaluations to make decision among $N(N$-1)/2 binary classifiers. Accordingly, for a four class problem, DAG decides the class label by three comparisons only out of six classifiers as shown in Fig. 4. Hence DAG method reduces the testing time of a sample than by OVO method.

3.5 Nonparallel Plane Proximal Classifier (NPPC)

Nonparallel plane proximal classifier (NPPC) is a recently developed kernel classifier that classifies a pattern by the proximity of it to one of the two nonparallel hyperplanes [21, 22]. The advantage of the NPPC is that its training can be accomplished by solving two systems of linear equations instead of solving a quadratic program as it requires for training standard SVM classifiers [17, 18] and its performance is comparable to that of the SVM classifier. This fact motivated us to evaluate the performance of multiclass-NPPC in tea quality prediction.

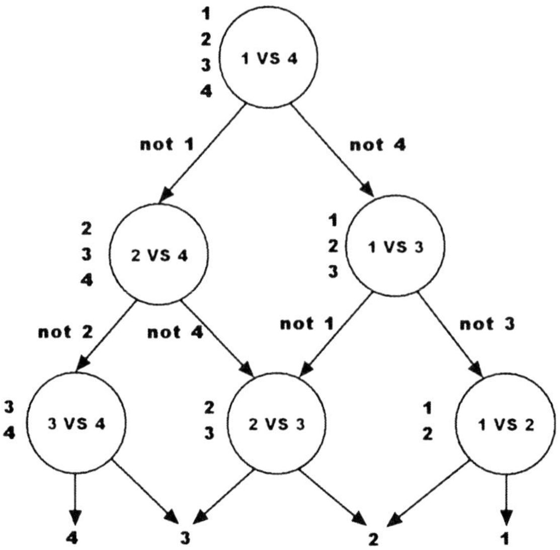

Fig. 4 Testing of a sample by DAG method with four different classes

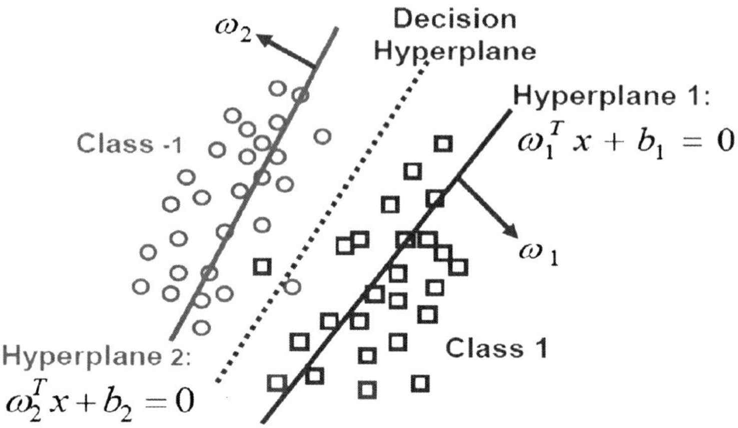

Fig. 5 Graphical representation of NPPC for a two dimensional data classification problem

NPPC [22] is a binary classifier and it classifies a pattern by the proximity of a test pattern to one of the two planes as shown in Fig. 5. The two planes are obtained by solving two nonlinear programming problems (NPP) with a quadratic form of loss function. Each plane is clustered around a particular class of data by minimizing sum squared distances of patterns from it and considering the patterns of the others class at a distance of 1 with soft errors. Thus, the objective of NPPC is to find two hyperplanes:

$$\omega_1^T x + b_1 = 0 \text{ and } \omega_2^T x + b_2 = 0 \tag{11}$$

where $\omega_1, \omega_2 \in \Re^n$ and $b_1, b_2 \in \Re$ are normal vectors and bias terms of the hyperplanes 1 and 2, respectively. To obtain the above two planes, NPPC solves the following pair of problems:

$$\begin{aligned} &\underset{(\omega_1, b_1, \xi_2)}{\text{Min}} \ \tfrac{1}{2} \left\| A\omega_1 + e_1 b_1 \right\|^2 + c_1 e_2^T \xi_2 + \tfrac{c_2}{2} \left\| \xi_2 \right\|^2 \\ &\text{s.t.} \quad -(B\omega_1 + e_2 b_1) + \xi_2 = e_2 \text{ and } \left\| \omega_1 \right\| = 1 \end{aligned} \tag{12}$$

$$\begin{aligned} &\underset{(\omega_2, b_2, \xi_1)}{\text{Min}} \ \tfrac{1}{2} \left\| B\omega_2 + e_2 b_2 \right\|^2 + c_3 e_1^T \xi_1 + \tfrac{c_4}{2} \left\| \xi_1 \right\|^2 \\ &\text{s.t.} \quad (A\omega_2 + e_1 b_2) + \xi_1 = e_1 \text{ and } \left\| \omega_2 \right\| = 1 \end{aligned} \tag{13}$$

Here, matrices $A \in \Re^{m_1 \times n}$ and $B \in \Re^{m_2 \times n}$ contain the m_1 and m_2 training patterns of class 1 and class -1 respectively in n dimensional space, $e_1 \in \Re^{m_1}$ and $e_2 \in \Re^{m_2}$ are vectors of ones and $\xi_1 \in \Re^{m_1}, \xi_2 \in \Re^{m_2}$ are the error variable vectors due to class 1 and class -1 data respectively and $c_1, c_2, c_3, c_4 > 0$ are four regularization parameters of the NPPC. The first term of each objective function minimizes the sum of the squared distances from the hyperplane to the patterns of respective class and the constraint requires that the patterns of opposite class are at a distance of 1 from the hyperplane with soft errors. The second and third terms of the objective functions constitute the general quadratic loss function. A new test sample $x \in \Re^n$ is assigned to a class l by comparing the following distance measure of it from the two hyperplanes given by (11), i.e.:

$$Class\, l = \underset{l=1,2}{\text{Min}} \left| \omega_l^T x + b_l \right| \tag{14}$$

The above formulation is known as linear NPPC. When the patterns are not linearly separable then one can use nonlinear NPPC. The linear NPPC can be extended to nonlinear classifiers by applying the kernel trick [18]. For nonlinearly separable patterns, the input data is first mapped into a higher dimensional feature space by some kernel function. In the feature space it implements a linear classifier which correspond a nonlinear separating surface in the input space. To apply this transformation, let $k(.,.)$ be any nonlinear kernel function and define the augmented matrix:

$$C = [A \quad B]^T \in \Re^{m \times n} \tag{15}$$

where $m_1 + m_2 = m$ the total patterns in the training set. Thus, nonlinear NPPC [22] finds two kernel generated surfaces:

$$k(x^T, C^T)\theta_1 + b_1 = 0 \quad and \quad k(x^T, C^T)\theta_2 + b_2 = 0 \tag{16}$$

by solving the following pair of objective functions:

$$\begin{aligned} &\underset{(\theta_1, b_1, \xi_2)}{\text{Min}} \quad \tfrac{1}{2}\|k(A, C^T)\theta_1 + e_1 b_1\|^2 + c_1 e_1^T \xi_2 + \tfrac{c_2}{2}\xi_2^T \xi_2 \\ &\text{s.t.} \quad -(k(B, C^T)\theta_1 + e_2 b_1) + \xi_2 = e_2 \\ &\mathit{and} \quad \|\theta_1\| = 1 \end{aligned} \tag{17}$$

and

$$\begin{aligned} &\underset{(\theta_2, b_2, \xi_1)}{\text{Min}} \quad \tfrac{1}{2}\|k(B, C^T)\theta_2 + e_2 b_2\|^2 + c_3 e_1^T \xi_1 + \tfrac{c_4}{2}\xi_1^T \xi_1 \\ &\text{s.t.} \quad -(k(A, C^T)\theta_2 + e_1 b_2) + \xi_1 = e_1 \\ &\mathit{and} \quad \|\theta_2\| = 1 \end{aligned} \tag{18}$$

Here, $\theta_1, \theta_2 \in \Re^m$ are the normal vectors to the kernel generated surfaces by the objective functions (17) and (18) respectively. A new test sample $x \in \Re^n$ is assigned to a class $j (j = 1, 2)$, by using the classification rule:

$$class(x) = \underset{j=1,2}{\arg\min}\ (|k(x^T, C^T)\theta_j + b_j|) \tag{19}$$

by comparing the Euclidean distance measure of it from the two kernel generated hypersurfaces given by (16).
This idea of classifying a pattern by the proximity to one of the two hyperplanes of binary NPPC can be extended to multiclass NPPC by using decomposition techniques as described in previous section for SVM classifiers. In [20], the authors have discussed these methods in details and therefore omitted here.

3.6 Vector-Valued Regularized Kernel Function Approximation

The problems of the decomposition techniques, as described above, with a large number of categories are: (i) large training time for large data set. (ii) Large prediction time of a test sample. (iii) Decomposition techniques often suffer from the data unbalance problem which causes a biased classifier. (iv) The parameter tuning task becomes tedious for large data set. In order to overcome these difficulties of decomposition techniques Ghorai et al. [23] proposed a regression based multiclass data classification technique, called VVRKFA, as described below.
Unlike other multiclass classifiers VVRKFA [23, 24], utilizes the regression or function approximation approach to map the feature vectors to its label vectors in a space whose dimension is equal to the number of categories present in the classification problems. The classification is performed in this label space by comparing the Mahalanobis distance of a test pattern from the respective class centroids. The idea of VVRKFA classifier is graphically illustrated in Fig. 6 for a three class classification problem. The performance of VVRKFA classifier is comparable to that of SVM classifier and it takes less time to train a large data set [23].

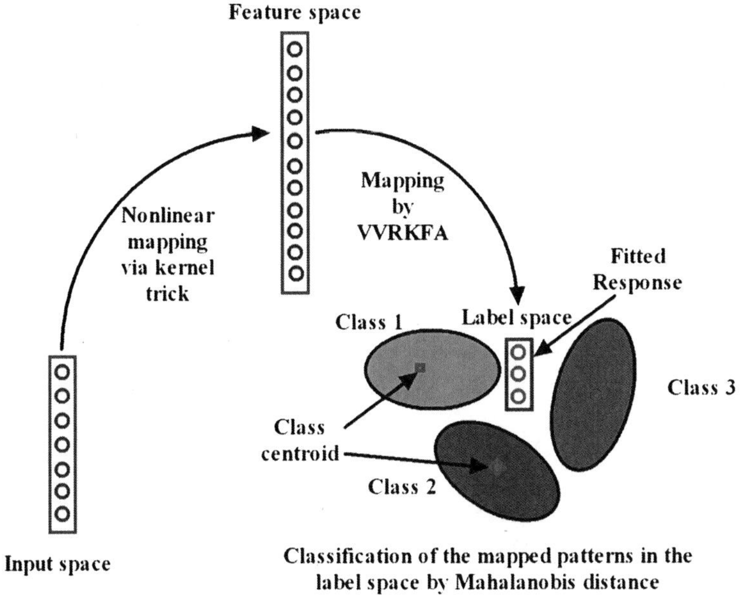

Fig. 6 VVRKFA method of classification with three different classes

Let us consider a Nclass data classification problem. The label vector y_i of a sample $x_i \in \Re^n$ of jth class is chosen as the indication vector of the classes according to the following rule:

$$y_i = [y_{1i}, y_{2i}, \ldots, y_{Ni}]^T \text{with } y_{ji} = 1 \text{and } y_{ki} = 0 \text{ for } k \neq j \in N \quad (20)$$

The VVRKFA is realized on these training samples $(x_i, y_i) \in (\Re^n, \Re^N)$ by solving the following optimization problem:

$$\begin{aligned} \text{Min} \quad & J(W, b, \xi) = \tfrac{C}{2}\text{tr}\left([\, W\ b]^T [\,\Theta\ b]\right) + \tfrac{1}{2}\sum_{i=1}^{m} \|\xi_i\|^2 \\ \text{s.t.} \quad & Wk(x_i^T, B^T)^T + b + \xi_i = y_i, \quad i = 1, 2, \ldots, m. \end{aligned} \quad (21)$$

In (21), $W \in \Re^{N \times \bar{m}}$ is the regression coefficient matrix that maps the inner products of the feature space to the label space and $b \in \Re^{N \times 1}$ is the bias vector, $\xi_i \in \Re^{N \times 1}$ is the slack or error variable vector and C is a regularization parameter. The matrix $B \in \Re^{\bar{m} \times n}$ is the kernel basis formed by randomly picking up $\bar{m}(\square\ m)$ training patterns from the training set prior to the training. Let, the matrix $A \in \Re^{m \times n}$ represents the training set. Then, the minimizing problem (12) reduces to an unconstrained minimization problem as:

$$\min J(W, b) = \frac{C}{2}\mathrm{tr}(\begin{bmatrix} W & b \end{bmatrix}^T \begin{bmatrix} W & b \end{bmatrix}) + \frac{1}{2}\left\| Wk(A, B^T)^T - be^T - Y \right\|_F^2 \tag{22}$$

where $Y \in \Re^{N\times m}$ is the matrix containing m label vectors of the training data and $\|\bullet\|_F$ is Frobenius norm of any matrix. Optimization problem (22) can be simplified as

$$\min J(\Theta) = \frac{C}{2}\mathrm{tr}(\Theta^T\Theta) + \frac{1}{2}\|\Theta P - Y\|_F^2 \tag{23}$$

where $\Theta = \begin{bmatrix} Wb \end{bmatrix} \in \Re^{N\times(\bar{m}+1)}$ and $P = \begin{bmatrix} k(A, B^T) & e \end{bmatrix}^T \in \Re^{(\bar{m}+1)\times m}$ is augmented matrix. The global solution of (23) can be obtained by differentiating it with respect to Θ and equating it to zero (NULL matrix), i.e.:

$$\begin{aligned} C\Theta + [\Theta P - Y]P^T &= \mathbf{0} \\ \Leftrightarrow \Theta &= YP^T[CI + PP^T]^{-1} \end{aligned} \tag{24}$$

Once Θ is found, the expression of $W \in \Re^{N\times\bar{m}}$ and $b \in \Re^{N\times 1}$ can also be obtained from it. Finally, the vector valued mapping function to map a feature vector into the low dimensional label space becomes:

$$\rho(x_i) = Wk(x_i^T, B^T)^T + b \tag{25}$$

A test pattern $x_t \in \Re^n$ is assigned to a class j $(j = 1, \ldots, N)$, by comparing its Mahalanobis distance from the respective class centroids $\bar{\rho}^{(j)} = \frac{1}{m_j}\sum_{i=1}^{m_j}\hat{\rho}(x_i)$, i.e.,

$$Class(x_t) = \arg\min_{1\le j\le N} d_M(\hat{\rho}(x_t), \bar{\rho}^{(x_j)}|\hat{\Sigma}) \tag{26}$$

where $\hat{\Sigma} = \sum_{j=1}^{N}(m_j - 1)\hat{\Sigma}^{(j)}/(m - N)$ is the pooled within class covariance matrix and d_M is the Mahalanobis distance.

4 Results and Discussions

4.1 Experimental Setup

Experimentations with e-nose have been performed with 174 tea samples that were collected from Khongea tea estate, Mateli tea estate, Gelburn tea estate and Fulbari tea estate of north-east India. The multi-sensor array in the e-nose system consists of five sensors and they produce five readings for eachsample. It has been observed that the taster's score varies within a range from 3 to 8. So, this black tea quality prediction problem becomes a six-category classification problem. We have computed the performances of multiclass SVM, NPPC and VVRKFA classifiers by using both polynomial and Gaussian kernel. To implement the multiclass SVM classifier we have used LIBSVM Toolbox [34]. VVRKFA and NPPC are implemented in MATLAB as described in [24]. The optimal value of the regularization parameter C for SVM and c_1, c_2, c_3 and c_4 for NPPC are selected from the set of values $\{2^i \,|i = -5, -4, \ldots\ldots, 15\}$, whereas C for VVRKFA is selected from the set $\{10^i \,|i = -6, -5, \ldots\ldots, -1\}$. The Gaussian kernel parameter γ for all the methods was selected from the set of values $\{2^i \,|i = -9, -8, \ldots\ldots, 9\}$ and the degree of polynomial kernel is selected from the set $\{2, 3, 4, 5, 6\}$. The parameters are selected by the performance on a tuning set comprising of 20 % of training data while the remaining 80 % training data are used to train the classifier [35]. Once the parameter set is selected the training and testing sets are merged to train the final classifier model using the selected parameters. We have evaluated the performance of e-nose system using both 10-fold cross validation (10-fold CV) and leave-one-out cross validation (LOOCV) method [35]. In the 10-fold CV method the data are divided into 10 subsets. Out of these 10 subsets, one subset is used as the test set and other 9 subsets are used as the training set. As a result, 10 train-test trials were conducted in this method. The classification rates are then averaged over these 10-folds for estimating the average classification accuracy. In LOOCV the number of folds is equal to the total number of training patterns in the data set. Therefore, only one pattern is used to test the classifier while the rest are used to train the classifier in each fold. Each training feature is normalized in the range [0 1]. In order to calculate the accuracy, random permutations of the data points are performed before proceeding to the tests.

4.2 Experimental Results

In Tables 3 and 4 we have shown the performances of 10-fold CV and LOOCV method, respectively, using OVR, OVO and DAG method of decomposition of multiclass SVM and NPPC as well as VVRKFA classifier considering both RBF and polynomial kernels. In each table we have reported average percentage testing accuracy and standard deviation with the selected parameters for each model. The best accuracy offered by a particular method in each table is made bold.

From Table 3 it is observed that all the models of SVM can produce an accuracy greater than 94 %. But the performances of different multiclass NPPC models are not as good as SVM classifier models. Whereas the performance of VVRKFA classifier is the best among all the models considering either RBF kernel (98.33 %) or polynomial kernel (98.27 %) with very low standard deviations.

Table 3 10-fold cross validation performance of different classifier models

Classifier	Kernel type	Selected parameters		Avg. % testing accuracy (std. dev.)
		C	γ or d	
OVR SVM	RBF	256	4	97.68 (3.87)
	Polynomial	**8192**	**2**	96.01 (4.38)
OVO SVM	RBF	4	32	95.36 (5.11)
	Polynomial	2	3	94.84 (5.94)
DAG SVM	RBF	512	2	97.68 (3.87)
	Polynomial	1024	2	96.01 (5.11)
OVR NPPC	RBF	$c_1 = c_3 = 2^{-7}, c_2 = c_4 = 2^{-3}$	32	94.18 (5.88)
	Polynomial	$c_1 = c_3 = 2^{-8}, c_2 = c_4 = 2^{-5}$	4	87.39 (8.57)
OVO NPPC	RBF	$c_1 = c_3 = 2^{-6}, c_2 = c_4 = 2^{-4}$	8	94.84 (6.07)
	Polynomial	$c_1 = c_3 = 2^{-5}, c_2 = c_4 = 2^{-3}$	4	93.17 (4.89)
DAG NPPC	RBF	$c_1 = c_3 = 2^{-8}, c_2 = c_4 = 2^{-5}$	4	95.42 (5.03)
	Polynomial	$c_1 = c_3 = 2^{-8}, c_2 = c_4 = 2^{-5}$	4	93.17 (5.55)
VVRKFA	RBF	10^{-6}	0.50	**98.33 (3.56)**
	Polynomial	10^{-5}	4	98.27 (2.65)

Table 4 LOOCV performance of different multiclass SVM models

Classifier	Kernel type	Selected parameters		Avg. percentage testing accuracy (std. dev.)
		C	γ or d	
OVR SVM	RBF	256	4	**99.43 (7.56)**
	Polynomial	8192	2	96.55 (18.25)
OVO SVM	RBF	4	32	95.40 (20.94)
	Polynomial	2	3	95.40 (20.94)
DAG SVM	RBF	512	2	98.28 (13.02)
	Polynomial	1024	2	97.13 (16.71)
OVR NPPC	RBF	$c_1 = c_3 = 2^{-7}, c_2 = c_4 = 2^{-3}$	32	93.68 (24.34)
	Polynomial	$c_1 = c_3 = 2^{-8}, c_2 = c_4 = 2^{-5}$	4	86.21 (34.48)
OVO NPPC	RBF	$c_1 = c_3 = 2^{-6}, c_2 = c_4 = 2^{-4}$	8	95.40 (20.94)
	Polynomial	$c_1 = c_3 = 2^{-5}, c_2 = c_4 = 2^{-3}$	4	92.53 (26.29)
DAG NPPC	RBF	$c_1 = c_3 = 2^{-8}, c_2 = c_4 = 2^{-5}$	4	95.40 (20.94)
	Polynomial	$c_1 = c_3 = 2^{-8}, c_2 = c_4 = 2^{-5}$	4	93.68 (24.34)
VVRKFA	RBF	10^{-6}	0.50	98.28 (13.02)
	Polynomial	10^{-5}	4	97.70 (14.29)

On the other hand, from Table 4 it is observed that the OVR multiclass SVM with RBF kernel achieves the highest classification accuracy of 99.43 % with a standard deviation of 7.56 %. This performance is slightly better that VVRKFA classifier with RBF kernel which is 98.28 % with a standard deviation of 13.02 %. The LOOCV performance of NPPC is again slightly less compared to SVM or VVRKFA classifier

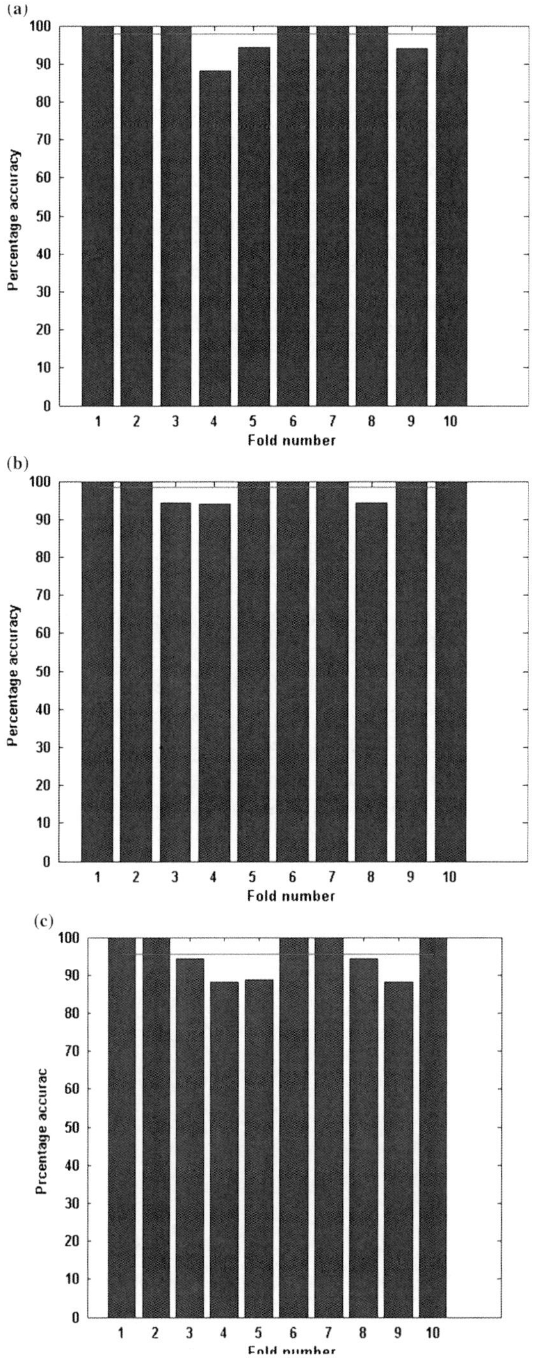

Fig. 7 Bargraphs showing the 10-fold CV performance of multiclass **a** OVR SVM **b** VVRKFA and **c** DAG-NPPC model

with large standard deviations. The percentage standard deviation of LOOCV is greater than the 10-fold CV. This is because in the first case there is only one test sample. As a result the testing accuracy in each fold is either 100 % when it is accurately classified or 0 % when it is misclassified. This is the reason for obtaining high standard deviation among the classification accuracies in different folds.
In Fig. 7 we have shown visually the best performances of multiclass OVR SVM, VVRKFA and DAG NPPC models obtained experimentally in the 10-fold CV method. The Fig. 7 shows that the OVR SVM and VVRKFA models with RBF kernels achieve high classification accuracy in all the 10 individual folds. In view of all these results, it is observed that the model trained with OVR multiclass SVM and VVRKFA with Gaussian RBF kernel may be suitable for black tea quality prediction application.

5 Conclusions

The objective of this study is to develop a taster-specific computational model for the prediction of tea quality scores using an electronic nose. This work evaluates the performance three different types of multiclass kernel classifiers, namely SVM, NPPC and VVRKFA. Among these SVM and NPPC employ decomposition techniques of multiclass data classification and VVRKFA uses regression approach of multiclass data classification. From the experimental results of 10-fold CV and LOOCV it is observed that VVRKFA performs well in both the cases, whereas the performance of OVR SVM is the best among all in LOOCV method. Multiclass data usually suffers from data unbalance problem, i.e., one of the classes may have many more samples than the other. As a result the decision functions are usually biased by the dominating sample class. But in this application it is seen that this problem has not affected the trained classifier as the data set is a balance one. The observed small value of standard deviation of testing accuracy for all the cases signifies the fact that the mean value of testing accuracy is independent of the training set. The results obtained using presented method were encouraging and also resulted merits like fast, highly generalized and user friendly. Therefore, the model may be potentially helpful in the tea tasting centers in order to predict black tea quality.

References

1. R. Dutta, E.L. Hines, J.W. Gardner, K.R. Kashwan, M. Bhuyan, Tea quality prediction using a tin oxide-based electronic nose: An artificial intelligence approach. Sens. Actuators B: Chem. **94**(2), 228–237 (2003)
2. M. Pardo, G. Sberveglieri, Coffee analysis with an electronic nose. IEEE Trans. Instrum. Meas. **51**(6), 1334–1339 (2002)
3. D.D.H. Boothe, J.W. Arnold, Electronic nose analysis of volatile compounds from poultry meat samples, fresh and after refrigerated storage. J. Sc. Food Agric. **82**(3), 315–322 (2002)

4. J. Lozano, J.P. Santos, M. Aleixandre, I. Sayago, J. Gutierrez, M.C. Horrillo, Identification of typical wine aromas by means of an electronic nose. IEEE Sensor J. **6**(1), 173–178 (2006)
5. M. O'Connel, G. Valdora, G. Peltzer, R. Martin Negri, A practical approach for fish freshness determinations using a portable electronic nose. Sens. Actuators, B: Chem. **80**(2), 149–154 (2001).
6. N. Bhattacharyya, R. Bandyopadhyay, M. Bhuyan, B. Tudu, D. Ghosh, A. Jana, Electronic nose for black tea classification and correlation of measurements with "Tea Taster" marks. IEEE Trans. Instrum. Meas. **57**(7), 1313–1321 (2008)
7. B. Tudu, A. Jana, A. Metla, D. Ghosh, N. Bhattacharyya, R. Bandopadhyay, Electronic nose for black tea quality evaluation by an incremental RBF network. Sens. Actuat. B: Chem. **138**, 90–95 (2009)
8. T. Hastie, R. Tibshirani and J. Friedman, *The Elemetns of Statistical Learning: Data mining, Inference, Prediction*, 2nd edn. (Springer, New York, 2009)
9. R. O. Duda, P. E. Hart and D. J. Stork, *Pattern Classification*, 2nd edn. (John Wiley & Sons, Inc., Singapore, 2006)
10. J. R. Quinlan, *C4.5: Programs for Machine Learning* (Morgan Kaufmann Inc., 1993)
11. L. Breiman, J. Friedman, C.J. Stone, R.A. Olshen, *Classification and Regression Trees* (Chapman and Hall, New York, 1993)
12. E.L. Allwein, R.E. Schapire, Y. Singer, Reducing multiclass to binary: a unifying approach for margin classifiers. J. Mach. Learn. Res. **1**, 113–141 (2000)
13. L. Bottou, C. Cortes, J. Denker, H. Drucker, I. Guyon, L. Jackel, Y. LeCun, U. Muller, E. Sackinger, P. Simard, V. Vapnik, Comparison of classifier methods: A case study in hand writing digit recognition. in *Proceedings of the International Conference Pattern Recognition*, 1994, pp. 77–87
14. U. Kreßel, Pairwise Classification and Support Vector Machines, in *Advances in Kernel Methods-Support Vector Learning*, ed. by B. Schölkopf, C. J. C. Burges, A. J. Smola (MIT Press, Cambridge, 1999), pp. 255–268
15. J.C. Platt, N. Cristianini, J. Shawe-Taylor, Large margin DAG's for multiclass classification. Adv. Neu. Inform. Process. Syst. **12**, 547–553 (2000)
16. K. Crammer and Y. Singer, On the learn ability and design of output codes for multiclass problems. Mach. Learning **47**(2–3), 201–233 (2002)
17. V.N. Vapnik, *The nature of statistical learning theory* (John Wiley & Sons, Inc., New York, 1998)
18. N. Cristianini and J. Shawe-Taylor, *An Introduction to Support Vector Machines*, (Cambridge University Press, Cambridge, 2000)
19. C.-W. Hsu, C.-J. Lin, A comparison of methods for multiclass support vector machines. IEEE Trans. N. Nets. **13**(2), 415–425 (2002)
20. S. Ghorai, A. Mukherjee, S. Sengupta, P. K. Dutta, Multicategory Cancer Classification from Gene Expression Data by Multiclass NPPC Ensemble. in *IEEE International Conference on Systems in Medicine and Biology*, (IIT-Kharagpur, 2010)
21. S. Ghorai, A. Mukherjee, P.K. Dutta, Nonparallel plane proximal classifier. Sign. Process. **89**(4), 510–522 (2009)
22. S. Ghorai, S.J. Hossain, A. Mukherjee, P.K. Dutta, Newton method for nonparallel plane proximal classifier with unity norm hyperplanes. Sign. Proces. **90**(1), 93–104 (2010)
23. S. Ghorai, A. Mukherjee, P.K. Dutta, Discriminant analysis for fast multiclass data classification through regularized kernel function approximation. IEEE Trans. Neu. Netw **21**(6), 1020–1029 (2010)
24. S. Ghorai, A. Mukherjee, P.K. Dutta, *Advances in Proximal Kernel Classifiers* (LAP LAMBERT Academic Publishing, Germany, 2012)
25. M. Pardo, G. Sberveglieri, Classification of electronic nose data with Support vector machines. Sens. Actuat. B: Chem. **107**, 730–737 (2005)
26. M. Gaudioso, W. Khalaf, On the Use of the SVM Approach in Analyzing an Electronic Nose, in *Proceedings of the 7th international Conference on Hybrid Intelligent Systems* (Kaiserslautern, 2007), pp. 42–46

27. L. Xie, X. Wang, Gas quantitative analysis with support vector machine. in *Proceedings of Chinese Control and Decision Conference*, (Jinhua, 2009), pp. 2362–2366
28. W. Khalaf, C.P.M. Gaudioso, Least square regression method for estimating gas concentration in an electronic nose system. Sensors **9**(3), 1678–1691 (2009)
29. X. Wang, J. Chang, K. Wang, M. Ye, Least square support vector machines in combination with principal component analysis for electronic nose data classification, in *Proceedings of 2nd International Symposium on Information Science and Engineering* (Shanghai, China, 2009), pp. 348–352
30. E. Phaisangittisagul, Improving sensor subset selection of machine olfaction using multiclass SVM, in *Proceedings of 3rd International Conference on Knowledge Discovery and Data Mining* (Bangkok, 2010), pp. 28–31
31. E. Phaisangittisagul, Approximating sensors' responses of odor mixture on machine olfaction, in *Proceedings of International Conference on Artificial Intelligence and Computational Intelligence*, vol. 2 (2009), pp. 60–64
32. C. Cortes, V.N. Vapnik, Support Vector Networks. Mach. Learn. **20**(3), 273–297 (1995)
33. J. Friedman, Another approach to polychotomous classification, Department of Statistics, Stanford University Stanford, Technical Report. Available at: http://www-stat.stanford.edu/reports/friedman/poly.ps.Z, 1996
34. C. C. Chang and C. J. Lin, LIBSVM: A library for support vector machines. Technical Reort, 2001. Available at: http://www.csie.ntu.edu.tw/~cjlin/libsvm
35. T. M. Mitchell, *Machine Learning*, (The McGraw-Hill Companies, Inc., Singapore, 1997), p 148 (Chapter 5)

Chapter 8
Electronic Nose Setup for Estimation of Rancidity in Cookies

D. Chatterjee, P. Bhattacharjee, H. Lechat, F. Ayouni, V. Vabre and N. Bhattacharyya

Abstract Rancidity is a major problem in food systems which occurs through lipid oxidation and significantly affects quality of foods. Owing to the drawbacks of conventional methods of estimating lipid oxidation in foods, alternative inexpensive technology such as electronic nose (e-nose) technology that can accurately and rapidly determine odor profile of foods, is gaining importance. First section of this chapter presents various applications of e-nose analysis of several food systems. Latter section of this chapter focusses on assessment of shelf-life of cookies formulated with un-encapsulated and encapsulated eugenol-rich clove extracts (obtained by supercritical carbon dioxide extraction from clove buds) as a source of natural antioxidant using ALPHA MOS e-nose system and a customized e-nose system. The e-nose results were further validated by conventional methods of rancidity determination. Shelf-life study for a storage period of 200 days revealed that normal cookies without any antioxidant have shelf-life of 100 days when stored in aluminium laminates at $23 \pm 2\,^{\circ}C$, while administration of encapsulated clove extract as antioxidant prevented rancidity in cookies for at least 200 days with a shelf-life lead of 100 days over control cookies. These observations were further affirmed by phytochemical analyses which indicated that the addition of encapsulated clove extract in cookies enhanced its nutraceutical potential and antioxidant activity for at least 200 days. Finally, linear regression equations were developed using which rancidity parameters of conventional biochemical assays of cookies can be directly derived from e-nose sensor responses.

D. Chatterjee · P. Bhattacharjee (✉)
Department of Food Technology and Biochemical Engineering, Jadavpur University, Kolkata 700032 , India
e-mail: pb@ftbe.jdvu.ac.in

H. Lechat, F. Ayouni · V. Vabre
Alpha-Mos, 20 rue Didier Daurat Avenue, 31200 Toulouse , France

N. Bhattacharyya
Centre for Development of Advanced Computing (C-DAC),
E-2/1 Block – GP Sector – V Salt Lake, Kolkata 700091, India

A. Mason et al. (eds.), *Sensing Technology: Current Status and Future Trends II*, Smart Sensors, Measurement and Instrumentation 8, DOI: 10.1007/978-3-319-02315-1_8,

Keywords Electronic nose · Cookies · Shelf-life study · Encapsulated and un-encapsulated clove extract as source of natural antioxidant

1 Introduction

Quality of a food can be defined as the degree of excellence which includes sensory properties, safety factors and its nutritional value. Lipid oxidation is the most important phenomenon in food systems causing rancidity which significantly affects quality of foods [1].

During lipid oxidation, the primary oxidation products that are formed by the autoxidation of unsaturated lipids are hydroperoxides, which have little or no direct impact on the sensory properties of foods. However, hydroperoxides are degraded to produce additional radicals which further accelerates the oxidation process and produce secondary oxidation products such as aldehydes, ketones, acids and alcohols, of which some are volatiles with very low sensory thresholds and have potentially significant impact on the sensory properties namely odor and flavor [2, 3]. Sensory analysis of food samples are performed by a panel of semi to highly trained personnel under specific quarantined conditions. Any chemical method used to determine lipid oxidation in food must be closely correlated with a sensory panel because the human nose is the most appropriate detector to monitor the odorants resulting from oxidative and non-oxidative degradation processes. The results obtained from sensory analyses provide the closest approximation to the consumers' approach. Sensory analyses of smell and taste has been developed in many studies of edible fats and oils and for fatty food quality estimation [1, 4, 5].

However, sensory analysis has some drawbacks originating from the subjectivity of human panels, which can give conflicting measurements when oxidized samples are analyzed. Moreover, sensory methods are usually time-consuming and characterized by poor reproducibility of the data. Therefore, oxidative changes in the food samples are assessed by either classical titrimetric methods or by instrumental analysis. Among the classical methods, the primary oxidation in foods are determined by estimation of acid value and peroxide value; whereas, the secondary oxidation products are estimated by *p*-anisidine value and from content of malonaldehyde and conjugated dienes. However, these titrimetric methods are time consuming, less reliable, less accurate and are of low sensitivity. Moreover, these procedures require additional lipid extraction steps from food samples that may produce unwanted interfering artifacts.

Lipid oxidation is also estimated by instrumental analyses such as by Rancimat test method and used for accelerated stability study [6]. Electron spin resonance can measure radical formation in many types of food matrices [7]. Analysis of volatiles with dynamic headspace/GC-MS is highly sensitive and can provide lot of information with regard to volatile lipid oxidation products and other volatiles with sensory impact in food samples. This type of data can also give information on possible reaction pathways of food deterioration [8]. Front-face fluorescence spectroscopy is another fast and nondestructive technique, reported to measure lipid oxidation in various types of poultry meat and meat loaves [9]. The basis for this method is that lipid oxidation products (hydroperoxides or aldehydes) can combine with primary amine groups in amino acids/ proteins/ peptides/ DNA and to reaction products that fluoresce when they are illuminated. The emitted fluorescent light is detected with a camera-type detector. Lipid oxidation products may also produce ultra-weak chemiluminescence. Sodium hypochlorite induced decomposition of hydroperoxides has been shown to give strong chemiluminescence [10] which has been used to determine the oxidative quality of refined fish oil [11]. However, it could not be successfully applied to any other food matrices. Although, these instrumental analyses are more reliable and accurate, they are very expensive and specific for particular food samples. Moreover, trained hands are required for analysis and interpretation of data. Hence, they cannot be used as a routine analysis technique for large batch size.

Since food industries demand a large number of samples to be analyzed with high sample throughput, there is a requirement of alternative inexpensive technology that would accurately determine the odor profile of foods in short time. In this context, the concept of electronic nose (e-nose) technology which works by mimicking the human olfactory system is gaining an edge over the techniques mentioned above.

2 Electronic Nose

Electronic nose is a chemosensory system consisting of a series of sensors that respond to the volatile components in the headspace above a tested sample. According to Gardner and Barlett [12], the term electronic nose refers to multisensory arrays, where responses of all sensors have partial specificity and an appropriate pattern recognition system capable of recognizing simple and complex odors. The working principle of e-nose is claimed to mimic the human olfactory system. The sensor array represents the sensors in the human nose. The circuitry represents the conversion of the chemical reaction on human sensors to electrical signals into the brain, while the software analysis mimics the brain itself [13]. A pattern recognition system is developed (either by statistical methods or by artificial neural network) to evaluate the responses of sensor arrays, which is used for identification and discrimination between samples [14].

Among the different types of sensors, the most commercially used ones in e-nose systems include metal oxide semiconductors, conducting polymers, bulk acoustic waves, surface acoustic waves, quartz-microbalance sensors and tin oxide sensors.

E-nose instrumentation has advanced rapidly over the last decade with several successful applications in food industries for effective and convenient process monitoring, shelf-life investigation, freshness evaluation, authenticity assessment and quality control for cooking and fermentation processes [15].

Table 1 Applications of e-nose analyses in monitoring food processing operations

Food systems	Parameters	Sensors
Wine-must [16]	Discrimination between fermentation stages	32 conducting polymers
Iberian hams (*Montanera*) [17]	Spoilage during curing process	16 tin-oxide thin films
Milk fermented with *Lactobacillus casei* strains for Gruyère cheese production [18]	Discrimination between aroma producing genotype strains	mass spectrometry based e-nose
Milk fermented with *Lactococcus lactis* strains [19]	Discrimination between aroma intensity scores	12 metal oxide sensors
Australian red wine [20]	Spoilage caused by *Brettanomyces* yeast	12 metal oxide sensors
Tomato cv. Cencara [21]	Dehydration process of tomato slices	10 metal oxide sensors
Mangoes (*Mangifera indica* L.) [22]	Discrimination between harvest maturities within a ripening stage	18 metal oxide sensors
	Discrimination between ripening stages within a maturity stage	
	Discrimination between fruit varieties	
Black tea (*Camellia sinensis*) [23]	Estimation of optimum fermentation time	8 metal oxide gas sensors
Danish blue cheese [24]	Monitoring the ripening process	14 conducting polymers
Pinklady apples [25]	Discrimination between ripening stages	21 metal oxide sensors
Apples [26]	Monitoring post-harvest ripening	12 quartz micro balance sensors

2.1 Food Process Monitoring

Successful applications of e-nose on monitoring of flavor and/or aroma components during a food production process have been reported by several authors and have been elaborated in Table 1.

2.2 Shelf-life Study

E-nose analysis has been successfully carried out to study shelf-life properties of different foods and beverages (Table 2).

Table 2 Applications of e-nose analysis in shelf-life studies

Food system	Type of study	Sensor
Mandarins '*ZaojinJiaogan*' [27]	Discrimination between shelf-lives under different storage conditions and prediction of fruit quality	10 metal oxide sensors
Heatwave treated tomatoes (*Lycopersicum esculentum*) [28]	Discrimination between shelf-lives at two different storage conditions	10 metal oxide sensors
Peaches (*Prunus persica* L.) [29]	Discrimination between shelf-lives of two cultivars and assessment of ripening stages	10 metal oxide sensors
Crescenza cheese [30]	Determination of thresholds of shelf lives at different storage temperatures	10 metal oxide semiconductor field-effect transistors and 12 metal oxide sensors
Pizza [31]	Analysis of sensory quality during storage	metal oxide sensors
Milk [32]	Determination of shelf-life and correlation with bacterial growth	18 metal oxide sensors
Extra-virgin olive oils [33]	Evaluation of oxidative status at different storage temperatures, time and other conditions	10 metal oxide semiconductor field-effect transistors and 12 metal oxide sensors
Refined rapeseed oil [1]	Evaluation of lipid autooxidation as a function of storage temperature and time	18 metal oxide sensors
Palm olein [13]	Determination of storage stability	surface acoustic wave sensors
Egg [34]	Determination of shelf-life	8 metal oxide sensors
Minced beef [35]	Determination of shelf life in high-oxygen modified atmosphere packaging at different storage temperatures	10 metal oxide sensors

2.3 Evaluation of Freshness in Food

Freshness is another important quality parameter in the food industry. E-noses have successfully predicted the freshness or spoilage and rancidity of different raw materials and food products by sensing different VOCs generated during storage of foods (Table 3).

Table 3 Applications of e-nose analysis in evaluation of freshness and rancidity in foods

Food system	Type of study	Sensor
Fresh Atlantic salmon (*Salmo salar*) [36]	Spoilage classification of salmon fillets during storage at different temperatures	32 conducting polymers
Cod-fish fillets [37]	Prediction of the state of freshness as a function of storage temperature and time	8 thickness shear mode sensors and 5 electrical conductivity sensors
Cured ham [38]	Predicion of the state of freshness	semiconductor gas sensors
Cold smoked Atlantic salmon (*Salmo salar*) [39]	Spoilage/freshness classification during storage at different temperatures	6 metal oxide sensors
Silver pomfret [40]	Predicion of the state of freshness	18 metal oxide sensors
Alaska pink salmon (*Oncorhynchus gorbuscha*) [41]	Spoilage classification under different storage temperatures	32 thin-film carbon-black polymer sensors
Sea food octopus [42]	Spoilage classification of octopus treated with different concentrations of formaldehyde and water	semiconductor gas sensors
Fresh tilapia fillets (*Oreochromis niloticus*) [43]	Discrimination between storage times of fillets treated with different concentration of sodium lactate	12 conducting polymer
Oysters (*Cassostrea virginica*) [44]	Prediction models for odor changes in shucked oysters	12 conducting polymer
Mushrooms (*Agaricus bisporus*) [45]	Prediction of the state of freshness	18 metal oxide sensor
Bread (India) [46]	Prediction of the state of freshness	4 tin-oxide sensors
Eggs [47]	Prediction of the state of freshness	4 tin-oxide sensors
Beef and sheep meats [48]	Spoilage classification and prediction of bacteriological parameters	6 tin-oxide sensors
Pork [3]	Analysis of lipid oxidation	hybrid gas-sensor
Milk [49]	Monitoring of rancidity	tin-oxide sensors
Virgin olive oil [50]	Detection of rancidity	32 conducting polymer
Olive oil [51]	Detection of rancidity	13 metal oxide sensor
Rapeseed oil [52]	Detection of rancidity	mass spectrometry based e-nose
Potato crisps [53]	Detection of rancidity	7 semiconductor gas sensors
Nuts mix (peanuts, almonds, cashew nuts, Brazil nuts, pecan nuts and hazelnuts) [54]	Detection of rancidity	mass spectrometry based e-nose

Table 4 Applications of e-nose analysis in food authenticity assessment

Food system	Type of study	Sensor
Tequila, whisky, vodka and red wine [55]	Discrimination between the four types of beverages	18 metal oxide sensors
Italian wines [56]	Recognition and quantification of adulterants	4 thin-film metal oxide sensors
Spanish wines [57]	Classification of wine varieties	16 tin-oxide sensors
Grape wine [58]	Characterization of aroma compounds	12 metal oxide sensors
Apricots (*Prunus armeniaca*) [59]	Discrimination between varieties	18 metal oxide sensor
Apples "Jina", "fuji", "Huaniu" [60]	Discrimination between cultivars	14 tin-oxide gas sensors
Virgin olive oils [61]	Detection of adulterants	12 metal oxide sensors
Extra virgin olive oils [62]	Discrimination between geographical origins	mass spectrometry based e-nose
Orange juices [63]	Discrimination between geographical origins	12 metal oxide sensors
Emmental cheese [64]	Discrimination between geographical origin	mass spectrometry based e-nose
Swiss unifloral honeys [65]	Discrimination between botanical origin of honey	mass spectrometry based e-nose
Aceto Balsamico Tradizionale di Modena' [66]	Classification of different aged products	mass spectrometry based e-nose
Cuminum cyminum L. [67]	Characterization of aroma active compounds	12 metal oxide sensors
Orange flavors [68]	Aromatic profiles of spray-dried encapsulated orange flavors	8 quartz micro balance sensors
Coffee [69]	Characterization of aroma compounds	18 metal oxide sensors
Spices (basil, cinnamon and garlic) [70]	Quantification of spice mixture compositions	12 conducting polymers
Milk powder [71]	Characterization of odor	28 conducting polymers

2.4 Food System Classification and Authenticity Assessment

For successful commercialization, accurate classification and labeling of foods and beverages are extremely important. E-nose analyses have successfully identified different alcoholic beverages and several other food matrices as shown in Table 4.

2.5 Food Quality Control Studies

E-nose analyses were also successfully applied in quality evaluation of different food products. In these studies, a clear discrimination among products was carried out on the basis of food quality, with an aim to separate the good products from the bad ones. This type of study is very important in any food industry prior to consumer usage. The food systems, whose quality control parameters were successfully ascertained by e-nose analyses, are listed in Table 5.

The most commonly employed sensors in e-nose analysis of food samples are metal oxide sensors (MOS). These sensors are composed of a film of metal oxide, a heating element and electrical contacts. MOS devices can be fabricated with either thick (sintering type) or thin oxide films. The most common oxide material is SnO_2, an n-type semiconductor; although zinc, titanium, and tungsten oxides have also been used. The n-type MOS devices display an increased conductivity in the presence of reducible gases such as H_2, CO and CH_4. The mechanism for the change in conductivity is based on the number of charged species adsorbed on the semiconductor surface. The sensitivities of MOS sensors can be manipulated by changing the operating temperature, using different semiconductor materials for the oxide films, and doping the semiconductor with metal catalysts such as lead, platinum, and palladium. Advantages of using MOS devices are that there are a diverse range of sensors readily available and are relatively inexpensive to produce [87].

Other sensors that are used during e-nose analyses of food systems are conducting polymers, surface acoustic wave sensors, quartz micro balance sensors, thickness shear mode sensors and tin oxide sensors. The major advantage of using sensors for detecting volatile rancidity markers in foods is that these sensors do not need any pre-treatment and do not use solvents to detect the presence of volatiles. Furthermore, their main advantages are their low cost and high commercial availability [88]. From the above cited literature it was found that there are no reports on the application of e-nose in estimating shelf-life of cookies, where rancidity is a major concern during its storage. Therefore, e-nose analysis of drop cookies were studied by the authors.

3 Cookies

Cookie, derived from the Dutch word koekie (little cake), is one of the most popular bakery products consumed globally due to its ready-to-eat nature and high nutritional value [89]. It contains high amount of fat which imparts desirable organoleptic quality and also contributes to texture and flavor quality of the product [90]. However, the fat present in the cookies has propensity to get oxidized during storage leading to problems of rancidity and deterioration of sensory value. Improving the shelf-life of a cookie is therefore a major challenge [91] and this could possibly be achieved by administration of antioxidants to prevent the cookie from lipid oxidation. Usually commercial synthetic antioxidants such as butylated hydroxytoluene (BHT) and butylated hydroxyanisole (BHA) are used to reduce the rate of oxidation processes. However, these antioxidants suffer from the drawbacks that they are volatile, readily decompose at high temperatures and are believed to be carcinogenic [92]. Currently there is an increasing demand for antioxidants derived from

Table 5 Applications of e-nose analysis in food quality control study

Food system	Type of study	Sensor
Virgin olive oils [72]	Discrimination between quality grades	12 metal oxide sensors
Italian dry red wines [73]	Prediction of sensorial descriptors	10 metal oxide sensors
Oranges and apples [74]	Evaluation of post-harvest quality	7 thickness shear mode sensors
Peaches and nectarines from several cultivars [75]	Evaluation of sensorial features typical of each class	7 thickness shear mode sensors
"Xueqing" pears [76]	Prediction of quality indices (firmness, soluble solids content and pH)	8 metal oxide sensor
Longjing green tea [77]	Discrimination between different quality grades	10 metal oxide sensor
Green tea [78]	Identification of grades of tea	semiconductor gas sensors
Onions (*Allium cepa*) [79]	Influence of edaphic factors on bulbs quality	32 conducting polymer
Hams [80]	Discrimination of different types of hams based on ham-flavor	16 tin-oxide thin film sensors
Chinese vinegars [81]	Identification of several commercial vinegars	9 doped nano-ZnO thick film sensors
Pineapples [82]	Discrimination between ripe and unripe pineapples based on VOC composition	8 micro-surface acoustic wave oscilators
Sugar [82]	Discrimination between sugar and sugar having off-flavors	8 micro-surface acoustic wave oscilators
Water [83]	Detection of different stages of differentiation of *Streptomyces* in potable water	14 conducting polymers
Wheat bread [84]	Discrimination of volatiles of refined and whole wheat bread	12 metal oxide sensors
Corylus avellana L. [85]	Identification of flavor of natural and roasted Turkish hazelnut varieties	8 quartz micro balance sensors and 8 metal oxide sensors
Milk [86]	Quality control of goat milk	metal oxide sensors

natural sources [93]. Many herbs and spices which are excellent sources of phenolic compounds have been reported to show appreciable antioxidant activities [94].

Clove (*Syzygium aromaticum* Linn) belonging to the family Myrtaceae, is rich in a phenolic compound known as eugenol (~88 %) which has considerable nutraceutical properties such as antioxidant, antimicrobial and anti-inflammatory activities [95, 96]. Therefore clove extracts could serve as a potential natural antioxidant in cookie. In our previous study [97], authors have successfully used eugenol-rich clove extracts obtained from clove buds (grown in India) as a source of natural antioxidant in cookies. Since the clove bud extracts were directly used in food applications, the green technology of supercritical carbon dioxide (SC-CO_2) extraction was employed to obtain the eugenol-rich clove extracts.

3.1 Formulation of Cookies Using Clove Extracts

Drop cookies were prepared as explained by Sakac et al. [98], with little modifications. Wheat flour (50 %), powdered sugar (20 %), hydrogenated vegetable fat (20 %), butter (2 %), milk powder (2 %), ammonium bicarbonate (0.8 %), table salt (0.6 %), sodium bicarbonate (0.4 %), lecithin (0.3 %), SC-CO_2 clove extract (0.5 %) and water (8 %), all on w/w of dough weight (wet weight), constituted the cookies. At first, the ingredients (wheat flour, sugar, milk powder, ammonium bicarbonate, table salt, sodium bicarbonate and lecithin) were mixed in a dough making machine for 6 min and remixed with hydrogenated vegetable fat, butter and water for another 15 min. The clove extract was separately mixed in hydrogenated vegetable fat and then added to the dough to ensure its complete homogeneous dispersion. The dough was then manually shaped into circular pieces (diameter: 50 mm; thickness: 10 mm), subsequently baked in a baking air oven at $190 \pm 2\,°C$ for 15 min and cooled to room temperature ($23 \pm 2\,°C$) to obtain ready-to-eat cookies. The cookies were packed in aluminum foil, placed in Ziploc pouches, flushed with nitrogen and stored at room temperature ($23 \pm 2\,°C$) for one month.

Three sets of cookies were prepared, namely ‘control cookies’ with no antioxidant added, ‘sample cookies’ with clove extract as antioxidant and ‘deliberately made-rancid cookies’. In order to detect the rancidity profile of the cookies, the e-nose system was trained with the ‘deliberately made-rancid cookies’. These training sets of cookies were prepared by keeping ‘control cookies’ in a rancidity chamber for 14, 21 and 28 days. The chamber was kept at ~ 40 °C and UV light was used to promote oxidation. On completion of the ageing treatment, the samples were removed from the rancidity chamber and stored in the conservation chamber to maintain the final rancidity stage [99].

3.2 E-nose Analysis of Cookies Using ALPHA MOS e-nose System

Among the different e-nose systems, the application of ALPHA MOS e-nose system (M/s ALPHA MOS, Toulouse, France) was vastly explored by several authors in a

Table 6 Volatile descriptors associated with the different sensors of e-nose systems

Sensors	Volatile description	Sensors	Volatile description
LY2/LG	Fluoride, chloride, oxynitride, sulphide	P30 /1	Hydrocarbons, ammonia, ethanol
LY2 /G	Ammonia, amines, carbon-oxygen compounds	T70 /2	Toluene, xylene, carbon monoxide
LY2 /AA	Alcohol, acetone, ammonia, aldehydes	T40 /1	Fluorine
LY2 /GH	Ammonia, amines	P40 /1	Fluorine, chlorine
P40 /2	Chlorine, hydrogen sulfide and fluoride	LY2 /gCTL	Hydrogen sulfide
P30 /2	Hydrogen sulphide, ketone	LY2 /gCT	Propane, butane
T30 /1	Polar compound, hydrogen chloride	T40 /2	Chlorine
P10 /1	Nonpolar compound: hydrocarbon, ammonia, chlorine	PA /2	Ethanol, ammonia, amine compounds
P10 /2	Nonpolar compound: methane, ethane	TA /2	Ethanol

variety of food matrices [19, 29, 32, 88]. The system is equipped with 18 metal oxide sensors inside three chambers. 6 sensors were un-doped metal oxide sensors, while remaining 12 sensors were doped with noble catalytic metals in order to shift the selectivity spectrum towards different chemical compounds. Details of 18 sensors are presented in Table 6. The determination of rancidity in cookies was carried out using two types of ALPHA MOS e-nose systems, namely Fox e-nose and Heracles e-nose. Both the systems analyze the volatile organic compounds (VOCs) and are used for faster qualitative and quantitative analysis of complex odorous matrices. Fox e-nose system is based on metal oxide sensor detection system. It detects rancidity from the responses of 18 e-nose sensors. Heracles e-nose system is based on gas chromatography principle which uses two capillary columns of different polarities. This system can develop chromatograms for easy comparisons of non-rancid samples with rancid ones.

3.2.1 Fox Electronic Nose System

The Fox e-nose used for determination of rancidity in cookies was ALPHA MOS Fox 4,000 system (Fig. 1). 5 g of cookie samples were heated at 50 °C inside a controlled thermostat-sampling chamber for a headspace generation time of 20 min and 2 ml of VOCs were injected to the Fox system with an autosampler HS100 from 10 ml sealed vials by means of a carrier gas (TOC grade synthetic air) at a flow rate of 2 ml/min. The gas sampling syringe was heated at 60 °C. Post injection, a valve was

Fig. 1 Experimental setup of Alpha MOS e-nose system

switched on and only carrier gas was blown into the sensor chambers to return to the baseline of the sensor signals. The acquisition time and time between subsequent analyses were 18 min and 7 min, respectively. These conditions ensured recovery of baseline prior to subsequent analysis [99].

3.2.2 Heracles Electronic Nose System

Odor profile analyses by Heracles e-nose (Flash e-Nose) was carried out on Alpha MOS GC system, equipped with two columns MXT-5 and MXT-1701 (both $l = 10$ m, i.d. $= 180\,\mu$m). The two columns have different polarities (non-polar MXT-5 and slightly polar MXT-1701) to mitigate possible co-elutions. 5 g of cookie samples were heated at 50 °C inside a controlled thermostat-sampling chamber for a headspace generation time of 20 min and 2 ml of volatiles were injected to GC system with an autosampler HS100 from 10 ml sealed vials by means of a carrier gas (N_2 at 25 KPa) at a flow-rate of 0.5 ml/min for 15 s. The gas sampling syringe was heated at 200 °C. The trap was initially kept at 40 °C and heated to 240 °C in 35 s. The GC column temperature was kept at 40 °C for 2 s and programmed to 280 °C at a rate of 3 °C/s and finally held at 280 °C for 18 s. The FID detector was maintained at 240 °C. The total acquisition time and time between subsequent analyses were 100 s and 8 min, respectively. The method of analysis was calibrated using an alkane solution (*n*-butane to *n*-hexadecane) to convert retention time in Kovat's indices and to identify the VOCs using AroChem Base. The data processing was carried out with inbuilt Alphasoft V12.4 software [99].

Both Fox and Heracles e-nose analyses were able to successfully detect the odor profile of rancid cookies. From the odor map generated on the basis of PCA (Principal Component Analysis), clear discrimination was observed between fresh and rancid cookies. The VOCs detected in cookies samples were confirmed using Kovats retention indices and it was found that the markers were aldehydes such as pentanal, hexanal, heptanal, octanal and nonanal; and alcohols namely *n*-butanol and pentanol, which are present in high amounts in 'deliberately made-rancid cookies'. Although these marker compounds were present in small amounts in 'control cookies' on storage, none of these were present in the 'samples cookies'. This affirmed that the cookies remained non-rancid for at least 30 days with addition of clove extract as natural antioxidant. However, high amount of eugenol was detected by e-nose sensors in all batches of 'sample cookies' which hindered accurate shelf-life estimation. Owing to the characteristic pungent smell of eugenol, the e-nose sensors (highly sensitive for eugenol) have recorded high intensity odor signal. Therefore, these 'sample cookies' were placed in a cluster, separate from control cookies in the odor map and could not be compared with 'control cookies' during shelf-life estimation [99].

3.3 Determination of Shelf-Life of Cookies Using a Customized E-nose System

Since SC-CO_2 clove extract was added as a source of natural antioxidant in improving the shelf-life of the cookies, it is imperative to ascertain whether the antioxidant potential of clove extract can prevent rancidity in the cookies and extend its shelf-life. Owing to the high sensitivity of ALPHA MOS e-nose system towards eugenol, it was difficult to ascertain the accurate shelf-life of cookies with and without antioxidants. The strong pungent smell of eugenol had to be masked, which was addressed by two different approaches. The first approach comprised development of a customized e-nose using C-DAC e-nose system (M/s C-DAC, Kolkata, India) (Fig. 2) with only four metal oxide gas sensors. The discrimination index obtained in PCA plots of ALPHA MOS e-nose system with 18 sensors was ~95 %. Many authors have reported that with large number of sensors, the discrimination indices are generally poor and better discrimination indices are obtained in PCA plots when the number of sensors was reduced to 4-6 [1, 51, 88]. The four sensors that were used in this customized e-nose are TGS 830, TGS 2600, TGS 2610 and TGS 2626 (M/s Figaro, USA). These sensors have reportedly detected alcohol and aldehyde such as 2-phenyl-ethanol and benzaldehyde and were not sensitive for eugenol [100, 101].

The second approach focused on improvisation of the product to improve sensor response by masking the pungency of eugenol in the cookies. Fresh cookies were formulated using encapsulated eugenol-rich clove extract in maltodextrin-gum arabic

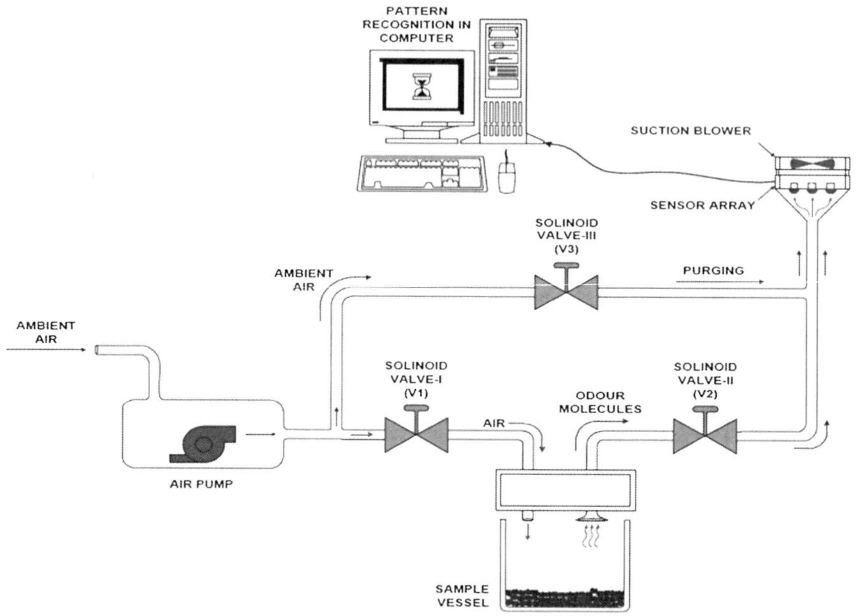

Fig. 2 Schematic diagram of customized C-DAC e-nose system for determination of rancidity in cookies

matrices. Microencapsulation of clove extract would also preserve the phytochemical properties and antioxidant activities in the cookies over longer period of time [102]. Application of encapsulated eugenol-rich clove extract as a natural source of antioxidant has been established in soybean oil by the authors, who found that it improved the shelf-life and frying stability of the oil [103].

Microencapsulation of eugenol-rich clove extract was carried out using Mini Spray Dryer B-290 (M/s Buchi, Switzerland). 2.5 g clove extract was added to 100 ml solution containing 12 g maltodextrin and 6 g gum arabic (1:4.8:2.4). The emulsion was created by mixing the solution in an Ultra-Turax homogenizer (M/s Ika, Germany) for 30 s. The inlet temperature was kept at 150 °C and outlet temperature at 86 °C. The spray gas, sample feed rate and atomization pressure were kept constant at 6 bar, 6.67 ml/min and −45 mbar respectively. The powder obtained was packed in aluminum foil and placed in Ziploc pouches (M/s Johnson, India), flushed with nitrogen and stored at 23 °C [103].

For determination of shelf-life of cookies, 'control cookies' and 'sample cookies' were prepared in accordance to the method, discussed in Sect. 3.1 with little modification and stored for a period of 200 days. The amount of encapsulated clove extract in the sample cookie formulation was 2.5 % of the dough weight. The e-nose system was trained with 'deliberately made-rancid cookies', prepared by storing 'control cookies' in a rancidity chamber for 1, 3, 10, 20 and 30 days.

Table 7 Phytochemical properties of cookies formulated with encapsulated clove extracts over a storage period of 200 days

Storage period (Days)	Eugenol content (mg/g cookie)	Phenolic content (mg gallic acid equivalent/g cookie)	IC_{50} of DPPH radical scavenging assay (μg/ml)
0	4.73a	2.22a	17.17c
20	4.27a	2.16a	18.39c
40	4.07c	2.16a	18.68bc
60	4.04a	2.15a	19.38abc
80	4.04a	2.14a	21.69abc
100	3.99a	1.97a	22.87abc
120	3.96a	1.92a	23.85abc
140	3.94a	1.88a	25.24abc
160	3.59a	1.68a	25.62abc
180	3.23a	1.58a	27.13ab
200	3.03a	1.54a	28.35a

Different letters in a column indicate significant difference at $p < 0.05$ level

3.3.1 Phytochemical Properties of Cookies Formulated with Encapsulated Clove Extracts

'Sample cookies' formulated with encapsulated clove extract were studied for their phytochemical properties such as eugenol content, total phenolic content and antioxidant activity. The eugenol content was determined by densitometric assay [104] as mg eugenol/g cookie; the phenolic content as mg gallic acid equivalent/g of cookie using Folin-Ciocalteu reagent [105] and antioxidant activity was determined by measuring the radical scavenging activity of DPPH [106] and expressed as IC_{50} values. These phytochemical assays were conducted at an interval of 20 days for a storage period of 200 days (Table 7). Since no antioxidant activity of the 'control cookies' was observed by DPPH assay and its phenolic content (0.11 mg gallic acid equivalent/g of cookie) was found to be significantly lower in comparison with cookies prepared with clove extracts, they were not subjected to further analyses.

Eugenol content of the freshly prepared 'sample cookies' was 4.73 mg/g cookie and showed nominal degradation to 3.04 mg/g cookie over a storage period of 200 days. The total phenolic content of the freshly prepared 'sample cookies' was 2.22 mg gallic acid equivalent/g cookie which degraded insignificantly to 1.54 mg gallic acid equivalent/g cookie during the storage period. Similar observation was also obtained for antioxidant activity, wherein, IC_{50} values of freshly prepared 'sample cookies' (17.17 μg/ml) has shown increase to 28.35μg/ml over storage. Therefore, from the phytochemical analyses, it could be concluded that the encapsulated clove extract does not degrade during preparation of cookies at high temperature and retains its antioxidant activity in appreciable amount even at the end of storage period of 200 days.

3.3.2 Estimation of Shelf-Life of Cookies

Shelf-lives of cookies formulated with and without antioxidants were studied by determining their rancidity profiles using both e-nose analysis and conventional biochemical assays over a storage period of 200 days.

3.3.3 E-nose Analysis for Determining Rancidity of the Cookies

The e-nose analyses of cookies (with and without antioxidants) were carried out at an interval of 20 days for a storage period of 200 days using a customized C-DAC e-nose system. Optimization of the e-nose process parameters such as batch size of cookie, granularity and heating time were fixed through preliminary trials. 50 g of ground cookie powder (dp = 0.5 mm) was charged into a 100 ml vial and heated for 450 s at 50 °C. The headspace time, sampling time and purging time was kept constant at 30, 50 and 300 s, respectively. The e-nose system was trained with a set of deliberately-made rancid cookies in an accelerated rancidity chamber.

The response of the four sensors of e-nose for a cookie sample was determined from the ($\Delta R/R$) value which is the change in the resistance of metal oxide gas sensor due to the VOCs of cookies (with or without clove extract) with respect to the base value [100]. The base value of control cookies is the resistance shown by the sensors due to VOCs of freshly prepared cookies (without antioxidant); whereas, base value of antioxidant-rich cookies is the resistance shown by the sensors due to VOCs of freshly prepared cookies with encapsulated clove extract. The representation of sensor responses in terms of ($\Delta R/R$) value is reported for different food matrices. While, Mildner-Szkudlarz et al. [1] used ($\Delta R/R$) value of e-nose sensors for monitoring the autoxidation of rapeseed oil, Bhattacharyya et al. [100] used the same in assessing optimum fermentation time of black tea.

The responses ($\Delta R/R$) of four metal oxide gas sensors of e-nose system for different sets of cookies have been presented in Table 8. From these data, an odor map was generated on the basis of PCA plot (Fig. 3) with discrimination index of 99.21 %, indicating the clusters to be distinctly separated. It was observed that all the 'sample cookies' with different storage periods have formed a well-defined cluster along with the 'control cookies' whose storage period was less than 100 days. The 'control cookies' with storage period greater than 100 days, formed a distinct cluster and was present in the same quadrant along with the cluster of 'deliberately-made rancid cookies'. However, none of the encapsulated clove extract formulated cookies were in the quadrant of 'deliberately-made rancid cookies'. Therefore, it could be concluded that 'control cookies' with storage period over 100 days had similar odor profile as that of the rancid cookies, indicating that cookies tend to become rancid after 100 days. Addition of encapsulated clove extract prevented rancidity in cookies for at least 200 days. This customized e-nose set up could detect rancidity in cookies after 100 days of storage. However, visible degradation of cookies was

Table 8 E-nose sensor responses ($\Delta R/R$) of 'control cookies' and 'sample cookies' during the storage period of 200 days along with deliberately made rancid cookies

Storage period (Days)	'Control cookies' (without antioxidant)				'Sample cookies' (with encapsulated clove extract)			
	TGS 830	TGS 2600	TGS 2610	TGS 2626	TGS 830	TGS 2600	TGS 2610	TGS 2626
0	0.0000a	0.0000a	0.0000a	0.0000a	0.0000a	0.0000a	0.0000a	0.0000a
20	0.0469a	0.1253a	0.0403a	0.0458a	0.0159a	0.0002a	0.0158a	0.0232a
40	0.0654a	0.1474a	0.0606a	0.0673a	0.0263a	0.1229a	0.0285a	0.0272a
60	0.0686a	0.1660a	0.0999a	0.0827a	0.0151a	0.0681a	0.0463a	0.0165a
80	0.1228a	0.3162a	0.1033a	0.1348a	0.0519a	0.1268a	0.0532a	0.0472a
100	0.1393a	0.3162a	0.1183a	0.1455a	0.0653a	0.1612a	0.0690a	0.0586a
120	0.1476a	0.3190a	0.1277a	0.1521a	0.0662a	0.1722a	0.0752a	0.0611a
140	0.1706a	0.3292a	0.1357a	0.1598a	0.0679a	0.2007a	0.0865a	0.0683a
160	0.1765a	0.3841a	0.1576a	0.1777a	0.0215a	0.1419a	0.0956a	0.0212a
180	0.1818a	0.4408a	0.1576a	0.1823a	0.0551a	0.1715a	0.0987a	0.0584a
200	0.2356a	0.4491a	0.1729a	0.2206a	0.0756a	0.2063a	0.0995a	0.0734a

Different letters in a column indicate significant difference at $p < 0.05$ level

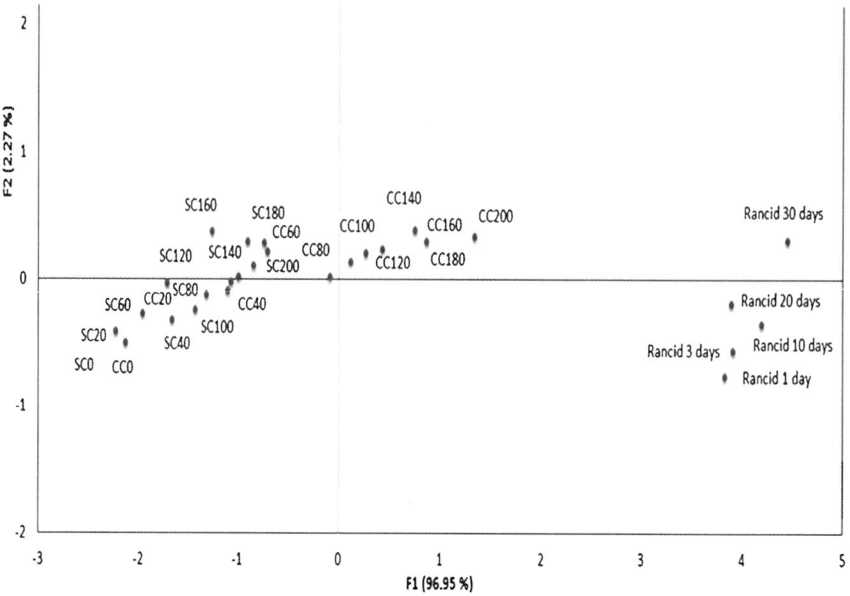

Fig. 3 Odor map of different cookie sets on the basis of PCA using four metal oxide sensors of C-DAC e-nose system

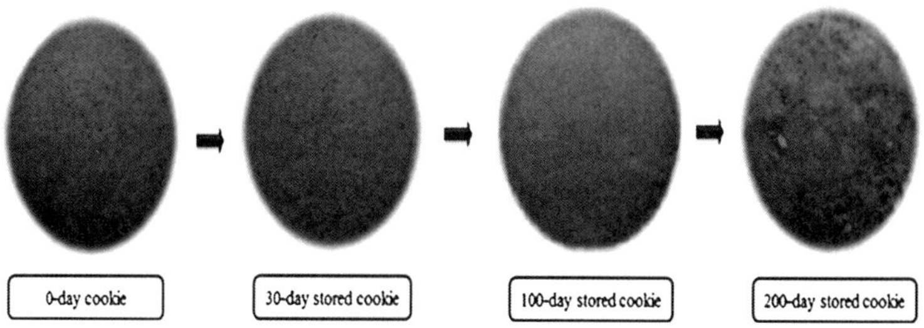

Fig. 4 Control cookies with different storage period

noticeable only at the end of the storage period of 200 days (Fig. 4). This established the reliability of the e-nose technique in assessment of rancid-spoilage of cookies.

3.3.4 Conventional Method for Determining Rancidity of the Cookies

For further affirmation of the data obtained from e-nose analysis of cookies, the conventional biochemical assays such as determination of free fatty acid (FFA) content, peroxide value (PV) and thiobarbituric acid value (TBA) were carried out to estimate the rancidity of different cookie sets during its storage (Table 9). The peroxide value (PV) and free fatty acid (FFA) content of the cookies were determined by alkali titration method according to the AOAC method [107]. PV was expressed as $meqKg^{-1}$ of cookie; whereas, FFA was expressed as % oleic acid. The TBA value of the cookies were determined in accordance to the method reported by Che-Man and Tan [108] and expressed as mmol malonaldehyde/g cookie from the standard curve prepared.

Although no significant differences exist in the FFA contents of 'control cookies' and 'sample cookies', the PV and TBA values of the 'control cookies' were significantly higher than the 'sample cookies', during the storage period. However, the rancidity parameter of both 'control cookies' and 'sample cookies' were significantly lower than that of 'deliberately made-rancid cookies'. The FFA content, PV value and TBA values of the freshly prepared 'control cookies' were 0.48 % oleic acid, 4.34 $meqKg^{-1}$ and 0.26 mmol malonaldehyde/g cookie respectively; while for the freshly prepared 'sample cookies', the respective values were 0.46 % oleic acid and 2.09 $meqKg^{-1}$, with no malonaldehyde detected. From the rancidity parameters of the biochemical assays, it was found that the 'control cookies' remained non-rancid until storage time of 100 days beyond which their rancidity parameters were found to be similar with those of 'deliberately made-rancid cookies'. However, no such trend was observed for 'sample cookies' set and it remained non-rancid till the end of storage period of 200 days. Therefore, from the biochemical assays, it could be concluded that addition of encapsulated clove extract as a source of natural antioxidant, enhanced shelf-life of cookies by 100 days.

These observations are in good agreement with the conclusion derived from e-nose analyses. Hence, from the shelf-life study of the cookies (conducted by both e-nose analyses and conventional biochemical assays), it could be concluded that normal cookies have shelf-life of 100 days when stored in aluminum foil, placed in Ziploc pouches, flushed with N_2 at $23 \pm 2\,^{\circ}C$. Administration of encapsulated clove extract as a source of natural antioxidant enhances its shelf-life by an additional 100 days.

3.3.5 Correlation Between the Rancidity Parameters of Cookies Determined by E-nose Analysis and Biochemical Assays

From the experimental data, a linear correlation was found between the responses of the different sensors of e-nose system and rancidity parameters obtained by biochemical assays. Therefore, multiple regression analyses were carried out to develop three linear regression equations to predict the FFA value, PV value and TBA value (individually) of the cookies as a function of responses of four metal oxide gas sensors (TGS 830, TGS 2600, TGS 2610 and TGS 2626), which are provided below.

Table 9 FFA, PV and TBA values of 'control cookies' and 'sample cookies' during the storage period of 200 days along with deliberately made rancid cookies

Storage period (Days)	'Control cookies' (without antioxidant)			'Sample cookies' (with encapsulated clove extract)		
	FFA content (% oleic acid)	PV($meqKg^{-1}$)	Malonaldehyde content (mmol /g cookie)	FFA content (% oleic acid)	PV($meqKg^{-1}$)	Malonaldehyde content (mmol /g cookie)
0	0.48a	4.34d	0.26b	0.46a	2.09b	0.00a
20	0.49a	10.71d	3.32ab	0.48a	2.62b	0.00a
40	0.73a	11.14d	4.97ab	0.71a	3.55b	0.08a
60	0.83a	11.53d	5.59ab	0.72a	5.31b	0.09a
80	0.85a	13.47cd	5.76ab	0.82a	8.28ab	0.09a
100	0.97a	21.36bc	6.19ab	0.86a	10.95ab	0.26a
120	0.98a	21.78bc	6.23ab	0.89a	10.98ab	0.29a
140	0.99a	22.11bc	6.46ab	0.96a	11.39ab	0.35a
160	1.22a	22.95b	8.38ab	1.20a	14.75a	0.87a
180	3.70a	45.01a	10.21a	1.20a	15.80a	1.57a
200	3.96a	49.5a	10.99a	1.45a	16.87a	1.83a

Different letters in a column indicate significant difference at $p < 0.05$ level

$$Y_1 = -0.2145 + 19.7517 \times X_1 - 1.2596 \times X_2 + 24.7720 \times X_3 - 24.4989 \times X_4 \quad (1)$$

$$Y_2 = -0.9471 + 230.5910 \times X_1 + 27.0360 \times X_2 + 384.8341 \times X_3 - 385.5990 \times X_4 \quad (2)$$

$$Y_3 = -1.3581 + 46.6934 \times X_1 + 4.6030 \times X_2 + 12.7123 \times X_3 - 10.9291 \times X_4 \quad (3)$$

where, Y_1 is the FFA value of the cookies (% oleic acid), Y_2 is the PV value of the cookies (meqKg^{-1}) and Y_3 is the TBA value of the cookies (mmol malonaldehyde/g cookie). X_1, X_2, X_3 and X_4 are the e-nose responses obtained from sensor TGS 830, TGS 2600, TGS 2610 and TGS 2626 respectively. The correlation coefficients of Eqs. 1–3 are 0.85, 0.95 and 0.97 respectively. The validity of the fitted models were indicated by the insignificant lack of fit ($p = 1.00$) for these equations and by F-values of 12.38, 43.83 and 91.63 corresponding to Eqs. 1, 2 and 3, respectively. Therefore, these equations could be used to predict the conventional rancidity parameters of the cookies (with or without antioxidant) from the sensor responses of the customized e-nose.

4 Conclusion

In this chapter, authors have discussed the major applications of electronic noses in various food systems where this technology has been successfully applied in process monitoring, shelf-life study, detection of rancidity, assessment of authenticity and quality control of the food products. Metal oxide sensors are found to be the most responsive sensors for food matrices. Authors have reported for the first time on the usage of e- nose in assessment of shelf-life of cookies, where lipid oxidation is a major problem. Comparison of odor profiles of cookies formulated with and without SC-CO_2 eugenol-rich clove extract (as a source of natural antioxidant) was carried out by Fox e-nose and Heracles e-nose systems. However, owing to high sensitivity of these systems towards eugenol, it was difficult to ascertain accurate shelf-life of cookies using the same. A customized e-nose system was therefore designed using 4 metal oxide gas sensors to ascertain the shelf-life of cookies formulated with and without antioxidant (encapsulated clove extract). Rancidity parameters adjudged by e-nose analyses and biochemical assays were found to be in close agreement. This study established that cookies without administered antioxidants have shelf-life of 100 days when stored in Al laminates at $23 \pm 2\,°C$, while administration of encapsulated clove extract as a source of natural antioxidant prevented rancidity for at least 200 days and thereby, enhanced its shelf-life by 100 days. Phytochemical analyses also affirmed that the cookies formulated with encapsulated clove extract as antioxidant, enhanced its nutraceutical potential for at least 200 days. Linear regression equations were developed from the correlated data set (rancidity parameters from e-nose analysis and biochemical assays), which can successfully predict FFA, PV and TBA values of cookies from the sensor responses of our customized e-nose system.

References

1. S. Mildner-Szkudlarz, H.H. Jelen, R. Zawirska-Wojtasiak, The use of electronic and human nose for monitoring rapeseed oil autoxidation. Eur. J. Lipid Sci. Technol. **110**, 61–72 (2008)
2. L.J. Dugan, Lipids, in *Principles of Food Science, Part I, Food Chemistry*, ed. by O.W. Fenema (Marcel Dekker Inc., New York, 1976), p. 183
3. E. Olsen, G. Vogt, D. Ekeberg, M. Sandbakk, J. Pettersen, A. Nilsson, Analysis of the early stages of lipid oxidation in freeze-stored pork back fat and mechanically recovered poultry meat. J. Agric. Food Chem. **53**, 338–348 (2005)
4. F. Angerosa, L. Di Giacinto, R. Vito, S. Cumitini, Sensory evaluation of virgin olive oils by artificial neural network processing of dynamic head-space gas chromatographic data. J. Sci. Food Agric. **72**, 323–328 (1996)
5. N. Shen, S. Moizuddin, L. Wilson, S. Duvick, P. White, L. Pollak, Relationship of electronic nose analyses and sensory evaluation of vegetable oils during storage. J. Am. Oil Chem. Soc. **78**, 937–940 (2001)
6. E.N. Frankel, *Lipid Oxidation* (The Oily Press Dundee, Scotland, 1998)
7. E. Hansen, L. Lauridsen, L.H. Skibsted, R.K. Moawad, M.L. Andersen, Oxidative stability of frozen pork patties: Effect on fluctuating temperature on lipid oxidation. Meat Sci. **68**, 479–484 (2004)
8. D.S. Mottram, Flavour formation in meat and meat products: a review. Food Chem. **62**, 415–424 (1998)
9. J.P. Wold, M. Mielnik, Non-destructive assessment of lipid oxidation in minced poultry meat by autofluorescence spectroscopy. J. Food Sci. **65**, 87–95 (2000)
10. Y. Yamamoto, E. Niki, R. Tanimura, Y. Kamiya, Study of oxidation by chemiluminescence. IV. Detection of low levels of lipid hydroperoxides by chemiluminescence. J. Am. Oil Chem. Soc. **62**, 1248–1250 (1985)
11. J. Pettersen, Chemiluminescence of fish oils and its flavour quality. J. Sci. Food Agric. **65**, 307–313 (1994)
12. J.W. Gardner, P.N. Bartlett, Intelligent Chem-SADs for artificial odour-sensing of coffees and lager beers. in *Proceeding of the 11th Symposium on Olfaction and Taste* (Tokyo, 1994), pp. 660–693
13. H.L. Gan, C.P. Tan, Y.B. Che Man, I. NorAini, S.A.H. Nazimah, Monitoring the storage stability of RBD palm olein using the electronic nose. Food Chem. **89**, 271–282 (2005)
14. B. Innawong, P. Mallikarjunan, J.E. Marcy, The determination of frying oil quality using a chemosensory system. LWT Food Sci. Technol. **37**, 35–41 (2004)
15. M. Peris, L. Escuder-Gilabert, A 21st century technique for food control: Electronic noses. Anal. Chim. Acta **638**, 1–15 (2009)
16. C. Pinheiro, C.M. Rodrigues, T. Schäfer, J.G. Crespo, Monitoring the aroma production during wine-must fermentation with an electronic nose. Biotechnol. Bioeng. **77**, 632–640 (2002)
17. M. García, M. Aleixandre, M.C. Horrillo, Electronic nose for the identification of spoiled Iberian hams. in *Proceedings of Spanish on Conference Electron Devices* (Tarragona, 2005), pp. 537–540
18. L. Marilley, S. Ampuero, T. Zesiger, M.G. Casey, Screening of aroma-producing lactic acid bacteria with an electronic nose. Int. Dairy J. **14**, 849–856 (2004)
19. N. Gutiérrez-Méndez, B. Vallejo-Cordoba, A.F. González-Córdova, G.V. Nevárez- Moorillón, B. Rivera-Chavira, Evaluation of aroma generation of *Lactococcus lactis* with an electronic nose and sensory analysis. J. Dairy Sci. **91**, 49–57 (2008)

20. W. Cynkar, D. Cozzolino, B. Dambergs, L. Janik, M. Gishen, Feasibility study on the use of a head space mass spectrometry electronic nose (MS e-nose) to monitor red wine spoilage induced by *Brettanomyces* yeast. Sensors Actuators B Chem. **124**, 167–171 (2007)
21. P. Pani, A.A. Leva, M. Riva, A. Maestrelli, D. Torreggiani, Influence of an osmotic pre-treatment on structure-property relationships of air-dehydrated tomato slices. J. Food Eng. **86**, 105–112 (2008)
22. M. Lebrun, M.N. Ducamp, A. Plotto, K. Goodner, E. Baldwin, Fox + GC/MS development of electronic nose measurements for mango (*Mangifera indica*) homogenate and whole fruit. Proc. Florida State Hortic. Soc. **117**, 421–425 (2004)
23. N. Bhattacharya, B. Tudu, A. Jana, D. Ghosh, R. Bandhopadhyay, M. Bhuyan, Preemptive identification of optimum fermentation time for black tea using electronic nose. Sensors Actuators B Chem. **131**, 110–116 (2008)
24. J. Trihaas, P.V. Nielsen, Electronic nose technology in quality assessment: Monitoring the ripening process of Danish blue cheese. J. Food Sci. **70**, E44–E49 (2005)
25. J. Brezmes, E. Llobet, X. Vilanova, J. Orts, G. Saiz, X. Correig, Correlation between electronic nose signals and fruit quality indicators on shelf-life measurements with pinklady apples". Sensors Actuators B Chem. **80**, 41–50 (2001)
26. U. Herrmann, T. Jonischkeit, J. Bargon, U. Hahn, Q.Y. Li, C.A. Schalley, E. Vogel, F. Vögtle, Monitoring apple flavor by use of quartz microbalances. Anal. Bioanal. Chem. **372**, 611–614 (2002)
27. A. Hernández-Gómez, J. Wang, G. Hu, P. García-Pereira, Discrimination of storage shelf-life for mandarin by electronic nose technique. LWT Food Sci. Technol. **40**, 681–689 (2007)
28. A. Hernández-Gómez, J. Wang, G. Hu, A. García-Pereira, Monitoring storage shelf life of tomato using electronic nose technique. J. Food Eng. **85**, 625–631 (2008)
29. S. Benedetti, S. Buratti, A. Spinardi, S. Manninoa, I. Mignani, Electronic nose as a non-destructive tool to characterise peach cultivars and to monitor their ripening stage during shelf-life. Postharvest Biol. Technol. **47**, 181–188 (2008)
30. S. Benedetti, N. Sinelli, S. Buratti, M. Riva, Shelf life of Crescenza cheese as measured by electronic nose. J. Dairy Sci. **88**, 3044–3051 (2005)
31. J.S. Vestergaard, M. Martens, P. Turkki, Application of an electronic nose system for prediction of sensory quality changes of a meat product (pizza topping) during storage. LWT-Food Sci. Technol. **40**, 1095–1101 (2007)
32. S. Labreche, S. Bazzo, S. Cade, E. Chanie, Shelf life determination by electronic nose: application to milk. Sensors Actuators B Chem. **106**, 199–206 (2005)
33. M.S. Cosio, D. Ballabio, S. Benedetti, C. Gigliotti, Evaluation of different storage conditions of extra virgin olive oils with an innovative recognition tool built by means of electronic nose and electronic tongue. Food Chem. **101**, 485–491 (2007)
34. W. Yongwei, J. Wang, B. Zhou, Q. Lu, Monitoring storage time and quality attribute of egg based on electronic nose. Anal. Chimica Acta **650**, 183–188 (2009)
35. S. Limbo, L. Torri, N. Sinelli, L. Franzetti, E. Casiraghi, Evaluation and predictive modeling of shelf life of minced beef stored in high-oxygen modified atmosphere packaging at different temperatures. Meat Sci. **84**, 129–136 (2010)
36. W.X. Du, C.M. Lin, T. Huang, J. Kim, M. Marshall, C.I. Wei, Potential application of the electronic nose for quality assessment of salmon fillets under various storage conditions. J. Food Sci. **67**, 307–313 (2002)
37. D. Natale, G. Olafsdottir, S. Einarsson, E. Martinelli, R. Paolesse, A. D'Amico, Comparison and integration of different electronic noses for freshness evaluation of cod-fish fillets. Sensors Actuators B Chem. **77**, 572–578 (2001)
38. V. Rossi, C. Garcia, R. Talon, C. Denoyer, J.L. Berdagu'e, Rapid discrimination of meat products and bacterial strains using semiconductor gas sensors. Sensors Actuators B Chem. **37**, 43–48 (1996)
39. G. Olafsdottir, E. Chanie, F. Westad, R. Jonsdottir, C.R. Thalmann, S. Bazzo, S. Labreche, P. Marcq, F. Lundby, J.E. Haugen, Prediction of microbial and sensory quality of cold smoked Atlantic salmon (*Salmo salar*) by Electronic Nose. JFS S Sensory Nutritive Qualities Food **70**, S563–S574 (2005)

40. T. Yi, X. Jing, X. Hong, Y. Sheng-ping, Food-fish fox application of electronic nose in the prediction model of shelf-life of *Pampus argenteus*. J. Fish. China **1000**, 3–8 (2010)
41. J. Chantarachoti, A.C.M. Oliveira, B.H. Himelbloom, C.A. Crapo, D.G. Mclachlan, Portable electronic nose for detection of spoiling Alaska pink salmon (*Oncorhynchus gorbuscha*). J. Food Sci. **71**, S414–S421 (2006)
42. S. Zhang, C. Xie, Z. Bai, M. Hub, H. Li, D. Zeng, Spoiling and formaldehyde-containing detections in octopus with an e-nose. Food Chem. **113**, 1346–1350 (2009)
43. F. Korel, D.A. Luzuriaga, M.Ö. Balaban, Objective quality assessment of raw tilapia (*Oreochromis niloticus*) fillets using electronic nose and machine vision. J. Food Sci. Sensory Nutritive Qualities Food **66**, 1018–1024 (2001)
44. O. Tokusoglu, M. Balaban, Correlation of odor and color profiles of oysters with electronic nose and color machine vision. J. Shellfish Res. **23**, 143–148 (2004)
45. K. Tamaki, T. Tamaki, T. Yamakasi, Fox studies on the deodorization by mushroom (*Agaricus bisporus*) extract of garlic extract-induced oral malodor. J. Nutr. Sci. Vitaminol. **53**, 277–286 (2007)
46. C. Botre, D. Gharpure, Analysis of volatile bread aroma for evaluation of bread freshness using an electronic nose (e-nose). Mater. Manufact. Process. **21**, 279–283 (2006)
47. R. Dutta, E.L. Hines, J.W. Gardner, D.D. Udrea, P. Boilot, Non-destructive egg freshness determination: an electronic nose based approach. Measur. Sci. Technol. **14**, 190–198 (2003)
48. N.E. Barbri, E. Llobet, N.E. Bari, X. Correig, B. Bouchikhi, Electronic nose based on metal oxide semiconductor sensors as an alternative technique for the spoilage classification of red meat. Sensors **8**, 142–156 (2008)
49. S. Capone, M. Epifani, F. Quaranta, P. Siciliano, A. Taurino, L. Vasanelli, Monitoring of rancidity of milk by means of an electronic nose and a dynamic PCA analysis. Sensors Actuators B Chem. **78**, 174–179 (2001)
50. R. Aparicio, S.M. Rocha, I. Delgadillo, M.T. Morales, Detection of rancid defect in virgin olive oil by the electronic nose. J. Agric. Food Chem. **48**, 853–860 (2000)
51. A. Apetrei, I.M. Apetrei, S. Villanueva, J.A. de Saja, F. Gutierrez-Rosales, M.L. Rodriguez-Mendez, Combination of an e-nose, an e-tongue and an e-eye for the characterisation of olive oils with different degree of bitterness. Anal. Chim. Acta **663**, 91–97 (2010)
52. J. Hong, C.L. Lim, H.J. Son, J.Y. Choi, N.B. Soo, Rancidity analysis of rapeseed oil under different storage conditions using mass spectrometry-based electronic nose. Korean J. Food Sci. Technol. **42**, 699–704 (2010)
53. M. Vinaixa, A. Vergara, C. Duran, E. Llobet, C. Badia, J. Brezmes, X. Vilanova, X. Correig, Fast detection of rancidity in potato crisps using e-noses based on mass spectrometry or gas sensors. Sensors Actuators B **106**, 67–75 (2005)
54. K. Yoshida, E. Ishikawa, M. Joshi, H. Lechat, F. Ayouni, M. Bonnefille, Quality control and rancidity tendency of nut mix using an electronic nose. in *Proceedings of the First Indo-Japan Conference on Perception and Machine Intelligence* (Heidelberg, New York, 2012), pp. 163–170
55. J.A. Ragazzo, P. Chalier, J. Crouzet, C. Ghommidh, Identification of alcoholic beverages by coupling gas chromatography and electronic nose. in ed. by A.M. Spanier, F. Shahidi, T.H. Parliament, C. Mussinan, C.T. Ho, E.T. Contis *Food Flavors and Chemistry* (Royal Society of Chemistry, Cambridge, 2001), pp. 404–411
56. M. Penza, G. Cassano, Recognition of adulteration of Italian wines by thin-film multisensor array and artificial neural networks. Anal. Chimica Acta **509**, 159–177 (2004)
57. J. Lozano, J.P. Santos, J. Gutiérrez, M.C. Horrillo, Comparative study of sampling systems combined with gas sensors for wine discrimination. Sensors Actuators B Chem. **126**, 616–623 (2007)
58. R.C. McKellar, H.P.V. Rupasinghe, X. Lu, K.P. Knight, The electronic nose as a tool for the classification of fruit and grape wines from different Ontario wineries. J. Sci. Food Agric. **85**, 2391–2396 (2005)

59. H.M. Solis-Solis, M. Calderon-Santoyo, P. Gutierrez-Martinez, S. Schorr-Galindo, J.A. Ragazzo-Sanchez, Discrimination of eight varieties of apricot (*Prunus armeniaca*) by electronic nose, LLE and SPME using GC-MS and multivariate analysis. Sensors Actuators B Chem. **125**, 415–421 (2007)
60. Z. Xiaobo, Z. Jiewen, Comparative analyses of apple aroma by a tin-oxide gas sensor array device and GC/MS. Food Chem. **107**, 120–128 (2008)
61. M.C.C. Oliveros, J.L.P. Pavón, C.G. Pinto, M.E.F. Laespada, B.M. Cordero, M. Forina, Electronic nose based on metal oxide semiconductor sensors as a fast alternative for the detection of adulteration of virgin olive oils. Anal. Chimica Acta **459**, 219–228 (2002)
62. M.S. Cosio, D. Ballabio, S. Benedetti, C. Gigliotti, Geographical origin and authentication of extra virgin olive oils by an electronic nose in combination with artificial neural networks. Anal. Chimica Acta **567**, 202–210 (2006)
63. C. Steine, F. Beaucousin, C. Siv, G. Peiffer, Potential of semiconductor sensor arrays for the origin authentication of pure Valencia orange juices. J. Agric. Food Chem. **49**, 3151–3160 (2001)
64. L. Pillonel, S. Ampuero, R. Tabacchi, J.O. Bosset, Analytical methods for the determination of the geographic origin of Emmental cheese, volatile compounds by GC-MS-FID and electronic nose. Eur. Food Res. Technol. **216**, 179–183 (2003)
65. S. Ampuero, S. Bogdanov, J.O. Bosset, Classification of unifloral honeys with an MS-based electronic nose using different sampling modes: SHS, SPME, and INDEX. Eur. Food Res. Technol. **218**, 198–207 (2004)
66. M. Cocchi, C. Durante, A. Marchetti, C. Armanino, M. Casale, Characterization and discrimination of different aged 'Aceto balsamico tradizionale di modena' products by head space mass spectrometry and chemometrics. Anal. Chimica Acta **589**, 96–104 (2007)
67. R. Ravi, M. Prakash, K.K. Bhat, Characterization of aroma active compounds of cumin (*Cuminum cyminum L.*) by GC-MS, e-Nose, and sensory techniques. Int. J. Food Prop. **16**, 1048–1058 (2013)
68. M.V. Galmarini, M.C. Zamora, R. Baby, J. Chirife, V. Mesina, Aromatic profiles of spray-dried encapsulated orange flavours: influence of matrix composition on the aroma retention evaluated by sensory analysis and electronic nose techniques. Int. J. Food Sci. Technol. **43**, 1569–1576 (2008)
69. T. Michishita, M. Akiyama, Y. Hirano, M. Ikeda, Y. Sagara, T. Araki, Gas chromatography/olfactometry and electronic nose analyses of retronasal aroma of espresso and correlation with sensory evaluation by an artificial neural network. J. Food Sci. **75**, S477–S489 (2010)
70. H. Zhang, M.Ö. Balaban, J.C. Principe, K. Portier, Quantification of spice mixture compositions by electronic nose: part I. Experimental design and data analysis using neural networks. J. Food Sci. E Food Eng. Phys. Prop. **70**, E253–258 (2005)
71. A. Biolatto, G. Grigioni, M. Irurueta, A.M. Sancho, M. Taverna, N. Pensel, Seasonal variation in the odour characteristics of whole milk powder. Food Chem. **103**, 960–967 (2007)
72. F. Lacoste, F. Bosque, R. Raoux, Developments in analytical methods and management: Is it possible to use an "electronic nose" for the detection of sensorial defects in virgin olive oil? OCL-Oléagineux Corps Gras Lipides **8**, 78 (2001)
73. S. Buratti, D. Ballabio, S. Benedetti, M.S. Cosio, Prediction of Italian red wine sensorial descriptors from electronic nose, electronic tongue and spectrophotometric measurements by means of genetic algorithm regression models. Food Chem. **100**, 211–218 (2007)
74. C.D. Natale, A. Macagnano, E. Martinelli, R. Paolesse, E. Proiettia, A. D'Amico, The evaluation of quality of post-harvest oranges and apples by means of an electronic nose. Sensors Actuators B Chem. **78**, 26–31 (2001)
75. C.D. Natale, A. Macagnano, E. Martinelli, E. Proietti, R. Paolesse, L. Castellari, S. Campani, A. D'Amico, Electronic nose based investigation of the sensorial properties of peaches and nectarines. Sensors Actuators B Chem. **77**, 561–566 (2001)
76. G. Zhang, J. Wang, Y. Sheng, Predictions of acidity, soluble solids and firmness of pear using electronic nose technique. J. Food Eng. **86**, 370–378 (2008)

77. Y. Yu, J. Wang, H. Zhang, Y. Yu, C. Yao, Identification of green tea grade using different feature of response signal from e-nose sensors. Sensors Actuators B Chem. **128**, 455–461 (2008)
78. Y. Yu, J. Wang, C. Yao, H. Zhang, Y. Yu, Quality grade identification of green tea using e-nose by CA and ANN. LWT-Food Sci. Technol. **41**, 1268–1273 (2008)
79. L. Abbey, D.C. Joyce, J. Aked, B. Smith, C. Marshall, Electronic nose evaluation of onion headspace volatiles and bulb quality as affected by nitrogen, sulphur and soil type. Ann. Appl. Biol. **145**, 41–50 (2004)
80. M. García, M. Aleixandre, J. Gutiérrez, M.C. Horrillo, Electronic nose for ham discrimination. Sensors Actuators B Chem. **114**, 418–422 (2006)
81. Q. Zhang, S. Zhang, C. Xie, C. Fan, Z. Bai, Sensory analysis' of Chinese vinegars using an electronic nose. Sensors Actuators B Chem. **128**, 586–593 (2008)
82. N. Barié, M. Bücking, M. Rapp, A novel electronic nose based on miniaturized SAW sensor arrays coupled with SPME enhanced headspace-analysis and its use for rapid determination of volatile organic compounds in food quality monitoring. Sensors Actuators B Chem. **114**, 482–488 (2006)
83. A.C. Bastos, N. Magan, Potential of an electronic nose for the early detection and differentiation between *Streptomyces* in potable water. Sensors Actuators B Chem. **116**, 151–155 (2006)
84. H.D. Sapirstein, S. Siddhu, M. Aliani, Discrimination of volatiles of refined and whole wheat bread containing red and white wheat bran using an electronic nose. J. Food Sci. **77**, S399–406 (2012)
85. C. Alasalvar, E. Pelvan, B. Bahar, F. Korel, H. Olmez, Flavour of natural and roasted Turkish hazelnut varieties (*Corylus avellana L.*) by descriptive sensory analysis, electronic nose and chemometrics. Int. J. Food Sci. Technol. **47**, 122–131 (2012)
86. A. Amari, N.E. Beri, B. Bouchikhi, Quality control of goat milk by a portable electronic nose based on metal oxide semiconductor sensors. Instrum. Measur. Metrologie **11**, 31–49 (2011)
87. S.E. Stitzel, M.J. Aernecke, D.R. Walt, Artificial Noses. Ann. Rev. Biomed. Eng. **13**, 1–25 (2011)
88. D.L. Garcia-Gonzalez, R. Aparicio, Classification of different quality virgin olive oils by metal-oxide sensors. Eur. Food Res. Technol. **218**, 484–487 (2004)
89. C.M. Ajila, K. Leelavathi, U.J.S.P. Rao, Improvement of dietary fiber content and antioxidative properties in soft dough biscuits with the incorporation of mango peel powder. J. Cereal Sci. **48**, 319–326 (2008)
90. A. Waheed, G. Rasool, A. Asghar, Effect of interesterified palm and cottonseed oil blends on cookie quality. Agric. Biol. J. North Am. **1**, 402–406 (2010)
91. V. Reddy, A. Urooj, A. Kumar, Evaluation of antioxidant activity of some plant extracts and their application in biscuits. Food Chem. **90**, 317–321 (2005)
92. M. Martinez-Tome, A.M. Jimenez, S. Ruggieri, N. Frega, R. Strabbioli, M.A. Murcia, Antioxidant properties of Mediterranean spices compared with common food additives. J. Food Prot. **64**, 1412–1419 (2001)
93. H.L. Madsen, G. Bertelsen, Spices as antioxidants. Trends Food Sci. Technol. **6**, 271–277 (1995)
94. W. Zheng, S.Y. Wang, Antioxidant activity and phenolic compounds in selected herbs. J. Agric. Food Chem. **49**, 5165–5170 (2011)
95. L. Hernández-Ochoa, B. Aguirre-Prieto, G.V. Nevárez-Moorillón, N. Gutierrez-Mendez, E. Salas-Muñoz, Use of essential oil and extracts from spices in meat protection. J. Food Sci. Technol. Article in press DOI: http://dx.doi.org/10.1007/s13197-011-0598-3, (2011)
96. M. Ogata, M. Hoshi, S. Urano, T. Endo, Antioxidant activity of eugenol and related monomeric and dimeric compounds. Chem. Pharm. Bull. **48**, 1467–1469 (2000)
97. D. Chatterjee, P. Bhattacharjee, A. Banerjee, Phytochemical analyses and food applications on clove bud extracts obtained by liquid and supercritical carbon dioxide extraction technologies, in *Proceedings of 2012 International Conference on Engineering and Applied Science*, ISSN 2227-0299, ISBN 978-986-87417-1-3 (Beijing, 2012), pp. 392–409

98. M.B. Sakac, J.F. Gyura, A.C. Misan, Z.I. Seres, B.S. Pajin, D.M. Soronja-Simovicm, Antioxidant activity of cookies supplemented with sugarbeet dietary fiber. Sugar Ind. **136**, 151–157 (2010)
99. D. Chatterjee, P. Bhattacharjee, H. Lechat, F. Ayouni, V. Vabre, Assessment of shelf-life of cookies formulated with clove extracts using electronic nose: Estimation of rancidity in cookies. in *Proceedings of the 2012 Sixth International Conference on Sensing Technology*, ISBN 978-1-4673-2245-4, DOI: http://dx.doi.org/10.1109/ICSensT.2012.6461709 (Kolkata, 2012), pp. 404–409
100. N. Bhattacharyya, S. Seth, B. Tudu, P. Tamuly, A. Jana, D. Ghosh, R. Bandyopadhyay, M. Bhuyan, S. Sabhapandit, Detection of optimum fermentation time for black tea manufacturing using electronic nose. Sensors Actuators B **122**, 627–634 (2007)
101. M. Palit, B. Tudu, N. Bhattacharyya, A. Dutta, P.K. Dutta, A. Jana, R. Bhandyopadhyay, A. Chatterjee, Comparison of multivariate preprocessing techniques as applied to electronic tounge based pattern classification for black tea. Anal. Chimica Acta **675**, 8–15 (2010)
102. J. Shaikh, R. Bhosale, R. Singhal, Microencapsulation of black pepper oleoresin. Food Chem. **94**, 105–110 (2006)
103. D. Chatterjee, P. Bhattacharjee, Comparative evaluation of the antioxidant efficacy of encapsulated and un-encapsulated eugenol-rich clove extracts in soybean oil: Shelf-life and frying stability of soybean oil. J. Food Eng. **117**, 545–550 (2013)
104. D. Chatterjee, P. Bhattacharjee, Supercritical carbon dioxide extraction of eugenol from clove buds: Process optimization and packed bed characterization. Food Bioprocess Technol. **6**, 2587–2599 (2013)
105. G.A. Spanos, R.E. Wrolstad, Influence of processing and storage on the phenolic composition of Thompson seedless grape juice. J. Agric. Food Chem. **38**, 1565–1571 (1990)
106. A. Aiyegoro, A.I. Okoh, Preliminary phytochemical screening and in vitro antioxidant activities of the aqueous extract of *Helichrysum longifolium* DC. BMC Complement. Altern. Med. **10**, 2–8 (2010)
107. Association of official analytical chemists (AOAC), *Official Methods of Analysis of AOAC International Methods 28.030*, 11th *Ed. AOAC International* (Gaithersburg, MD, 1970)
108. Y.B. Che-Man, C.P. Tan, Effects of natural and synthetic antioxidants on changes in refined, bleached, and deodorized palm olein during deep-fat frying of potato chips. J. Am. Oil Chem. Soc. **76**, 331–339 (1999)

Chapter 9
Optimization of Sensor Array in Electronic Nose by Combinational Feature Selection Method

P. Saha, S. Ghorai, B. Tudu, R. Bandyopadhyay and N. Bhattacharyya

Abstract Electronic nose (e-nose) is a machine olfaction system and the sensor array is an essential part of the electronic olfaction process. A pattern recognition unit is necessary in electronic nose system to efficiently decide about the output of the test using the responses of all the sensors in the array. The output of a pattern recognition algorithm depends on the quality of the feature set used for training and testing. Relevant and independent feature set improves the performance of a pattern classification algorithm. In some applications of electronic nose, the responses of few sensors are highly corrupted with noise and are either irrelevant or are redundant to the process. These sensors should be identified and eliminated from the sensor system for better accuracy. This work addresses the selection of sensors in an e-nose system by different feature selection methods and then integrates them to achieve improved classification performance. We have used three types of feature selection methods namely, t-statistics, Fisher's criterion and minimum redundancy maximum relevance (MRMR) technique to select the most informative features. We have tested the proposed method on data obtained from the major aroma producing chemicals of black tea. Multi-class support vector machine (SVM) has been used as a pattern

P. Saha (✉) · S. Ghorai
Department of Applied Electronics and Instrumentation Engineering, Heritage Institute of Technology, Kolkata 700107, India
e-mail: pradip.saha@heritageit.edu

S. Ghorai
e-mail: santanu.ghorai@heritageit.edu

B. Tudu · R. Bandyopadhyay
Department of Instrumentation Electronics Engineering, Jadavpur University, Kolkata 700098, India
e-mail: bt@iee.jusl.ac.in

R. Bandyopadhyay
e-mail: rb@iee.jusl.ac.in

N. Bhattacharyya
Centre for Development of Advanced Computing (CDAC), Kolkata 700091, India
e-mail: nabarun.bhattacharyya@cdac.in

A. Mason et al. (eds.), *Sensing Technology: Current Status and Future Trends II*, Smart Sensors, Measurement and Instrumentation 8, DOI: 10.1007/978-3-319-02315-1_9,

classifier in an electronic nose with black tea samples. The experimental results show that the performance of the e-nose system increased by 3–7 % with the use of the proposed combinational feature selection technique.

Keywords Black tea · Electronic nose · Feature selection methods · Multiclass support vector machine

1 Introduction

Recent applications of electronic nose (e-nose) [1] technologies have come through advances in sensor design, material improvements, software innovations and progress in improved systems integration. As a result e-nose systems are used increasingly in many industries in different steps of production. Prediction of black tea quality is such an example where e-nose systems are being used now-a-days. E-nose system is designed to detect and discriminate complex odors of black tea. The sensor array in an electronic nose consists of metal oxide semiconductor gas sensors with non-specific sensitivity towards aroma components in black tea. Thus, there may be significant correlation between the response of two different sensors and such kind of information is known as redundant information. In some applications of electronic nose, the responses of few sensors are highly corrupted with noise making it irrelevant to the system. The performance of pattern recognition depends on feature set but the presence of redundant or irrelevant features degraded the quality of the discovered rules for pattern classification. Therefore, it is necessary to reduce the irrelevant and redundant features for the improvement of classification accuracy. Consequently, feature selection or sensor selection is an important part for the success of application specific instruments.

In last decades, many methods have been proposed for feature selection and these methods can be categorized into three groups: filter [2–4], wrapper [5] and embedded method [6]. Filter methods select the best subset of features using some predefined evaluation criteria that are independent of the learning algorithm. These are known as classifier independent methods. Wrapper methods are classifier-specific in that they utilize the learning algorithm as the evaluation function and search for the best subset of features that optimizes the generalization performance. Embedded methods combine feature selection with the learning algorithm. The method is strongly coupled with a specific learning algorithm and thus, there is a limitation to its application to other learning algorithms. Rank ordering feature selection procedure has been proposed by Wilson et al. [7] on samples of breath alcohol mixtures. Feature extraction based on eigenvectors of principle component analysis (PCA) technique has been reported in [8]. However, the PCA-based technique does not guarantee the highest classification rate as it ignores the contribution of a feature on classification. Pardo et al. [9] considered PCA plot exhaustive search using cross validated K-nearest neighbor algorithm (K-NN) to find a sensor subset of three sensors from an array of 19 sensors. GA-based stochastic search algorithm is proposed by Phaisangittlsagul et al. [10] for sensor selection and multiclass support vector machine (SVM) was used as the classifier. The filter based methods operate in isolation for ranking the features and do not consider the correlation among the features. Thus, the redundancy among the selected features is not used. To overcome this difficulty, Ding and Peng [11] proposed minimum redundancy maximum relevance (MRMR) method of feature selection where they used the mutual information criteria for selecting a set of most informative features. Rough set theory has been applied by the authors in [12] for the classification and feature optimization procedure.

In this work, we have selected relevant features by using combination of three different feature selection methods to increase the prediction accuracy. Three different multi-class SVM classifiers have been employed for the evaluation of the feature sets for this application. The experimental results show an improvement of the classification accuracy by 10 % using the proposed technique of feature selection and hence sensor optimization.

The rest of the paper is organized as follows. The data collection procedure by e-nose is described in Sect. 2. Section 3 describes the three feature selection methods and SVM classifier is described in Sect. 4. Sections 5 and 6 together describe the proposed combinatorial feature selection technique along with experimental results and associated discussions. Finally Sect. 7 concludes the work.

2 Data Collection by E-nose

2.1 E-nose Set Up

Prediction of tea quality is not an easy job due to the presence of innumerable compounds and their multi-dimensional contribution in determining the quality of tea. In practice, tea samples are tested by human sensory panel called "Tea Tasters," who assign quality scores in the scale of 1–10 according to the quality of tea. This method is highly subjective, and the scores vary from taster to taster [1]. The prediction of black tea quality by experimental means, co-relation of sensor array response with Tea Tasters' scores have been established by the author in [13] with good accuracy.

2.2 Customized Electronic Nose Setup for Black Tea

A customized electronic nose setup has been developed for quality evaluation of tea aroma, the details of which are presented in [13]. The sensor array of e-nose is one of the most important parts. Here, the sensor array contains eight gas sensors from Figaro, Japan i.e., TGS-2610, TGS-2620, TGS-2611, TGS-2600, TGS-816, TGS-831, TGS-832 and TGS-823. The experimental conditions for black tea quality evaluation have been optimized on the basis of repeated trials and sustained experimentation under the conditions as given in Table 1. During the experiments, dry tea samples have been used to avoid the effect of humidity.

Table 1 Experimental setup

Amount of black tea sample	50 gm
Temperature	$60 \pm 3\,^{\circ}\mathrm{C}$
Headspace generation time	30 s
Collection time	100 s
Purging time	100 s
Air-flow rate	5 mL/s

Table 2 Sample tea taster's score sheet

Sample code	Scores (1–10)				
	Leaf quality	Infusion	Liquor	Aroma	Leaf quality
VIKRAM240806-01	7	5	3	5	7
VIKRAM260806-03	6	5	5	8	6
VIKRAM200806-06	6	4	4	6	6
VIKRAM290806-09	5	5	6	7	5

2.3 Sample Collection

Tea samples have been used for the experiments from the tea gardens of M/s. Vikram India Limited. Samples of tea produced in their gardens are sent regularly to the tea-tasting centers for quality assessment by the tea tasters. For our experiments, one expert tea taster was deputed by the concerned industry to provide taster's score to each of the samples and these scores considered for the correlation study with the computational model. A sample tea taster score sheet is given in Table 2. Aroma scores basically indicate the smell and flavor of the tea samples whereas liquor scores indicate the combined perception of taste, briskness and astringency of the samples. The scores assigned to leaf quality and infusions are based on visual inspection of the samples by the experts or testers. The aroma scores of Table 2 are used for the training of the computational models of multi-class SVM classifiers as described in the next section.

3 Feature Importance Ranking

In this section we have described briefly the three feature selection algorithms used in this application.

3.1 Fisher Criterion for Feature Selection

Filter-based feature selection methods rank the features as a pre-processing step prior to the learning algorithm, and select those features with high ranking scores and for this purpose a number of performance criteria have been reported earlier, such as mutual information [14], Fisher score [15], Relief F [16], Laplacian score [17] and Trace ratio criterion [18], among which Fisher score is one of the most widely used criteria for supervised feature selection that selects each feature independently according to their scores using the Fisher criterion. Here, to select the discriminant features, we calculated the Fisher criterion of each feature. Consider, n_i as the number of samples in the ith class, $\overline{X_i}$ is the mean of samples in the ith class, X_{ij} is the jth sample in the ith class, X is the mean of total samples and c is the number of class. The Fisher score of kth feature is defined as:

$$F^k = \frac{SB^k}{SW^k} \tag{1}$$

where $SB^k = \sum_{i=1}^{c} n_i(\overline{X_i} - X)^2$ is the between class variance and is the within class variance of the feature. A larger F-ratio indicates a greater ratio of the inter-class variability to the intra-class variability and therefore, can be used to measure the discriminant capability of the features. So, we have arranged the features in descending order according to the F values. We have selected the features with higher Fisher scores as the single-feature Fisher criterion is able to retain all the relevant features for which this ratio is maximum.

3.2 T-Test

The t-score [19, 20] is nothing but a t-statistic between the centroid of a specific class and the overall centroid of all the classes. The t-score (TS) of feature i is defined by the following equation:

$$TS_i = \max\left\{\left|\frac{\bar{X}_{\mathrm{ik}} - \bar{X}_i}{m_k S_i}\right|,\ k = 1, 2, \ldots., K\right\}, \tag{2}$$

where

$$\bar{X}_{ik} = \sum_{j \in c_k} \frac{\bar{X}_{\mathrm{ij}}}{n_k} \tag{3}$$

$$\bar{X}_i = \sum_{j=1}^{n} \frac{X_{ij}}{n} \tag{4}$$

$$S_i^2 = \frac{1}{n-K} \sum_k \sum_{j \in c_k} (X_{ij} - \bar{X}_{ik})^2 \tag{5}$$

$$m_k = \sqrt{\frac{1}{n_k} - \frac{1}{n}}. \tag{6}$$

Here, k is the no. of classes and $\max\{y_k, k = 1, 2, \ldots, K\}$ is the maximum of all y_k, c_k refers to class k that includes n_k number of samples. X_{ij} is the expression value of feature i in sample j. $\overline{X}_{ik}$ is the mean expression value in class k for feature i. n is the total number of samples. S_i is the pooled within-class standard deviation of ith feature. Another possible model for T-score could be a t-statistic between the centroid of a specific class and the centroid of all the other classes.

3.3 MRMR Method of Feature Selection

The filter-based methods operate in isolation for ranking the features and do not consider the correlation among the features. Thus, the redundancy among the selected features is not used. To overcome this problem, MRMR method have been used that takes into account both minimum redundancy and maximum relevance criteria to select the additional features that are maximally dissimilar to the already identified features.

Let S be the subset of features to be selected by MRMR. Then the minimum redundancy condition between two features F_i and F_j is:

$$\min W_I = \frac{1}{|S|^2} \sum_{i,j \in S} I(F_i, F_j) \tag{7}$$

where $|S|$ indicates the number of features in S and $I(F_i, F_j)$ is the mutual information of two features F_i and F_j. The mutual information between two variables depend on their joint probabilistic distribution $P(F_i, F_j)$ and the respective marginal probabilities $P(F_i)$ and $P(F_j)$ as:

$$I(F_i, F_j) = \sum_{i,j} P(F_i, F_j) \log \frac{P(F_i, F_j)}{P(F_i)\, P(F_j)} \tag{8}$$

Now, we have to maximize the total relevance of all the features in S to obtain the maximum relevance condition for selecting the ith feature by the relation:

$$\max V_I = \sum_{i \in S} I(h, F_i) \tag{9}$$

where $h = \{h_1, h_2, ..., h_k\}$ represent the target class of the classification problem and the mutual information determines the relevance of the ith feature F_i for the classification task. Now by combining the conditions (7) and (9) into a single optimization problem we got the MRMR feature set where $\max(V_I - W_I)$ known as the mutual information difference (MID) criterion and $\max \frac{V_I}{W_I}$ known as the mutual information quotient criterion (MIQ) of the MRMR feature selection method. In this work, we have used MIQ condition of MRMR method as it performs better than MID scheme consistently shown by Ding and Peng [11].

4 Support Vector Machine

Support vector machine (SVM) is a widely used machine learning algorithm for binary data classification based on the principle of structural risk minimization (SRM) [21, 22] unlike the traditional empirical risk minimization (ERM) of artificial neural network. For a two class problem SVM finds a separating hyperplane that maximizes the width of separation of between the convex hulls of the two classes. To find the expression of the hyperplane SVM minimizes a quadratic optimization problem as follows:

$$\begin{aligned} &\text{Min } \frac{1}{2}||w||^2 + C\sum_{i=1}^{m}\xi_i \\ &\text{s.t. } y_i\left(w^T x_i + b\right) \geq 1 - \xi_i \; for \; i = 1, 2, ..., m \end{aligned} \tag{10}$$

where, $x_i \in \Re^n$ is the ith training pattern in n-dimensions, $y_i \in \{-1, 1\}$ is the class label of the ith training pattern, $w \in \Re^n$ and $b \in \Re$, respectively, are the normal to the hyperplane and bias term and m is the number of training patterns. $\xi_i (\geq 0)$ is the non-negative slack variables or soft margin errors of the ith training sample and $C(\geq 0)$ is a penalty or regularization parameter that determines the trade-off between the maximization of the margin i.e., small $||w||^2$ and minimization of the classification error i.e., small $\sum_{i=1}^{m}\xi_i$. The parameter C is a user defined parameter and it regulates the trade off between the margin width and the generalization performance. This formulation of SVM is known as the soft SVM. The solution of the quadratic programming (QP) problem (10) is obtained by forming its Lagrangian dual as follows:

$$\begin{aligned} &L(\alpha) = \sum_{i=1}^{m}\alpha_i - \frac{1}{2}\sum_{i,j=1}^{m}\alpha_i\alpha_j y_i y_j (x_i^T x_j) \\ &\text{s. t. } \sum_{i=1}^{m}\alpha_i y_i = 0, \quad c \geq \alpha_i \geq 0 \text{ for } i = 1, 2, ..., m \end{aligned} \tag{11}$$

where $\alpha_i(0 < \alpha_i < C)$ is the Lagrange multiplier. Finally, solving the above dual problem the expression of the decision hyperplane is obtained as:

$$f(x) = sgn(w^T x + b) = sgn(\sum_{i=1}^{m} \alpha_i y_i (x_i^T x) + b) \tag{12}$$

This linear SVM classifier is usually used for the linearly separable data. But in most of the cases the data are not linearly separable. To overcome this difficulty, kernel trick is employed to learn nonlinear SVM classifier. By this procedure the patterns are first mapped from the input space to the high dimensional feature space by using a transformation function $\Phi(.,.)$. Then a linear SVM classifier is implemented in that feature space with the transformed data. The linear separating hyperplane corresponds to the nonlinear separating hyperplane in the input space. To implement such a classifier again leads to solving a QP problem similar to (10) with all x_i replaced by $\Phi(x_i)$. Then of course the training algorithm would depend only on the dot product of the patterns in the feature space. This dot product can be computed by using some kernel function:

$$k(x_i, x_j) = \Phi(x_i)^T \Phi(x_j) \tag{13}$$

Thus, using the kernel trick, the Lagrangian dual problem in the feature space will maximize the following problem:

$$\begin{aligned} &\text{Max } L(\alpha) = \sum_{i=1}^{m} \alpha_i - \frac{1}{2} \sum_{i,j=1}^{m} \alpha_i \alpha_j y_i y_j k(x_i, x_j) \\ &\text{s. t. } \sum_{i=1}^{m} \alpha_i y_i = 0, \quad c \geq \alpha_i \geq 0 \ \text{for} i = 1, 2, ..., m \end{aligned} \tag{14}$$

The corresponding decision function becomes:

$$\begin{aligned} f(x) &= sgn\left(\sum_{i=1}^{m} y_i \alpha_i (\Phi(x_i)^T \Phi(x)) + b\right) \\ &= sgn\left(\sum_{i=1}^{m} y_i \alpha_i k(x_i, x) + b\right) \end{aligned} \tag{15}$$

There are many possible kernels, such as linear, Gaussian, polynomial and multi-layer perceptron etc. In this study, we have used Gaussian (RBF) kernel function of the form as given below:

$$k(x_i, x_j) = \exp(-\gamma \left\| x_i - x_j \right\|^2) \tag{16}$$

The extensive application of binary SVM classifier provoked several researchers to discover the efficient way of extending binary classifier to multi-class classifier [23–26]. In the literature there are one-versus-rest (OVR), one-versus-one (OVO), directed acyclic graph (DAG), all at once, error correcting code etc. Among these we have used three most commonly used methods of multi-class SVM classifier in this application. A brief description of the OVR, OVO and DAG method is explained below.

4.1 One-Versus-Rest (OVR) SVM Classifier

In OVR method k numbers of binary classifier models are constructed for a k class problem [23]. In doing so the patterns of ith class considered as the positive samples and the patterns of all other classes are considered as the negative samples for the ith binary classifier model. This decomposition method has been shown in Fig. 1 for a three class problem which shows that it builds three classifiers: (a) class 1 versus classes 2 and 3, (b) class 2 versus classes 1 and 3, and (c) class 3 versus classes 1 and 2. The combined OVR decision function chooses the class of a sample that corresponds to the maximum value of the argument of k binary decision functions specified by the furthest "positive" hyperplane.

4.2 One-Versus-One (OVO) SVM Classifier

In contrast to OVR, OVO method trains ${}^{N}C_2 = N(N-1)/2$ binary classifier models considering each possible pair for a N class problem. Figure 2 illustrates how six binary classifiers are trained in OVO method for a four class problem. The testing of a sample in OVO method is performed by max win strategy [24, 26]. By this technique each of the trained binary classifier can deliver one vote for its favored class and the class with maximum votes specifies the class label of the sample. Thus as the number of classes increases the training and testing time also increase in this method.

4.3 Directed Acyclic Graph (DAG) SVM Classifier

To reduce the testing time of OVO method Platt introduced DAG method [25, 26]. The training part of the DAG method is same as that of the OVO method. But DAG method of testing follows a rooted binary directed acyclic graph to make a decision as shown in Fig. 3. The final decision is made when a test sample reaches to the leaf node. Thus, both OVO and DAG methods have the same training phase but during testing DAG uses only $(N-1)$ evaluations to make decision among $N(N-1)/2$

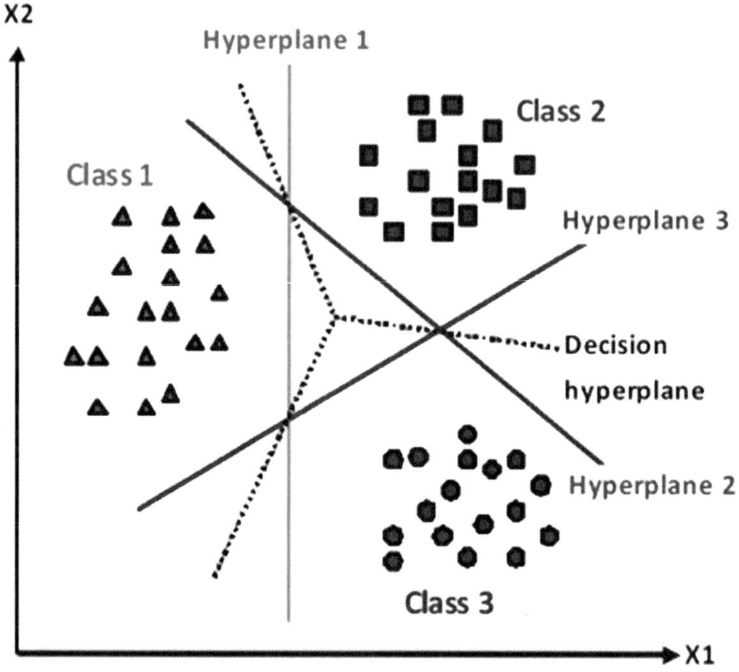

Fig. 1 One-versus-rest SVM classifier for a three class problem constructs three classifiers

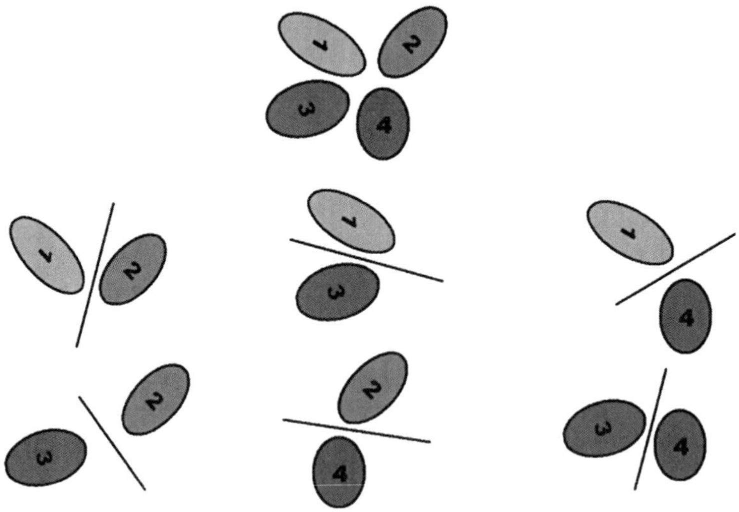

Fig. 2 OVO method of decomposition of a multiclass classification problem containing four different classes. For a four class problem it trains six binary classifiers with taking two different classes at a time

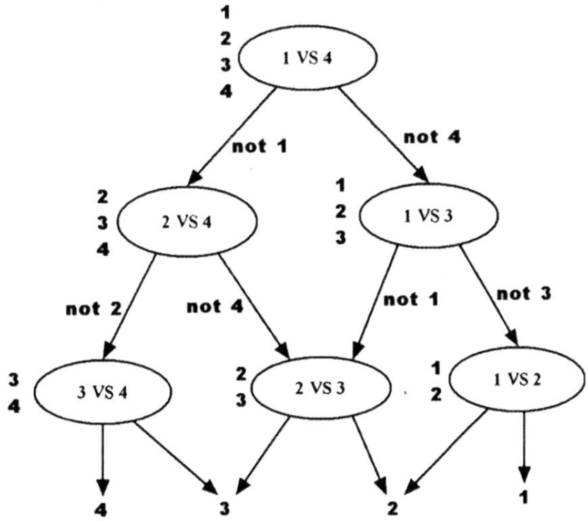

Fig. 3 Testing of a sample by DAG method with four different classes

binary classifiers. Accordingly, for a four class problem, DAG decides the class label by three comparisons only out of six classifiers as shown in Fig. 3. Hence DAG method reduces the testing time of a sample than by OVO method.

5 Experiments and Results

In this section we have described the experimental set up, experimental procedure, proposed method and results.

5.1 Experimental Set Up

The multi-sensor array in the e-nose system consists of eight sensors and they produce eight readings for each sample. It has been observed that the taster's score varies within a range from 3 to 6. So, this black tea quality prediction problem becomes a four-category classification problem. We have computed the performance of the OVR, OVO and DAG multi-class SVM classifiers by using Gaussian kernel with the help of LIBSVM Toolbox [27]. The optimal value of the regularization parameter C is selected from the set of value $\{2^i | i = -5, -4, ..., 11, 12\}$. The Gaussian kernel parameter γ is selected from the set of values $\{2^i | i = -9, -8, ..., 9, 10\}$. The parameters for each model are selected by the performance on a tuning set comprising of 20 % of training data while the remaining 80 % training data are used to train the classifier [28]. Once the parameter set is selected the training and testing sets are merged to train the final classifier model using the selected parameters. We have

Table 3 Feature ranking list by different algorithms

Feature ranking algorithm	Ranking list							
F-ratio	6	8	7	4	2	1	3	5
T-score	4	8	2	7	1	6	3	5
MRMR	4	6	2	7	8	5	3	1

evaluated the performance of the different feature sets using 10-fold cross validation (10-fold CV) method [28].

5.2 Experimental Procedure

We have first applied three feature ranking algorithms individually on the collected tea data obtained by e-nose responses. The feature ranking list as obtained by applying F-ratio, T-score and MRMR method is shown in the Table 3, where the numbers represents the sensor number as they are serially mentioned in Sect. 2.2. From the Table 3 it is observed that different feature ranking algorithms produce different ranking list. The experimental procedure that we have followed is explained with the help of a block diagram as shown in Fig. 4. It shows that each of the three feature selection algorithms is applied individually to rank the features. Then three different types of SVM models are trained by using first two features from a particular ranking and their 10-fold CV performances are observed. In this way, we have increased one

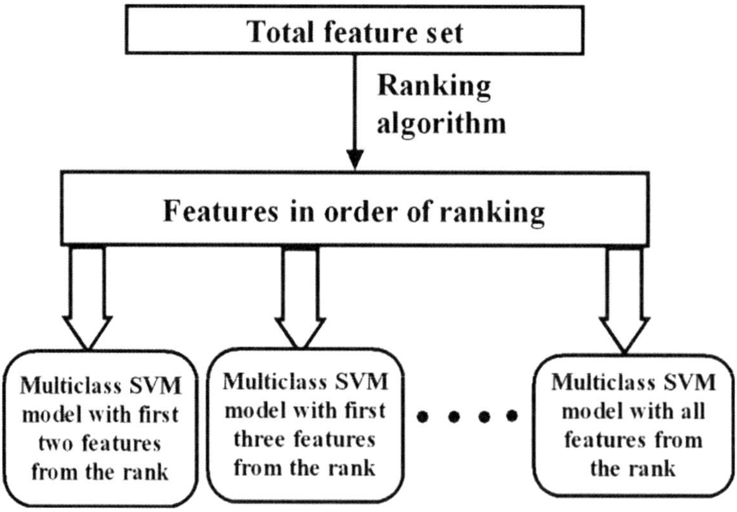

Fig. 4 Experimental procedure for performance evaluation of different feature sets

Table 4 10-fold CV performance of different feature sets by different multiclass SVM models

Classifier	No. of features from ranking	Average percentage of 10-fold CV accuracy obtained by ranking with		
		F-ratio	T-score	MRMR
OVR SVM	2	69.64 ± 11.17	**77.55 ± 15.20**	67.64 ± 17.10
	3	74.55 ± 15.52	75.55 ± 12.54	71.55 ± 11.36
	4	76.45 ± 14.93	74.55 ± 11.72	73.45 ± 9.10
	5	76.45 ± 11.09	73.45 ± 10.14	76.45 ± 11.09
	6	74.45 ± 10.24	74.45 ± 10.24	74.45 ± 10.24
	7	74.45 ± 10.24	74.45 ± 10.24	74.45 ± 10.24
	8	74.45 ± 10.24	74.45 ± 10.24	74.45 ± 10.24
OVO SVM	2	68.73 ± 13.53	68.55 ± 10.77	63.64 ± 19.86
	3	75.73 ± 12.86	70.45 ± 16.05	69.45 ± 13.21
	4	**78.36 ± 12.57**	67.55 ± 14.90	70.64 ± 14.64
	5	77.27 ± 12.84	72.55 ± 17.07	77.27 ± 12.84
	6	73.55 ± 17.86	73.55 ± 17.86	78.27 ± 20.48
	7	72.55 ± 19.87	72.55 ± 19.87	73.45 ± 19.05
	8	74.55 ± 19.00	74.55 ± 19.00	74.55 ± 19.00
DAG SVM	2	67.73 ± 14.47	69.55 ± 11.46	65.64 ± 16.30
	3	73.73 ± 12.81	70.64 ± 17.73	66.64 ± 12.79
	4	**78.36 ± 12.57**	66.55 ± 16.91	70.64 ± 14.64
	5	75.27 ± 16.43	72.55 ± 17.07	75.27 ± 16.43
	6	73.55 ± 17.86	73.55 ± 17.86	78.27 ± 20.48
	7	72.55 ± 19.87	72.55 ± 19.87	73.45 ± 19.05
	8	74.55 ± 19.00	74.55 ± 19.00	74.55 ± 19.00

feature at a time and observed the performances of different SVM models. Thus, we have tested the performance of seven OVR, OVO and DAG SVM models for each of the three feature ranking algorithms.

In Table 4, we have shown the average 10-fold CV accuracy and standard deviation of three different multi-class SVM classifiers on the feature sets obtained with the three ranking algorithms. The results show that the performances of different ranking algorithms with same number of features are different by all the three SVM models. Another important point observed from it is that performances by all the SVM models is poor considering all the features together than that of considering three or four features from the ranking list. This fact is better visualized from the Fig. 5, which shows the performances of OVR, OVO and DAG SVM models for each of the three ranking algorithms separately. For example, Fig. 5a shows that the performance OVR SVM on the features ranking with T-score is better compared to the ranking of both F-ratio and MRMR method considering up to 3 features. As the number of features gradually increases the performance of feature sets obtained by T-score decreases while the performance of feature sets obtained by F-ratio and MRMR method increases. Finally, it is observed that the performance with considering 6, 7 and 8 (i.e. all) number of features remains same. In other words these three features do

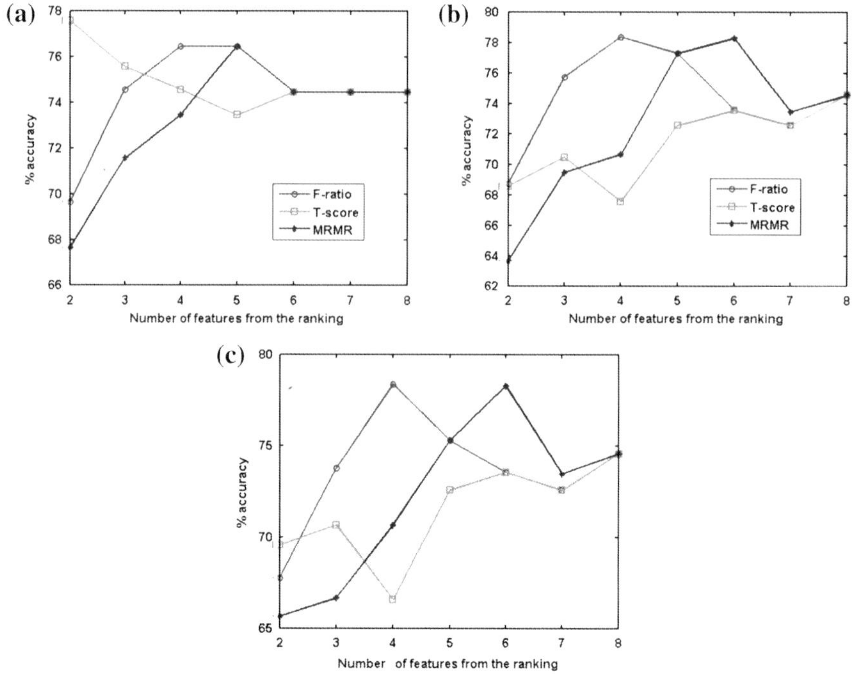

Fig. 5 Variation of performance of different feature sets for different ranking algorithm obtained with a OVR SVM b OVO SVM c DAG SVM classifiers

not contribute any important information to improve the performance of the classifier. Similar trends are observed from the Figs. 5b and c obtained by OVO and DAG SVM, respectively. Therefore, from these observations it is clear that not a single ranking algorithm can provide us a proper feature set which will be suitable for this application. For this purpose we have used a combinatorial feature selection technique as explained below.

6 Combinational Feature Selection Method

In Fig. 6 we have described the proposed combinational feature selection technique. The first two features of each individual ranking are taken together to form a feature set. It is observed from the Table 3 that the combined feature set consists of only three features, i.e., feature number 4, 6 and 8.

We have trained OVR, OVO and DAG SVM models with these three selected features and their performances are shown in Table 5. From the Table 5, it is observed that with these three features the average 10-fold CV accuracy becomes 81.36 % for OVR SVM and 81.27 % for both OVO and DAG SVM. This shows that an increase

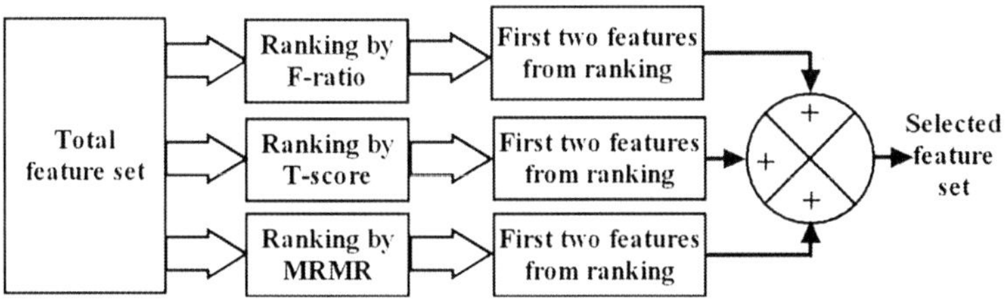

Fig. 6 Combinational feature selection technique

Table 5 Performance of different multiclass SVM models obtained by combinational feature selection technique

Selected features	Classifier type	Selected parameter set by tuning	Classification accuracy
4, 6, 8	OVR SVM	$C = 2^6, \quad \gamma = 2^9$	81.36 ± 15.77
	OVO SVM	$C = 2^9, \quad \gamma = 2^5$	81.27 ± 17.05
	DAG SVM	$C = 2^8, \quad \gamma = 2^5$	81.27 ± 13.81

of accuracy approximately 3–7 % than that of the performance of all the three types of SVM classifiers considering two to eight features in a set form individual ranking. Therefore, responses of sensors 4, 6 and 8 of the e-nose system, i.e., the sensor number TGS2600, TGS831 and TGS823, may be considered in this application for optimal performance.

7 Conclusions

An efficient method of feature selection and hence sensor optimization for an e-nose system is described in this work with the help of a problem of black tea quality prediction. This work shows that a feature set comprising of few features from the highest ranking of a feature selection algorithm will not necessarily produce the best classification performance. Since the performance of a classifier depends on the choice of the parameter also. Therefore, the feature and parameter of a classifier should be selected simultaneously to obtain the optimum performance. In our future work we shall look after this issue by using wrapper or embedded method of feature selection in this application.

References

1. R. Dutta, E.L. Hines, J.W. Gardner, K.R. Kashwan, M. Bhuyan, Tea quality prediction using a tin oxide-based electronic nose: An artificial intelligence approach. Sens. Actuators B, Chem. **94**(2), 228–237 (Sep. 2003)

2. N. Kwak, C.H. Choi, Input feature selection by mutual information based on Parzen window. IEEE Trans. Pattern Anal. Mach. Intell. **24**(12), 1667–1671 (2002)
3. M. Hall, Correlation-based feature selection for discrete and numeric class machine learning, in *Proceeding of the 17th International Conference on Machine Learning*, Stanford, pp. 259–266 (2000)
4. L. Yu, H. Liu, Efficient feature selection via analysis of relevance and redundancy. J. Mach. Learn. Res. **5**, 1205–1224 (2004)
5. M. Kudo, J. Sklansky, Comparison of algorithms that select features for pattern classifiers. Pattern Recogn. **33**(1), 25–41 (2000)
6. I. Guyon, A. Elisseeff, An introduction to variable and feature selection. J. Mach. Learn. Res. **3**, 1157–1182 (2003)
7. D.M. Willson, K. Dunman, T. Roppel, R. Kalim, Rank extraction in tin oxide sensor arrays. Sens. Actuators B **6**, 199–210 (2000)
8. P.W. Carey, K.R. Beebe, B.R. Kowalaski, D.L. Illman, T. Hirschfeld, Selection of adsorbates for chemical sensor arrays by pattern-recognition. Anal. Chem. **58**, 149–153 (1986)
9. M. Pardo, L.G. Kwong, G. Sberveglieri, K. Brubaker, J.F. Schneider, W.R. Penrose, J.R. Stetter, Data analysis for a hybrid sensor array. Sens. Actuators B **106**, 136–143 (2005)
10. E. Phaisangittisagul, H.T. Nagle, V. Areekul, Intelligent method for sensor subset selection for machine olfaction. Sens. Actuators B: Chem. **145**, 507–515 (2010)
11. H. Peng, F. Long, C. Ding, Feature selection on mutual information: Criteria of max-dependency, max-relevance, and min-redundancy. IEEE Trans. Pattern Anal. Mach. Intell. **27**(8), 1226–1238 (2005)
12. A. Bag, B. Tudu, J. Roy, N. Bhattacharyya, R. Bandyopadhyay, Optimization of sensor array in Electronic nose: A rough set based approach. IEEE Sens. J. **11**(11), 3001–3008 (November 2011)
13. N. Bhattacharyya, R. Bandyopadhyay, M. Bhuyan, B. Tudu, D. Ghosh, A. Jana, Electronic nose for black tea classification and correlation of measurements with "Tea Taster" marks. IEEE Trans. Instrum. Meas. **57**, 1313–1321 (2008)
14. D. Koller, M. Sahami, Toward optimal feature selection, in *ICML*, pp. 284–292 (1996)
15. P.E.H.R.O. Duda, D.G. Stork, *Pattern Classification* (Wiley-Interscience Publication, New York, 2001)
16. M. Robnik-Sikonja, I. Kononenko, Theoretical and empirical analysis of relieff and rrelieff. Mach. Learn. **53**, 23–69 (2003)
17. X. He, D. Cai, P. Niyogi, Laplacian score for feature selection, in *NIPS* (2005)
18. F. Nie, S. Xiang, Y. Jia, C. Zhang, S. Yan, Trace ratio criterion for feature selection, in *AAAI*, pp. 671–676 (2008)
19. R. Tibshirani, T. Hastie, B. Narashiman, G. Chu, Diagnosis of multiple cancer types by shrunken centroids of gene expression. Proc. Nat'l Acad. Sci. USA **99**, 6567–6572 (2002)
20. J. Devore, R. Peck, *Statistics: The Exploration and Analysis of Data*, 3rd edn. (Duxbury Press, Pacific Grove, 1997)
21. C. Cortes, V.N. Vapnik, Support vector networks. Mach. Learn. **20**(3), 273–297 (1995)
22. N. Cristianini, J. Shawe-Taylor, *An Introduction to Support Vector Machines* (Cambridge University Press, Cambridge, 2000), pp. 113–145
23. L. Bottou, C. Cortes, J. Denker, H. Drucker, I. Guyon, L. Jackel, Y. LeCun, U. Muller, E. Sackinger, P. Simard, V. Vapnik, Comparison Of classifier methods: A case study in hand writing digit recognition, in *Proceedings of the International Conference on Pattern Recognition*, 1994, pp. 77–87
24. U. Kreßel, Pairwise classification and support vector machines, in *Advances in Kernel Methods–Support Vector Learning*, ed. by B. Schölkopf, C.J.C. Burges, A.J. Smola (MIT Press, Cambridge, 1999), pp. 255–268
25. J.C. Platt, N. Cristianini, J. Shawe-Taylor, Large margin DAG's for multi-class classification. Adv. Neural Inf. Proc. Syst. **12**, 547–553 (2000)
26. C.-W. Hsu, C.-J. Lin, A comparison of methods for multi-class support vector machines. IEEE Trans. Neural Networks **13**(2), 415–425 (2002)

27. C.C. Chang, C.J. Lin, LIBSVM: A library for support vector machines. Technical Report, 2001. [Online]. Available: http://www.csie.ntu.edu.tw/~cjlin/libsvm
28. T.M. Mitchell, *Machine Learning* (The McGraw-Hill Companies, Inc., Singapore, 1997), pp. 148

Chapter 10
Exploratory Study on Aroma Profile of Cardamom by GC-MS and Electronic Nose

D. Ghosh, S. Mukherjee, S. Sarkar, N. K. Leela, V. K. Murthy, N. Bhattacharyya, P. Chopra and A. M. Muneeb

Abstract Cardamom is known as "Queen of Spices". It is one of the most highly priced spices in the world. The commercial part of the cardamom is the fruit (Capsule) of the plant that is used as a spice and a flavouring agent. The major quality measurement parameter of the cardamom is freshness, size, colour, aroma etc. Aroma is one of the main and crucial quality parameter for cardamom. The present practice of aroma quality estimation is GC, GC-MS, where different volatile oil and chemicals qualitative and quantitative tests are done. The present practice is laborious, time consuming and skilled manpower demanding process. In our present study an effort has been made to develop an Electronic Nose for rapid aroma determination of cardamom. Centres for Development of Advanced Computing (C-DAC), Kolkata has indigenously developed the Electronic Nose (E-Nose) to estimate the quality of food and agro produces. Three-clone specific cardamom samples ware tested using this system as an exploratory study to determine the quality of cardamom and found the system is able to differentiate the samples. The principal component analysis (PCA) shows distinct three clusters with principal component, PC1 (91.6 %) and PC2 (6.8 %). This paper demonstrates the quality estimation of cardamom by E-Nose.

Keywords Cardamom · Electronic nose · Aroma · Principal component analysis

D. Ghosh (✉) · S. Mukherjee · S. Sarkar · N. Bhattacharyya
Centres for Development of Advanced Computing (C-DAC), Saltlake, Kolkata Kol-91, India
e-mail: devdulal.ghosh@cdac.in

N. K. Leela · A. M. Muneeb
Indian Institute of Spice Research, ICAR, Calicut 673012, India

V. K. Murthy
PES Institute of Technology, 100-ft Ring Road, BSK III Stage, Bangalore 560085, India

P. Chopra
Electronics System Development and Application (ESDA), Department of Electronics and Information Technology (DeitY), MCIT, Electronics Niketan, New Delhi 110003, India

A. Mason et al. (eds.), *Sensing Technology: Current Status and Future Trends II*, Smart Sensors, Measurement and Instrumentation 8, DOI: 10.1007/978-3-319-02315-1_10,

1 Introduction

Small cardamom (*Elettaria cardamomum Maton*), one of the world's most ancient spices, was already mentioned in approximately 3000 BC in Sanskrit texts in India [1]. This "Queen of Spices" belongs to the ginger family (*Zingiberaceae*) and is the third most expensive spice in the world, after saffron and vanilla. It grows as a native in the southern Indian forests of the Western Ghats. Owing to its sensitiveness to wind, drought, and waterlogging, optimum yield is obtained on warm (10–35 °C) and humid (with >1,500 mm of well-distributed rainfall) mountain slopes at 600–1,500 m elevation, under a canopy of evergreen trees. Cardamom has been commercially cultivated in the Western Ghats for 150 years, and India has had a virtual trade monopoly until recently. Cardamom and black pepper were also the primary reason for establishing the sea route from Europe to the Far East. At present, the largest producers of true cardamom are Guatemala and India, and smaller producers include Tanzania, Sri Lanka, Papua New Guinea, El Salvador, Laos, and Vietnam. Cheaper substitutes to real cardamom (*Amomum spp.* and *Aframomum spp.*) are grown and used in some Asian countries [2]. India and Saudi Arabia consume more than half of the world's total cardamom production. In Arab countries and India, cardamom is a common flavoring ingredient for coffee and tea. In Scandinavia, as well as in Germany and Russia, it is used to flavor cakes, pastries, and sausages. It is popular in Indian and South Asian cooking and used to make spice blends, such as curries and garam masala. Chewing cardamom after a meal is recommended to aid digestion and to clean teeth. In Eastern medicinal practices it is used for curing ailments such as influenza, infections, asthma, bronchitis, cardiac disorders, diarrhea, nausea, cataracts, and for strengthening the nervous system [3]. It is also said to have a cooling effect in hot climates. Ancient Greeks and Romans already used its delicate aroma to make perfume [1].

1.1 About Indian Spice: Cardamom

Small cardamom (*Elettaria cardamomum Maton*) is perennial tropical herb plant of the ginger family and grows from thick rootstalk up to around 6–10 ft, indigenously grown in the evergreen forest of the Western Ghats in South India. Cardamom of commerce is the dried capsule of this plant indigenous to the evergreen forests of Western Ghats of South India. India is one of the largest producers of cardamom in the world and has major export market also in the world. Cultivated species are grouped in to three main groups primarily based on nature of panicle. These are 'Malabar', 'Mysore' and 'Vazhukka'. 'Vazhukka' is a natural hybrid between 'Malabar' and 'Mysore' [4]. The quality of cardamom depends on the essential oil content and composition. Malabar and Mysore types differ in the composition of essential oil. Malabar is camphory in aroma due to higher content of 1,8-cineole. Mysore contains more α-terpinyl acetate, which contributes to mild spicy flavor. In Mysore linalool and linalyl acetate are markedly higher. Detailed analysis of the variability in the composition of essential oil is described by Chempakam and Sindhu [3].

1.2 Brief Description of Electronic Nose

An electronic nose (Fig. 1) uses an array of non-specific broadly tuned sensors to discriminate odours [5–11]. The odours are analyzed by sensor array data with pattern recognition methods [4, 13]. A customized electronic nose set-up has been developed such that the same can be used in quality estimation of cardamom for monitoring of volatile emission pattern. The electronic nose consists of (a) sensor array, (b) micro-pump with programmable sequence control, (c) PC-based data acquisition and (d) olfaction software as illustrated in Fig. 2. The same system is used for study of cardamom.

An array of metal oxide semiconductor (MOS) sensors has been used for assessment of volatiles in the set-up. A series of experiments were carried out using a number of commercially available MOS sensors. From the response sensitivity of individual sensors, a set of eight gas sensors from Figaro, Japan (TGS-832, TGS-823, TGS-831, TGS-816, TGS-2600, TGS-2610, TGS-2611 and TGS-2620) (Fig. 3) has been selected for odour capture.

The outputs of the sensors are acquired in the PC through PCI data acquisition cards. The MOS sensors are conductometric in nature, and their resistance decreases when subjected to the odour vapor molecules. The change in resistance with respect

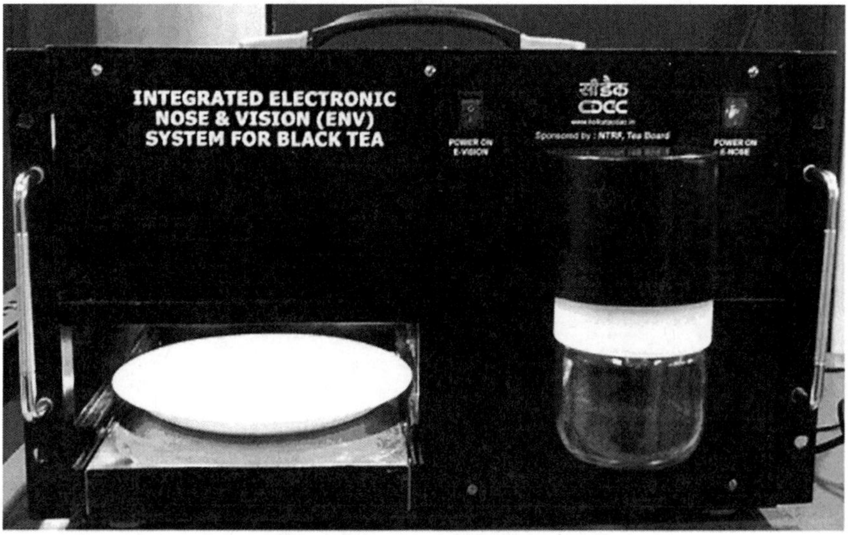

Fig. 1 Electronic nose and vision system

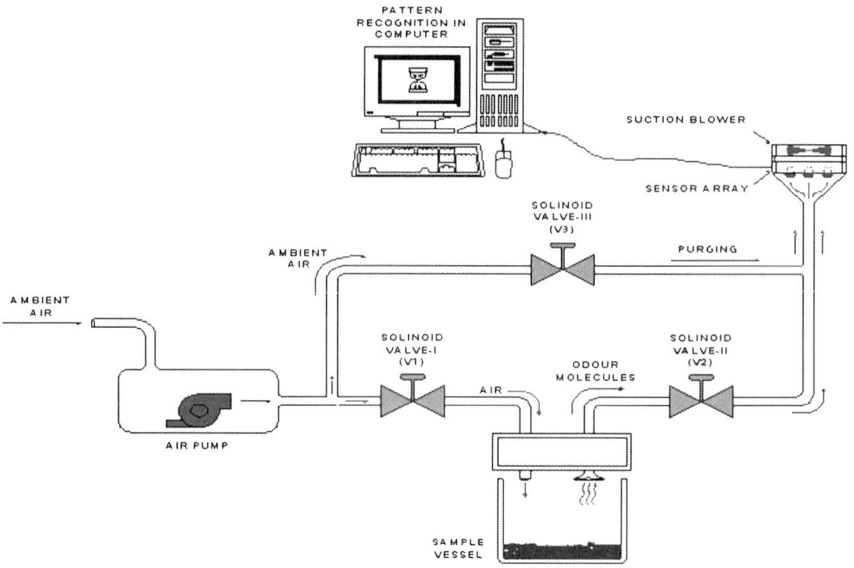

Fig. 2 Customized electronic nose set-up

Fig. 3 Sensor array

to their original values ($\Delta R/R$) is converted into voltage and then taken to the PC through analog to digital converter cards for subsequent analysis in the computational models.

The experimental sniffing cycle consists of automated sequence of internal operations: (i) headspace generation, (ii) sampling, (iii) purging and (iv) dormancy before the start of the next sniffing cycle (Fig. 2). Headspace generation ensures adequate concentration of volatiles released by sample within the sample holder by blowing regulated flow of air on the sample. During sampling, the sensor array is exposed to a constant flow of volatiles through pipelines inside the electronic nose system. During purging operation, sensor heads are cleared with blow of fresh air so that the sensors can go back to their baseline values. The programmable time dormancy cycle is the suspended mode of the electronic nose between two consecutive sniffing cycles.

The specially designed software, called olfaction software, controls the PC-based data acquisition and automated operation of all these cycles. The software has got features like programmable sequence control, dynamic fermentation profile display, data logging, alarm annunciation, data archival, etc. The software has been developed in LabVIEW® of National Instruments. Experimental conditions are as follows:

Heating time = 65 s
Amount of cardamom sample = 10 g,
Headspace generation time = 30 s,
Data collection time = 50 s,
Purging time = 100 s,
Airflow rate = 5 ml/s.

2 Experimental Procedure

2.1 Extraction of Essential Oil

Dried cardamom capsules (25 g) of four varieties, namely, CCS-1, GG, NKE-12 and RR-1 were crushed and the seeds were separated. The decorticated seeds were hydrodistilled for 3 h in Clevenger apparatus to extract the essential oil. The essential oil separated was collected, dried over anhydrous sodium sulphate and stored in a refrigerator until the analysis was carried out. The percent essential oil content was calculated. The varieties GG recorded maximum essential oil yield (6 %) per capsule weight basis and superior quality with lower 1,8-cineole and high terpinyl acetate contents.

2.2 Gas Chromatography Analysis

GC analysis was performed on a Shimadzu (GC-2010) Gas chromatograph fitted with FID detector and RTX-5 column (30 m × 0.25 mm, film thickness 0.25 μm).

Nitrogen was used as carrier gas at a flow rate of 1 ml/min the oven was programmed as follows: at 600–2,000 °C at the rate of 30 °C/min, again up to 2,200 °C at the rate of 50 °C/min, at which the column was maintained for 5 min. The injection port temperature was maintained at 2,500 °C, the detector temperature was 2,500 °C. The split ratio was 1:40. 0.1 μl sample was injected. The percentage composition component of the oil was determined by area normalization.

2.3 *Gas Chromatography: Mass Spectrometry*

The oil was analyzed using a Shimadzu GC-2010 Gas chromatograph equipped with QP 2010 mass spectrometer. RTX-5 column (30 m × 0.25 mm, film thickness 0.25 μm) coated with polyethylene glycol was used. Helium was used as the carrier gas at a flow rate of 1 ml/min. The oven was programmed as follows: at 600 °C for 5 min and then increased to 1,100 °C @ 50 °C/min, then up to 2,000 °C @ 30 °C/min, again up to 2,200 °C @ 50 °C/min, at which the column was maintained for 5 min. The injection port temperature was maintained at 2,600 °C, the detector temperature was 2,500 °C. The split ratio was 1:40 and ionization voltage was maintained at 70 eV and a 0.1 μl sample was injected. The retention indices were calculated relative to C8–C20 n-alkanes.

2.4 *E-Nose Analysis*

The experiment was conducted at CDAC, Kolkata. Indian Institute of Spice Research, Calucut, supplied the samples. Three cardamom verities were taken for E-Nose experimentation namely, CCS-1, GG, NKE-12. There were total three samples in number containing 100 g of cardamom sample of different clones. From three sample 10 g were taken for the E-Nose experiment and the experiment was conducted as above-mentioned condition i.e. heating time = 65 s, headspace time = 30 s, sampling time = 50 s, purging time = 100 s. For each sample three repetition ware done and also three-replication ware done. The same samples ware tested in QC lab of Indian Institute of Spice Research, Calicut, under ICAR.

3 Data Analysis and Results

3.1 *Essential Oil Yield and Composition*

The essential oil yield and composition of four cardamom varieties was shown in Tables 1 and 2 respectively.

Table 1 Essential oil yield of cardamom varieties

Serial number	Sample description	Oil yield (ml per 100 g capsules)
1	CCS-1	5.0
2	GG	6.0
3	NKE-12	5.0
4	RR-1	5.0

Table 2 Essential oil composition of cardamom varieties

Compound	RI	CCS-1	GG	NKE-12	RR-1
α-Pinene	934	2.1	2.0	2.3	1.8
Sabinene	974	3.9	4.2	4.0	3.8
β-Pinene	977	0.7	0.6	0.7	0.5
Myrcene	993	3.0	3.4	3.1	3.1
δ-3-Carene	1008	0.5	0.4	0.5	0.3
1,8-Cineole	1037	32.1	27.4	34.3	26.9
β- Z-Ocimene	1039	0.2	0.29	0.2	0.2
γ-Terpinene	1055	1.0	1.0	1.0	1.3
Terpinolene	1088	0.5	0.5	0.6	0.9
Linalool	1106	2.0	1.4	1.4	1.6
4-Terpineol	1180	2.9	2.7	2.8	3.5
α-Terpineol	1196	3.7	3.4	3.4	5.2
Neral	1229	0.7	0.9	0.3	1.0
Linalyl acetate	1254	1.4	0.9	1.0	1.0
Geraniol	1254	2.9	1.9	4.4	2.9
Geranial	1264	1.1	1.5	0.4	1.6
α-Terpinyl acetate	1365	34.0	32.0	33.3	37.4
Geranyl acetate	1379	0.9	0.5	1.0	1.1
α-Terpinyl propionate	1429	0.2	0.1	0.3	0.5
Nerolidol	1567	1.2	1.5	1.1	1.6

3.2 Data Normalization

In this data set we obtained one instantaneous response and one purging response for each cardamom sample. Instantaneous response is a matrix of 800×8 (since headspace time 30 s and sampling time 50 s, so total time is 80 s $(30 + 50)$. In 1 s 10 data are captured and there are eight MOS sensors, so total data set is (800×8) and purging response is a vector of 1×8 (when the process shifts from headspace to sampling, i.e. the response of 30th second). If V_I is the instantaneous response and V_P is the purging response then the normalization formula for all sensors is:

$$\frac{V_I - V_P}{V_P} \text{ for all sensors} \tag{1}$$

3.3 Data Matrix Preparation

Form the normalized matrix the maximum value is selected. For eight sensors eight values obtained. So finally it is vector of 1 × 8 and for three replications of three samples the matrix is 9 × 8.

3.4 Data Analysis

After collecting the data it was found that the norm aroma index (NAI) of the same sample is gradually decreasing. The bar graph shows that with increase of repetition the norm aroma index is decreasing (Fig. 4). This may be because at the time of headspace generation the air is blowing. For this the aroma of cardamom seeds is evaporating. For this in the time of second repetition the aroma emission is less and the norm aroma index is also less. Same tradition is observed in the third repetition also.

For this study three replications of three samples were selected at the time of experimentation. Thus the 9 × 8 data set is obtained and used for data analysis.

3.5 Cluster Analysis

For each cardamom samples we have one set of Electronic Nose data. From this data set the sensor wise maximum value is selected. Thus one the vector is derived. Thus for three replications three data set is made for one sample. For three samples three data sets will be obtained. Then these data sets were applied in principal component

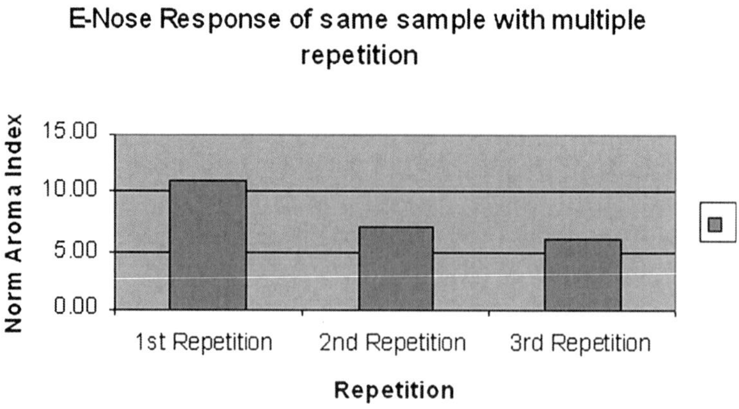

Fig. 4 Bar plot of NAI with repeated data

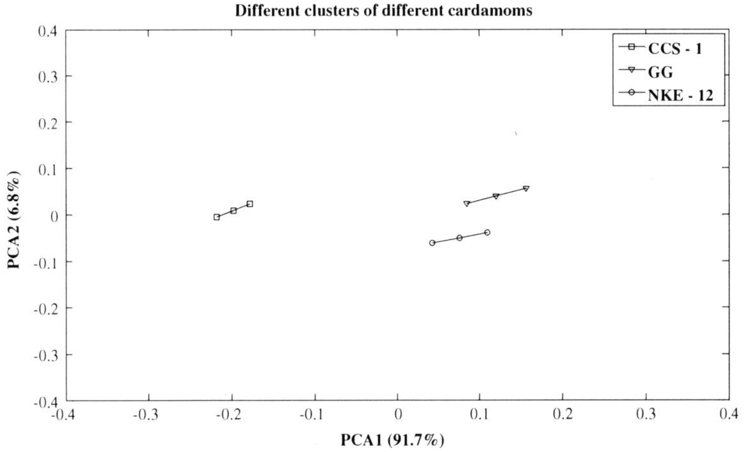

Fig. 5 Clustering analysis of E-Nose data based of three different samples

analysis (PCA) to visualize whether there are true three clusters. The clustering plot is given in Fig. 5 where three distinct groups are present. Principal component analysis (PCA) [6, 14] involves a mathematical procedure that transforms a number of possibly correlated variables into a lesser number of uncorrelated variables called principal components and reduces the number of dimensions also. Here first two dimensions (principal components) are plotted to visualize the clusters. The PC1 and PC2 contain 91.7 and 6.8 % information respectively.

4 Discussion

It is demonstrated that, in clustering analysis three distinct clusters are present for three samples. Using these data the classification model can be made and may be used for prediction of quality estimation of cardamom. Also the volatile chemical constituents responsible for aroma of these samples are available. So the chemical composition of cardamom can be predicted by the data set generated by E-Nose.

Acknowledgments This work is financially supported by the Department of Electronics and Information technology (DeitY), Government of India through a national program, titled, "Application of Electronics for Agriculture and Environment (eAgriEn)".

References

1. P.N. Ravindran, K.J. Madhusoodanan (eds.), *Cardamom: The Genus Elettaria (Medicinal and Aromatic Plants—Industrial Profiles)* (Taylor and Francis, London, 2002)

2. T.K.R. Nair, M.G. Kutty, Cardamom (*Elettaria cardamomum*) in Kerala, India, in *Forest Products, Livelihoods and Conservation: Case Studies of Non-timber Forest Product Systems*, vol. 1, Asia, ed. by K. Kusters, B. Belcher (CIFOR, Bogor, 2004), pp. 133–150
3. B. Chempakam, S. Sindhu, Cardamom, in *Chemistry of Spices*, ed. by V.A. Parthasarathy, B. Chempakam, T.J. Zachariah (CABI, London, 2008), pp. 41–58
4. K.V. Peter, E.V. Nybe, N. Miniraj (eds.), *Spices* (New India Publishing agency, New Delhi, 2007), pp. 43–66
5. N. Bhattacharya, B. Tudu, A. Jana, D. Ghosh, R. Bandhopadhyaya, M. Bhuyan, Preemptive identification of optimum fermentation time for black tea using electronic nose. Sens. Actuators B **131**(1), 110–116 (2008)
6. N. Bhattacharyya, R. Bandyopadhyay, M. Bhuyan, B. Tudu, D. Ghosh, A. Jana, Electronic nose for black tea classification and correlation of measurements with tea taster marks. IEEE Trans. Instrum. Measur. **57**(7), 1313–1321 (2008)
7. B.S. Mamatha, M. Prakash, S. Nagarajan, K.K. Bhat, Evaluation of the flavor quality of pepper (*Piper nigrum* L.) cultivars by GC-MS, electronic nose and sensory analysis techniques. J. Sens. Stud. **23**(4), 498–513 (2008)
8. U. Banach, C. Tiebe, Th Hübert, Application of electronic nose and ion mobility spectrometer to quality control of spice mixtures (Olfaction and Electronic Nose: Proceedings of the 13th International Symposium on Olfaction and Electronic Nose). AIP Conf. Proc. **1137**, 301–302 (2009)
9. R. Zawirska-Wojtasiak, S. Mildner-Szkudlarz, E. Wąsowicz, M. Pacyński, Gas chromatography, sensory analysis and electronic nose in the evaluation of black cumin (*Nigella sativa* L.) aroma quality. Herba Polonica **56**(4), 20–30 (2010)
10. H. Zhang, M. Balban, K. Portier, C.A. Sims, Quantification of spice mixture compositions by electronic nose, part II: comparison with GC and sensory methods. J. Food Sci. **70**(4), E259–E264 (2006)
11. L. Maggi, M. Carmona, C.P. del Campo, C.D. Kanakis, E. Anastasaki, P.A. Tarantilis, M.G. Polissioub, G.L. Alonsoa, Worldwide market screening of saffron volatile composition. J. Sci. Food Agric. **89**, 1950–1954 (2009)
12. K.R. Kashwan, M. Bhuyan, Robust electronic-nose system with temperature and humidity drift compensation for tea and spice flavour discrimination, in *Proceedings of the Sensors and the International Conference on New Techniques in Pharmaceutical and Biomedical Research*, 5–7 Sept 2005, pp. 154–158
13. H. Zhang, M.Ö. Balaban, J.C. Principe, Improving pattern recognition of electronic nose data with time-delay neural networks. Sens. Actuators B **96**, 385–389 (2003)
14. N. Bhattacharyya, S. Seth, B. Tudu, P. Tamuly, A. Jana, D. Ghosh, R. Bandyopadhyay, M. Bhuyan, S. Sabhapandit, Detection of optimum fermentation tim for black tea manufacturing using electronic nose. Sens. Actuators B **122**(2), 627–634 (2007)

Chapter 11
High Frequency Surface Acoustic Wave (SAW) Device for Toxic Vapor Detection: Prospects and Challenges

T. Islam, U. Mittal, A. T. Nimal and M. U. Sharma

Abstract Surface Acoustic Wave (SAW) based device is very much suitable for detecting very small quantity of vapors of explosive chemicals. There are three possible kinds of SAW devices which can be used for sensor fabrication such as delay lines, resonator and filter. Choice among them depends on individual preferences and sometimes it is the matter of chance. In this report, a high frequency 70 MHz SAW device has been developed for the detection of chemical warfare (CW) agents. SAW device is fabricated on ST-quartz substrate as it has negligible temperature coefficient at room temperature. The device is having dual oscillator circuit configuration for compensating the effects of temperature, humidity and pressure. Device is coated with suitable polymer and the applicability of the sensor for dimethyl methylphosphonate (DMMP) detection has been demonstrated.

Keywords SAW · Oscillator · Polymer coating · DMMP sensor

1 Introduction

Surface Acoustic Waves (SAW) can be generated at the free surface of an elastic solid [1]. When the solid is a piezoelectric the waves are coupled to potential fields as well. Through interactions with metal electrodes deposited on the solid surface and through digital sampling of these waves, a variety of signal processing operations can be realized. This field is broadly referred as acousto-electronics or ultrasonic-electronics. The foundation for this field was laid way back in 1885 by Lord Raleigh

T. Islam (✉)
Department of Electrical Engineering, F/O Engineering and Technology, Jamia Millia Islamia (Central University), New Delhi 110025, India
e-mail: tislam@jmi.ac.in

U. Mittal · A. T. Nimal · M. U. Sharma
Solid State Physics Laboratory, Lucknow Road, Timarpur, Delhi 110054, India

A. Mason et al. (eds.), *Sensing Technology: Current Status and Future Trends II*, Smart Sensors, Measurement and Instrumentation 8, DOI: 10.1007/978-3-319-02315-1_11,

[2] who stated that the behavior of waves upon the plane free surface of infinite homogeneous isotropic elastic solid is such that the disturbance is confined to a superficial region of thickness comparable with the wavelength. Surface acoustic waves are mechanical (acoustic) rather than electromagnetic in nature. However it was only in 1965 when White and Voltmer [1, 2] first exploited the phenomenon of SAW propagation for its application to electronic device. With the invention of interdigital transducer (IDT) that real development of this field began in right earnest [2, 3]. IDT is a metallic comb like structure carved on metallic film deposited on the piezoelectric substrate. SAW can be generated by applying RF voltage to IDT placed on the surface of the piezoelectric substrate. In a SAW device two IDTs are required and are called as input and output IDT. Input IDT converts the high frequency voltage signal into mechanical surface acoustic waves and output IDT converts mechanical SAW vibrations into electrical signal. Surface outside the IDT regions should be elastic, not necessarily being piezoelectric. In SAW devices the piezoelectric material is used to provide the electromechanical coupling. Velocity of SAW waves is of the order of about 105 times less than electromagnetic waves and wavelength is of the order 3 μm at the frequency of 1 GHz. Because of slow speed the acoustic waves enable one to delay a time varying signal by using very small space. Most common type of SAW devices are delay lines, band pass filters and resonators [4].

SAW propagation is sensitive to a number of different parameters such as temperature, humidity, pressure, etc. These parameters change the electrical characteristics of the SAW devices. Hence SAW devices are widely used in sensing applications. The word sensor is derived from sentire meaning 'to perceive'. 'Sensor' is a device that detects a change in a physical or chemical stimulus and turns it in to signal, which can be measured or recorded. It is like a transducer but the difference between sensor and transducer is to use 'sensor' for the sensing element and 'transducer' for sensing element plus any associated circuitry. The use of acoustic wave devices for sensor applications has been investigated since the early 1960s [3]. The first acoustic wave sensors were based on bulk acoustic waves. SAW sensors have the potential to offer fully integrated low cost, high performance, solid-state devices for a variety of physical, chemical and biological sensing applications. SAW physical sensors are used to measure the physical changes like acceleration, stress, strain, temperature, pressure, humidity, vacuum and motion [5]. SAW chemical sensors are used to sense the presence of almost any chemical, with detection limits extending down to few parts per trillion [5]. SAW biological sensors are used in medicines, food product industry, environment and defense. Biosensors are sensitive to monitor ingredients of food and drink, food additives, toxic components and contaminants [5]. In SAW devices the whole operation takes place on the surface, the characteristics of the device is susceptible to environmental perturbations on or near the surface.

A SAW device can be imparted sensitivity for a particular chemical species by coating the device surface with a selective adsorbent film. Absorption of chemical species into the sensor film causes change in its properties such as (mass or viscoelasticity). As a result SAW velocity and amplitude are changed. This process of chemical transduction is translated into shift in phase and amplitude of the signal. SAW chemical sensors can be further used in

- Detection of environmental pollutants [5].
- Detection of chemical warfare agents in trace level [5].
- Detection in explosive and narcotics of extremely low amount [5].
- Process parameter monitoring and control [5].
- Detection of hazardous chemicals in industrial environment [5].
- Passive remote sensing at hazardous and inaccessible locations [5].

Various techniques are used to measure the change in SAW properties via phase and amplitude. Wiring the SAW device in an oscillator loop as a frequency-determining element is the simplest and effective method for sensing [6]. In this case the change in phase is visualized as shift in frequency of the oscillator. The oscillator should have low noise, low power consumption, high stability, capable of sustaining mass loading and easy to handle. Electronic oscillators realized using LC tank circuit or RC network in the feedback loop of an amplifier have low Q factor in the range of $\sim 10^3$ and hence stability is less. Oscillator employing piezoelectric crystals have high Q factor of the order of $\sim 10^5$ and hence it shows higher stability and can be used in applications where demand on stability is more [5]. Schematic of SAW sensor has been shown in Fig. 1. It consists of two identical SAW devices one is used for sensing toxic agent and another is used for reference to compensate the temperature and pressure error.

1.1 Acoustic Wave Sensors

Piezoelectric effect allows transduction of electrical energy into acoustic energy and vice versa [7]. This is exploited to realize a variety of device configurations on piezoelectric crystals for sensor applications. An acoustic wave device in general

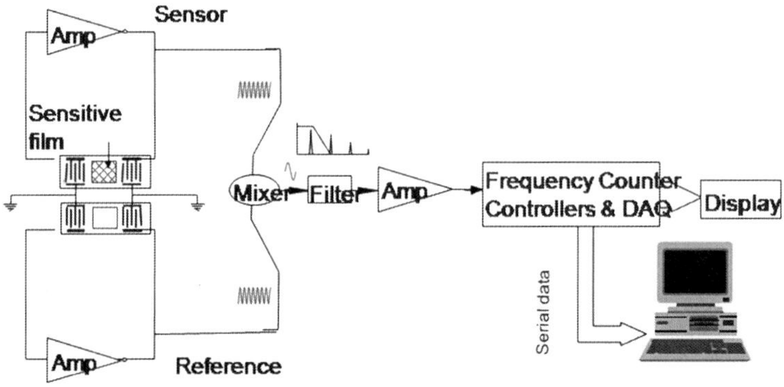

Fig. 1 Schematic diagram of interface electronics circuit for SAW sensor

is comprised of electro-acoustic transducers fabricated on crystal surface. Typically two transducers one for input and other for output are used. The characteristic of acoustic waves (velocity, power) propagating between two transducers is affected by electrical/physical boundary conditions on the surface and the surrounding environment [8]. This is precisely what makes the basis for realizing an acoustic wave sensor. Changes or a perturbation in surface conditions or immediate environment under physical, chemical or biological stimulus causes changes in acoustic wave properties, which can be monitored as electrical signals [5].

1.1.1 Physical Sensor

All acoustic wave sensors are sensitive, to varying degrees, to perturbations from many different physical parameters. As a matter of fact all acoustic wave devices manufactured for the telecommunication industry must be hermetically sealed to prevent any disturbances because they will be sensed by the device and cause an unwanted change in output [5]. But for sensor application requirement is just reversed. Acoustic wave sensors become pressure, torque, shock, strain and force etc. detectors under an applied stress that changes the dynamics of the propagating medium and called as physical sensor.

1.1.2 Chemical Sensor

An acoustic wave sensor becomes chemical sensor when a thin film is applied on piezoelectric substrate that adsorbs only specific chemical. Other techniques involve detection of change in thermal conductivity, electrical conductivity, etc. due to chemical interaction by SAW devices [5].

1.1.3 Biological Sensor

Biosensors are those devices that include as a "front end" bio recognition element that responds to only one biological parameter. In a biosensor the scope for a very high selectivity arises from 3-D structures of the biomolecules and the specific arrangement and location of functional groups. The functional groups located in 3-D structure engage in selective (mostly polar) interactions with molecules having complementary functional groups in proper orientation. Sensitive and selective detection of biochemically active compounds can be achieved by employing antigen–antibody, enzyme–substrate and other receptor–proteins pairs. When coating on acoustic wave sensor absorbs specific biological chemicals in liquid, they become a biosensor [5].

1.2 Surface Acoustic Wave Device

Besides discrete surface acoustic wave sensors which exploit the SAW fundamental principle, SAW resonators can be used as sensing devices. This device consists of a pair of electrodes positioned in a resonant cavity formed by two distributive SAW reflector arrays patterned on crystal surface. These reflector arrays consist of half—wavelength wide metal strips on grooves, when the surface wave is incident on these periodic structures, small amount of wave energy reflected from each discontinuity adds to a given nearly complete reflection (99 %). The high Q and low insertion loss that results make these devices extremely stable when incorporated in an oscillator circuit, resulting in lower detection limits. However reflector arrays add significantly to substrate area occupied by sensor and complicate coating of device surface.

1.3 Advantages of SAW Sensor

SAWs are well suited for either fixed or wireless sensor applications as the specific sensor constituent to be measured typically varies one or more of the SAW's physical parameters. Surface acoustic wave (SAW) based sensors and sensor systems have following advantages:

Sensitivity(S) of SAW sensor are very high as SAW devices can operate at very high frequency(f) hence we can achieve very high sensitivity ($S \alpha f^2$). SAW sensors:

- have low noise operation hence we can achieve low detection limit
- fabrication is based on IC planar technology, hence it can be mass producible and low cost [4]
- are in small size, rugged, reproducible, and reliable
- have fast response time and they are suitable for individual use
- can be placed in confined and inaccessible places
- are small and light weight
- operate in severe environments such as explosive, corrosive, radiated environments
- are hermetically sealed so there is no effect of the environmental condition.

1.4 Development of SAW Device for Vapor Sensing

Vapor sensor based on SAW devices were first reported in 1979, most of them rely on the mass sensitivity of the detector, in conjunction with a chemically selective coating that adsorbs the vapor of interest and result in an increased mass loading of the device [5]. In its simplest form SAW sensor consists of a delay line, which consists of an IDT at each end of an appropriate piezoelectric substrate. One IDT acts as transmitter and other acts as the receiver of acoustic energy which travels along the substrate surface. The propagating acoustic wave is free to interact with matter in contact with delay line surface. This wave film interaction results in alteration of wave characteristic such as amplitude, phase, velocity, harmonic contents etc. which are the basis of SAW chemical sensors. SAW delay line sensors are most easily configured to monitor either changes in SAW amplitude or SAW velocity. Amplitude measurements are made using an apparatus in which the Rayleigh wave is excited using an RF power source and the power at receiving end of SAW delay line is measured with RF power meter [5]. Measurement of SAW velocity is made indirectly by using SAW delay line in oscillator configuration. The resonant frequency of device is altered by changes in velocity of acoustic wave and can be measured accurately using a digital frequency counter. Experience has shown that measurement of SAW velocity provides a superior precision compared to SAW amplitude perturbation so we concentrate more on velocity measurement techniques.

1.5 Issues and Approaches for SAW Sensor Realization

The following issues are to be approached for the realization of SAW vapor sensor

- Stability
- Sensitivity
- Selectivity
- Response time
- Reversibility
- Reliability
- Temperature and pressure effects.

Sensing device should be fairly stable so that change in frequency can easily be distinguished due to chemical interaction. For sensing application highly stable systems are not required [5]. Several design considerations must be satisfied when selecting and applying the chemically sorptive coating. Ideally, the coating film should be completely reversible, meaning that it will absorb and then completely desorbs the vapor when purged with clean air. The rate at which the coating absorbs and desorbs should be fairly quick, <1 s, for instance. The coating should be robust enough to withstand corrosive vapors. It should be selective, absorbing only very specific vapors while rejecting others. The coating film must operate over a practical temperature range. It should be stable, reproducible, and sensitive. And finally, its thickness and uniformity are very important [9]. Temperature, pressure and humidity affect the performance of SAW sensor. To tackle temperature and pressure effects we can use one device as sensor and another device as reference, which effectively minimizing the effects of temperature and pressure variations. For humidity control we incorporate a filter, through which the desired vapor passes and which absorbs the moisture from the sample.

1.6 Application of SAW Sensor

SAW sensors are extremely versatile devices that are just beginning to realize their commercial potential. They are competitively priced, inherently rugged, very sensitive, and intrinsically reliable, and can be interrogated passively and wirelessly. The range of phenomenon that can be detected by SAW devices can be greatly expanded by coating the devices with materials that undergo changes in their mass, elasticity, or conductivity upon exposure to some physical or chemical stimulus. These sensors become pressure, torque, shock, and force detectors under an applied stress that changes the dynamics of the propagating medium. They become mass or gravimetric sensors when particles are allowed to contact the propagation medium, changing the stress on it. They become vapor sensors when a coating is applied that absorbs only specific chemical vapors. These devices work by effectively measuring the mass of the absorbed vapor. If the coating absorbs specific biological chemicals in liquids, the detector becomes a biosensor.

2 Surface Acoustic Waves

Surface acoustic wave is guided along the surface, with its amplitude decaying exponentially with depth. The wave is strongly confined, with typically 90 % of energy propagating within one wavelength of the surface. It is non-dispersive, with a velocity of typically 3,000 m/s [10]. A bounded medium also supports many other type of acoustic waves, and the boundary condition can substantially affects the nature of the waves. On the other hand, a medium with dimensions much larger than the wavelength can support waves with characteristic similar to those of wave in an infinite medium. The term bulk wave is often used to describe waves, which are not bounded to the medium. Surface acoustic waves can be generated at the free surface of elastic solid. In the SAW devices the generation of such waves is achieved by application of a voltage to metal film interdigital transducer (IDT) deposited on the surface of a piezoelectric solid [5, 9].

2.1 Interdigital Transducer

The most efficient and widely used means of generating and detecting surface elastic waves for electronic applications is the interdigital transducer shown in Fig. 2. In its simplest form it consists of a series of parallel metal plate electrodes periodically spaced on the surface of a piezoelectric substrate. The transducer is a three-port device: one electrical and two acoustic. When an alternating voltage is applied to its electrical port, time varying electric fields are set up within the substrate, and these excite alternating stress pattern via the piezoelectric effect. The electric field

Fig. 2 Inter digital transducer (IDT)

is reversed at each electrode, and therefore at frequencies for which the periodic length lambda of the array is equal to the wavelengths of elastic surface waves (synchronous), the waves are launched most efficiently in both directions normal to the electrodes [10].

The space between them known is the wavelength of the mechanical wave (λ). We can express the operating frequency f_0 of the device with SAW velocity (v_p) and wavelength (λ) as:

$$f_o = \frac{V_p}{\lambda} \tag{1}$$

3 Fabrication of SAW Device for Toxic Gas Detection

A SAW device is fabricated on a ST (42.75° Y-X) piezoelectric quartz substrate. The quartz substrate is known to have high electromechanical coupling coefficient (K^2) and low-temperature coefficient of frequency (TCF) [2]. A SAW device was comprised of an input interdigital transducer (IDT), a mass loading area and an output IDT, as shown in Fig. 1. The dimension of the one electrode is 11.26 μm and wave length is 45.06 μm. An Al thin film with a thickness of 200 nm was evaporated on a cleaned ST-quartz substrate by thermal evaporation technique and then, the IDT pattern was formed by standard photolithography process and steps for the SAW device fabrication are shown in Fig. 3. Once the IDTs are formed on the substrate, sensitive film of the desired gas has been deposited on the mass loading area. Terminals are connected for the testing of the device. Two identical devices have been fabricated.

3.1 Deposition of Sensing Layer on SAW Device

In SAW sensor, for obtaining significant sensitivity and selectivity for the measurement of a given analyte requires a chemical interface, henceforth referred to as "the coating". The coating, which should be physically or chemically bound to the sensor surface, may consist of a solid adsorbent, a chemical reagent, or a sorptive liquid or polymer [5, 11]. The coating acts as a chemically sensitive and selective element that immobilizes a finite mass of some chemical species from the environment. Resultant changes in physical and/or chemical properties of the coating, in turn perturb the underlying acoustic wave device. Perturbations of acoustic waves resulting from

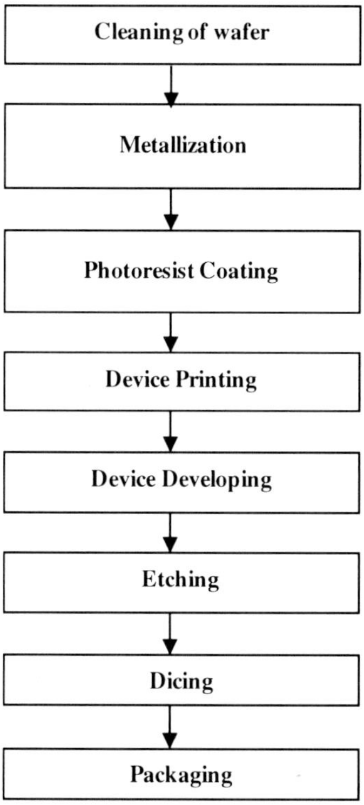

Fig. 3 SAW device fabrication flow chart

interactions of the coating with one or more analytes constitute the basis for detection and quantification [12].

3.2 General Coating Requirements

An attractive feature of acoustic-wave-based chemical sensors is that they impose relatively few constraints on the materials that can be used as chemically selective coatings. The film should have following properties to work as an excellent chemical interface:

- Uniformity
- Adherent
- Thin chemically and physically stable when in contact with its working medium (gas or liquid)
- Non conductive.

Typically, uniformity in film thickness is not crucial, but can be important in some circumstances. Assuming all parts of the film fall within the acceptable thickness limits discussed in the following, and that the particular film being examined has been calibrated, then under conditions of equilibrium between the analytes in the ambient phase and the film, variations in film thickness are unimportant. If, however, transient measurements are of interest- one means of identifying a chemical species is its rate of permeation through a given material-then uniformity becomes very important: non-uniformities in thickness will "smear out" the transient response, making identification difficult. Also, if device-to-device reproducibility is important, then the film must be quite uniform, unless aluminum films can be fabricated with the same set of non-uniformities. Uniform coverage of the acoustic wave path is of some importance as well. In the case of SAW delay line, the propagating wave front is typically fairly linear; a film that covers some parts of the wave path to a greater extent than others causes parts of the wave front adversely affects the signal-to-noise ratio. The selected materials must adhere to the device surface in such a manner that it moves synchronously with the acoustic wave and must maintain this adhesion in the presence of expected analytes and interferants. The adhesion of thin film to many types of surfaces, including those that are chemically very dissimilar to the coating material is also very crucial. Often, adhesion promoting inter layers have been developed for general classes of problems, such as securing a highly monopolymer film to a very polar substrate [13]. Highly conductive coatings (i.e., all metals and most semiconductors) must, of course, be electrically insulated from IDT electrodes in order to prevent shorting; this is not a concern with the planar electrodes of TSM quartz resonators [14]. This is readily accomplished by either deposition using a line-of-sight technique, e.g., vacuum evaporation, with a protective stencil interposed between the vice and the source; or depositing over the entire surface, and then chemically removing the conductive film from the IDTs (the film's chemical nature differ considerably from that of the IDT material in this case [15]. A final constraint is that the coating be acoustically thin. A somewhat standard rule of thumb is that thickness of the film less than 1 % of the acoustic wavelength are appropriate [5, 16]. In fact, whether or not a particular coating thickness is acoustically thin depends critically upon the acoustic properties of that material under the particular set of conditions (temperature, nature and concentration of contacting gaseous species, etc.) being evaluated. In other words, it is the acoustic wavelength in the coating film that is relevant; this can differ appreciably from the acoustic wavelength in the device substrate, particularly in materials such as rubbery polymers, which have vastly different sound velocities than, for example, quartz. In practice, coating thickness from a few angstroms to several μm have been utilized in sensing applications [17].

3.3 Coating Methods

Numerous methods have been described in the literature for depositing coatings onto piezoelectric acoustic sensors. They generally fall into three categories [5, 18]:

- Solvent casting techniques,
- Vacuum deposition techniques, and
- Vapor-phase deposition techniques.

Solvent Casting Techniques: Solvent casting is perhaps the simplest coating method. It requires that the coating material be soluble in a solvent that does not chemically attack the piezoelectric sensor device and its transducers. Once the coating material is dissolved the solution can be spread over the device and the solvent evaporated to leave the desired coating material. Popular techniques in this category include [5, 19]:

- Syringe deposition,
- Painting with small brushes or Q-tips,
- Dipping,
- Spraying, and
- Spin casting.

The coating reproducibility that is achievable with syringe deposition and painting can range from poor to good depending on the material used and the skill of the person applying the coating.

In any of the solvent casting techniques, less than a few micrograms of coating material are sufficient to completely coat the sensor device with a film of the appropriate thickness. Thus, the solutions used are usually quite dilute, the sequence of which is that solvent purity and equipment cleanliness must be carefully considered. Spray coating is performed by aspirating a dilute solution of the coating through an atomizing nozzle using a compressed-gas propellant (an inexpensive tool available at art supply stores, the air brush, is often utilized for this process). The fine, atomized mist of solution droplets is propelled toward the device they impact, stick, and evaporates, thereby leaving the non-volatile coating. Like the syringe- and paintbrush-deposited films, the coatings formed by this procedure often have somewhat irregular texture and coverage, but good reproducibility in thickness is possible, particularly if the acoustic sensor is operating during deposition process: monitoring the sensor output signal during deposition allow the apparent thickness of the coating to be measured in real time [20]. Spin casting generally offers the highest degree of film uniformity; and the greatest film-to-film thickness reproducibility. A commercially available "spinner" of the type used in the deposition of photoresist films for lithography, holds the substrate face-up on a motor-driven vacuum chuck that can be spun at hundreds to several thousand rpm. The surface of the device is then "flooded" with a viscous solution of the coating. When the motor is turned on, centrifugal and aerodynamic forces cause all but a thin layer of the solution to "fly" off the device surface; spinning continues long enough (10–60 s) for the vast majority of the solvent to evaporate, inhibiting recoalescence of film droplets. The uniformity and thickness reproducibility of the resulting film are often excellent. Film thickness is controlled by varying spinning speed and the solution viscosity through choice of solvent and concentration of coating material [21]. Dip coating, particularly using the Langmuir-Blodgett (LB) technique can be very reproducible. However, this method does require that the coating have ambiphilic properties (i.e., the individual molecules must possess a polar end and a nonpolar end separated by an intervening chain or body of a few atoms) and be somewhat water insoluble in order to form a stable monolayer at the air/water interface of the LB deposition trough. The required trough is commercially available but relatively expensive. For coating materials that do form stable monolayers, the LB deposition technique affords exquisite control of film thickness, since only a single monolayer of coating molecules is transferred to the device surface with each dip through the air/water interface. LB film have the advantage over other materials that they are highly ordered in two dimensions, a result of the alignment of all the molecules at the air/liquid interface to film deposition. The consequence of this is more predictable, precisely controlled chemical and physical properties than is available with the molecularly disordered materials produced with many other deposition techniques [22].

3.4 Interface Coating on SAW Device

The device has been coated with a solution having polymer "bisphenol" which is sensitive for DMMP nerve agent. Before formation of thin coating film, the device was cleaned with de ionized (DI) water and acetone to remove any particulate if sitting on the device. The device was coated through a process of "drop coating" using a micro-syringe. The liquid solution of $\sim 1\,\mu l$ is deposited on the surface and the device is then baked for 5 h at 110 °C. Unfortunately, thin film is not uniformly coated on the SAW device, which is very necessary for desired electrical characteristics of the sensor. It is clearly observed in the Fig. 4 that the coating of polymer over the device is non-uniform. The non-uniform coating of sensing film over the SAW device may occur due to improper cleaning of the device surface. The uncleaned particles may reduce the surface energy of the device for proper adhering of the film. To improve the uniformity, the surface of the device was then plasma cleaned with the help of plasma cleaning machine (Make ROTH & RAU). Plasma is generated by ionizing argon gas using microwave. This ionized gas will strike the surface of the device with very

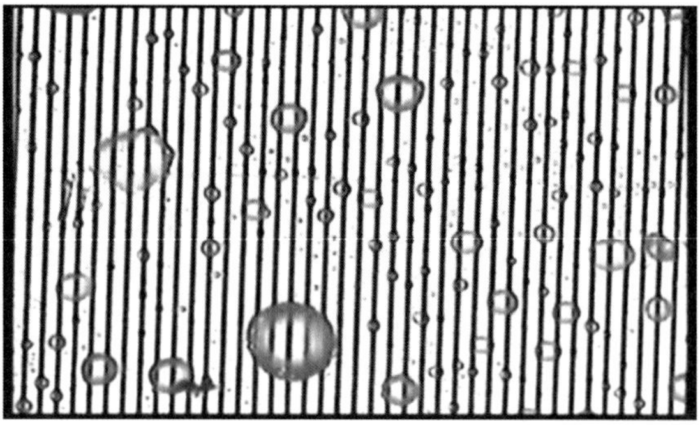

Fig. 4 Non uniform polymer film deposited on the SAW device

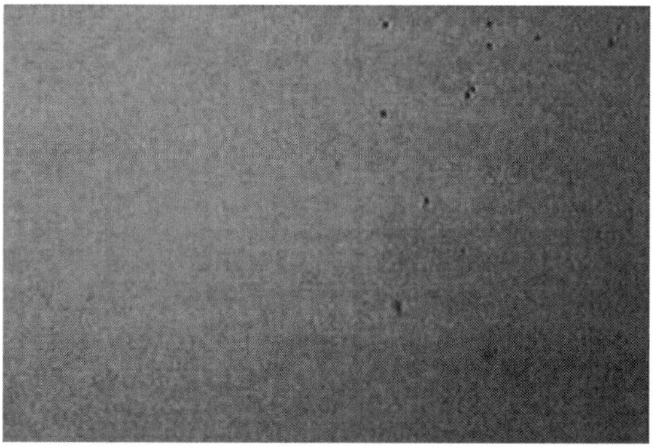

Fig. 5 Uniform polymer film deposited on the SAW device

high kinetic energy removing any particle on the surface. Immediately after plasma cleaning, we coat the polymer on SAW device and the film was baked at 110 °C for 5 h. The device surface with uniform coating suitable for sensing application is shown Fig. 5.

4 Characterization of SAW Device

The packaged device has been characterized by NA (Network Analyzer) model no 3577A and 35677A using S-parameter test set-up to determine the center frequency and insertion loss of the device which is the important parameter of SAW sensor and required for its oscillator design. The response of the packaged device is shown in Fig. 6. The central frequency is around 70 MHz. SAW oscillators have been successfully used in a number of applications where high frequency and low noise are important conditions. SAW oscillators are of high spectral purity, small power consumption and size for fundamental frequency of operation up to GHz range i.e. the oscillators can be realized in the range from about 20 MHz to 3 GHz. SAW oscillators have high value of quality factor (defined as reciprocal of fraction of energy stored to fraction of energy dissipated per radian). It is now possible to construct SAW oscillators with Q value of 1,00,000. For a fundamental frequency of 100 MHz it is possible to have $Q = 10^5 - 10^7$ (instead of only 4,000–5,000 for the simple delay line). The power dissipation is comparably very low in a SAW oscillator. SAW oscillator uses a SAW device in the feedback circuit. These oscillators are called SAW oscillators because SAW device in the feedback circuit is used for frequency control. The SAW device is connected to a wide band amplifier in a feedback loop and provides stability through frequency selective feedback. The sensor electronics includes the signal processing and the display units.

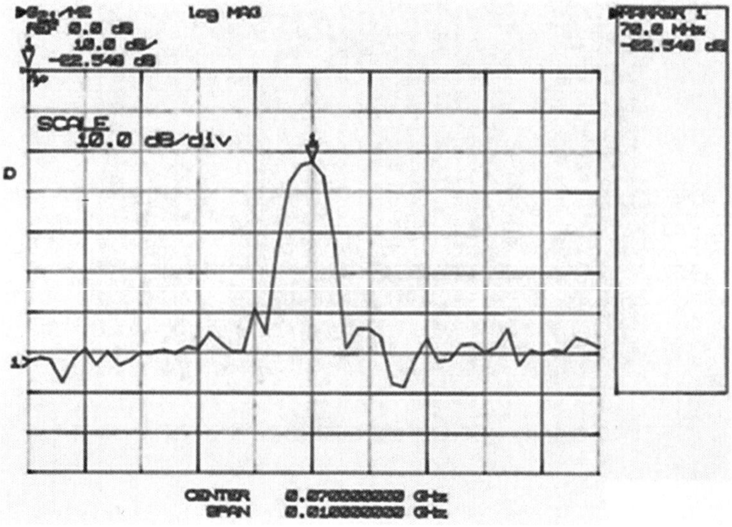

Fig. 6 Response of the SAW device for its central frequency

4.1 Condition for Oscillation

The two necessary conditions required to obtain oscillations are:

Total phase shift around the loop must be equal to a multiple of 2π. Amplifier gain must exceed the combined loss of the delay element. Schematic of SAW oscillator is shown below in Fig. 7. The output of the amplifier is connected to transmitter IDT and the input of amplifier is connected to receiver IDT. To maintain oscillation the product of amplifier gain 'A' and feedback gain 'β' should be equal to unity i.e.

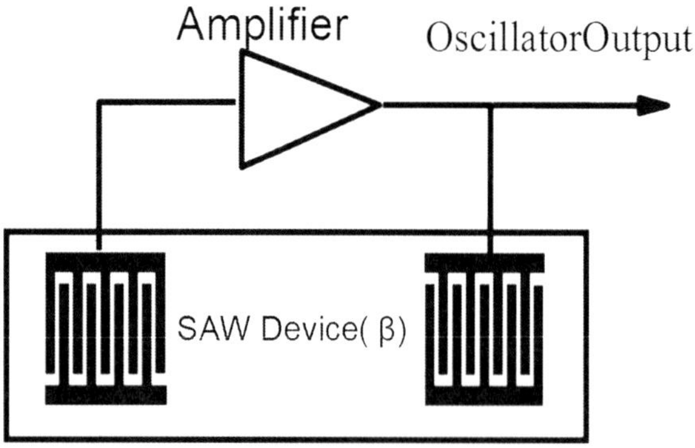

Fig. 7 SAW oscillator

$$|A\beta| = 1 \tag{2}$$

$$\angle A\beta = 2n\pi \tag{3}$$

These conditions must be satisfied simultaneously for obtaining the oscillations. For realizing the circuit, the methodology adopted has been shown in Fig. 7 and every aspect of that is discussed in the forth-coming headings.

4.2 Amplifier Design

The chosen SAW device work in RF range hence to make an oscillator, the role of the RF amplifier is critical. It needs to have enough gain over the required frequency range of operation and should be stable in itself [23]. An attempt has been made to design an amplifier, which may be capable of supporting oscillations with 70 MHz. SAW filter device is mounted in a SAW package. Keeping a speculation in mind that the total losses may not exceed 30 dB, the amplifier has been designed using INA12063 amplifier IC [24].

RF Amplifier: It is one of the major building blocks of modern RF circuits. The amplifier is usually required to provide low noise gain with low distortion at both small and large signal levels. It should also be stable, i.e. not generate unwanted spurious signals, and the performance should remain constant with time. The input and output match is also desirable. Thus for amplifier we require:

- Low noise
- Low distortion
- Stable operation
- Filtering of unwanted signal
- High bandwidth.

For SAW oscillator the gain of the amplifier should be greater than the loss of the device so that sustained oscillation should be generated. Amplifier should be properly biased as per application. Careful attention must be given to the design of the SAW filter and its RF amplifier. Optimal performance can be accomplished by having a large number of pairs of interdigital electrodes and series inductor accurately tuned with them to achieve maximum power transfer with the RF amplifier. The RF amplifier itself should be operated with moderate gain and should have low noise figure. Proper heat sinking of the RF amplifier will help to minimize thermally induced electronic phase shift variations [25]. By keeping above issues in the mind amplifier has been designed with INA12063 which is a 1.5 GHz low noise self-biased transistor amplifier. The salient features of the INA12063 have been listed below:

- Integrated active bias circuit
- Single positive supply voltage
- Adjustable current from 1 to 10 mA
- 2 dB noise figure at 900 MHz
- 25 dB gain at 100 MHz.

The amplifiers have tuning facility through an inductor and capacitors [26]. But for tuning, most of the times only experience plays an important role. For fixing R_{bis} the formula is used is:

$$R_{bis} = 10(V_d - 0.8)/I_c \tag{4}$$

where V_d is device voltage and I_c is the desired collector current.

For our circuit design we have chosen $V_d = 3V$, $I_c = 5\,mA$, Hence $R_{bis} = 4.4\,K\Omega$

Because 4.4 KΩ resistance was not available so we are using 4.6 KΩ resistance for biasing. The amplifier has provision to be tuned on both the input and output side. But tuning cannot be done simply with formula:

$$F = 1/2\pi\sqrt{LC} \tag{5}$$

It is also done with trial and error and experience.

4.3 Mixer

The difference in frequency between the sensor oscillator and the reference one is the signal of our interest, which carries information about chemical species. In order to find out the frequency difference between two oscillators, first both the frequencies are mixed and then the difference frequency is extracted using a low pass filter. Mixer is a non-linear resistance having two sets of inputs terminals and one set of output terminals. When two signals of different frequencies are applied to two terminals, the output terminal will have several frequencies including the difference frequency between the two input frequencies. The most common type of mixers utilized are bipolar transistors, FET, dual gate MOSFET and integrated circuits. Depending on the application, one can choose any of them [27].

4.4 Low Pass Filter

Out of various mixed frequencies, the information carrying frequency is the difference frequency. In order to separate it from the other frequencies available at the output of the mixer, a low pass filter is needed. It is expected that the difference frequency will be of the order of a few kHz only. In order to accommodate the default difference between the reference and sensor oscillator, the cutoff frequency of the filter has been kept at 1 MHz. A passive low pass filter of the form shown in Fig. 8 has been utilized for this purpose. This filter did not require any complicated design techniques. The components values have been chosen as per the formula:

$$f_c = 1/(2\pi RC) \tag{6}$$

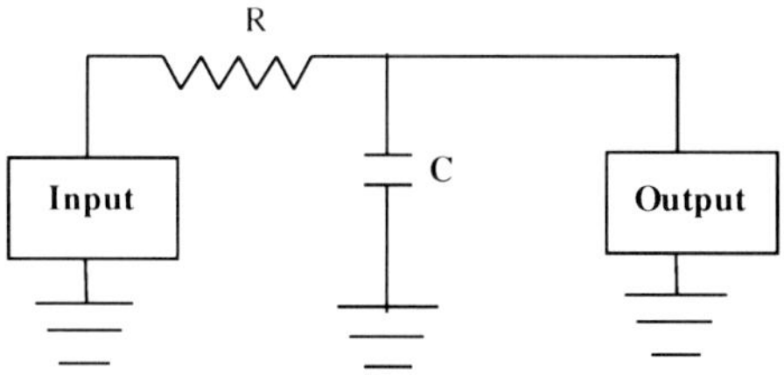

Fig. 8 First order low pass filter circuit

However, by using higher order filter response can be improved further. The loss in the passive filter is expected to be in limit. Otherwise also an amplifier following the filter can make up the thing. It is expected that the loss in the passive filter will be affordable as the mixer has an in-built amplifier. The cutoff frequency is around 1 MHz.

4.5 Testing of Amplifier

The RF amplifier has been designed which is capable of supporting oscillations at 70 MHz frequency with SAW filter device mounted in the package. Keeping a speculation in mind that the total losses may not exceed 30 dB, the amplifiers have been designed using INA12063 instrumentation type amplifier IC. INA12063 which is an 1.5 GHz low noise self-biased transistor amplifier have 25 dB gain at 100 MHz. The performance of the amplifier has been determined experimentally and the frequency response of the amplifier has been shown in Fig. 9. From the graph, we have observed that the peak is at around 70 MHz and gain is almost flat (23 dB) from 60 to 90 MHz range. Thus frequency range from 60 to 90 MHz has been used for analyzing the performance of the SAW device in presence of vapor molecules to be detected. 70 MHz SAW device is connected in feedback of amplifier circuit and we have recorded the oscillator signal on spectrum analyzer. The output of the spectrum analyzer is shown in Fig. 10.

5 Working of SAW Sensor

SAW device, when coated with polymer coating behaves as a SAW sensor. SAW devices as a vapor sensor are attractive because of their small size ($< 0.1\,cm^3$), ruggedness, low cost, electronic output, sensitivity and adaptability to a wide variety of vapor phase analytical problems. The presence of thin organic film on the SAW device surface can potentially cause attenuation of the wave through interaction with both longitudinal and the vertical shear components. Attenuation increases with the length of the SAW device and also with the inverse of the wavelength

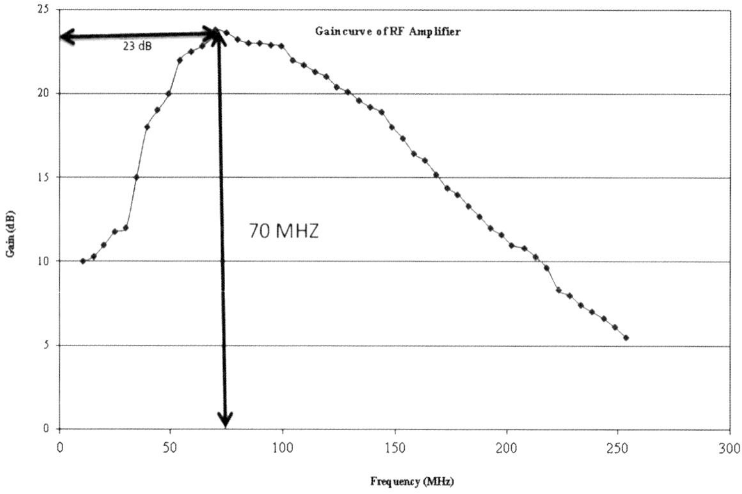

Fig. 9 Frequency response curve of the RF amplifier

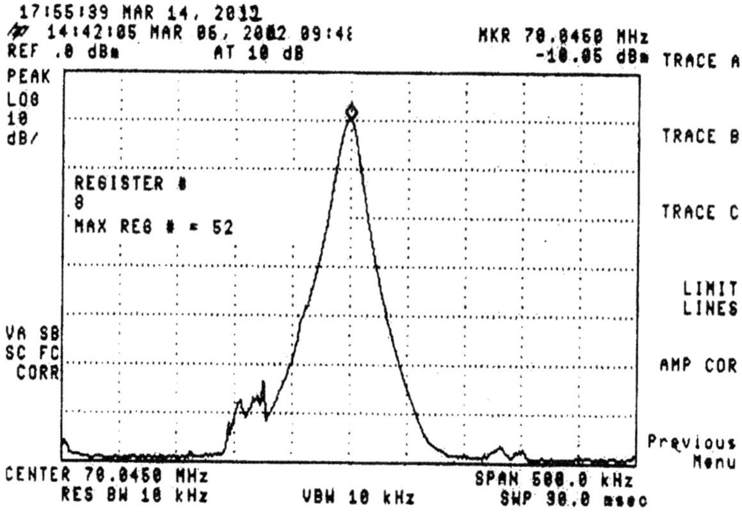

Fig. 10 70 MHz SAW oscillator output

(i.e. operating frequency) in relation with oscillator circuit performance. If the adjacent medium is thick enough (i.e. its thickness is greater then λ), then compressional wave is generated. Ambient gases easily meet this condition but for film, which is very thin (less than λ/100) compressional wave cannot exist and can therefore be neglected. Thus, the attenuation produced by thin organic coating film will be caused primarily through interactions with the shear component. In this situation the viscosity, elastic modulus and thickness of the organic coating will significantly affect the attenuation of the shear component. For organic film whose thickness is less than one percent of the acoustic wavelength attenuation is very small. This level of attenuation can be quite significant for SAW delay line oscillator since the loss introduced by the film make it more difficult to sustain oscillation in the device. The coating, through either physisorption or chemisorption, must interact selectively with the vapor of interest to cause a change in mass or mechanical property of the coating. While the SAW oscillator is easily capable of detecting monolayer adsorption on to its surface, greater sensitivity can be obtained by using a thicker coating film through which the vapor can diffuse and interact with greater number of sorption sites than on a smooth surface. Vapor diffusion rate and device response time is closely related. Clearly film that is very dense will not permit easy diffusion and the device will respond very slowly. For polymeric coating rapid diffusion is most easily obtained if the polymer is highly amorphous and well above its glass transition temperature. For such a film the shear modulus is probably small enough to make it mechanical contribution to SAW oscillator frequency a negligible quantity. Thus, incremental changes in film density or thickness (i.e., mass per unit area) will result in a corresponding resonant frequency change. Volatility of most liquids is high enough to cause a steady baseline drift due to evaporation. Interaction involving chemisorption will usually be quite selective but irreversible under normal conditions and will be most useful in dosimetry applications where an integral signal related to the concentration and time of exposure to a particular vapor is desired. The large dynamic range capability of SAW oscillator and the low cost of the device permit irreversible interaction to be considered as practical system. For reversible responses, interaction-involving physisorption will be required. Unfortunately the relatively low energies of adsorption associated with physisorption processes could permit several different vapors with similar properties to interact with the coating film. The sensitivity and size of SAW delay oscillator sensors is directly related to their resonant frequencies. Frequency shift for a given mass loading will increase with the square of the operating frequency. Frequency shift obtained for a given mass loading is independent of the length of the SAW delay line. Mass loading effects is given by Sauerbrey's equation [28, 29]:

$$\Delta f = -C \frac{f_0^2}{A} \Delta m \tag{7}$$

Where Δm is the mass change of quartz surface, Δf is the frequency change caused by Δm, f_0 is the fundamental resonant frequency of SAW device in the absence of the deposited mass, A is the geometric area of electrode and C is constant. The area of a SAW delay line with constant IDT impedance from device to device and constant number of wavelengths will vary inversely with the square of the operating frequency [30]. The reductions of the device area also have implication with regard to the minimum mass change that can be detected by a SAW delay line oscillator. Since the device is sensitive to mass per unit area, reductions in device area result in a corresponding reduction in the minimum detectable mass change.

6 Development of a Sensor Head

Sensor head basically a covering for the SAW devices so that they could be exposed only to the chemical intended. It is shown in Fig. 11. The 70 MHz SAW sensor is basically designed to test the presence of a nerve agent called 'Sarin'. Since it is very harmful and dangerous for testing at laboratory level, another material called DMMP (Di-methyl methyl phosponate) has been used for testing. DMMP is less harmful and has the same properties as Sarin.

7 Detection Mechanism

The detection of chemical analytes can be based on changes in one or more of physical characteristics of thin film or layer in contact with the device surface. Some of the intrinsic film properties that can be utilized for detection include mass/area, elastic stiffness (modulus) viscoelasticity, viscosity electrical conductivity, and permittivity. In addition, changes in extrinsic variables such as temperature and pressure also produce a sensor response [31, 32]. SAW sensor works as a mass sensor, when analytes of interest comes in contact of polymer film, which is coated, on SAW device. Desired toxic gas is exposed to the device through the carrier gas such as N_2 and then gas molecules are desorbed [5, 33, 34]. Device has been tested for determining the characteristics such as (i) frequency shift with different concentration of vapor (ii) transient response for response and recovery times (iii) transient response for reproducibility of the sensor. Since arrangement has been made to compensate the

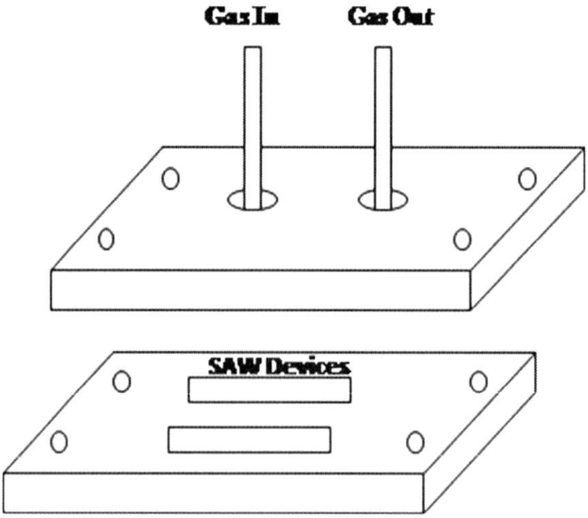

Fig. 11 Schematic diagram of the sensor cell

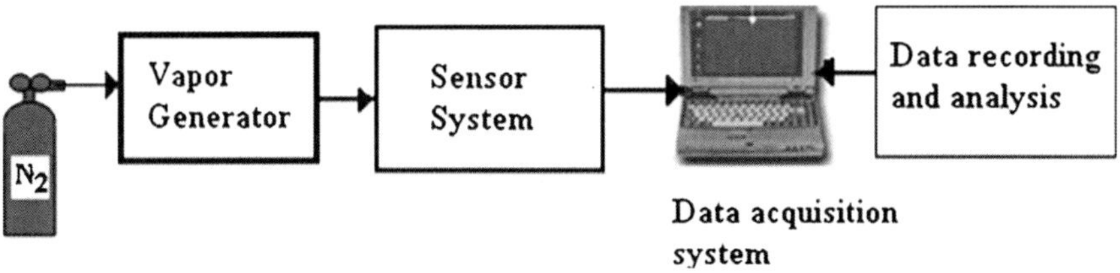

Fig. 12 Block diagram of the experimental set up

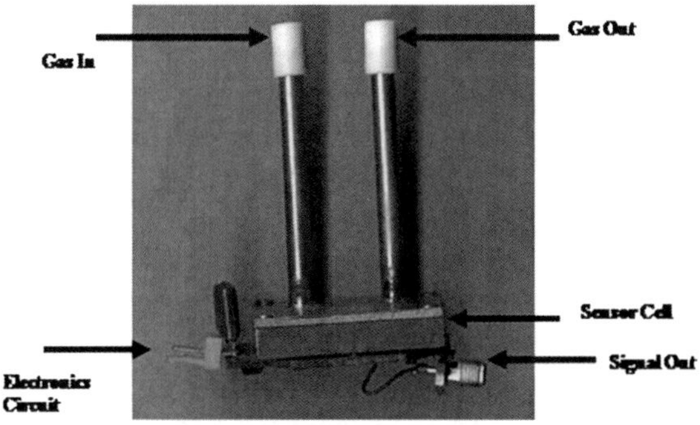

Fig. 13 Photograph of the sensor cell with detection electronics circuit

error due to temperature variation with signal conditioning circuit, effect of temperature variation has not been studied. Figure 12 shows the scheme of the experimental setup and Fig. 13 shows the photograph of the sensor head with sensor and signal conditioning circuit.

8 Results

Due to adsorption of chemical analytes on the device, the frequency of SAW oscillator is shifted. The frequency shift is then recorded in the PC through data acquisition system. Initially the sensor was refreshed by passing N_2 gas and then exposed to a certain concentration of analytes. The response curve for several cycles at fixed concentration of 0.5 ppm is shown in Fig. 14. During refreshing, the device output is returned to its initial oscillation frequency. The output of the device is highly reproducible. From the transient curve shown in Fig. 15 the response time of the sensor is 16 s, recovery time is 45 s and base line is very stable. There is almost 700 Hz frequency change due to change in vapor concentration of 0.5 ppm. Results show that there is significant change is frequency due to very small change is analytes. Thus the device has the potential for detecting chemical war fare agents. Response

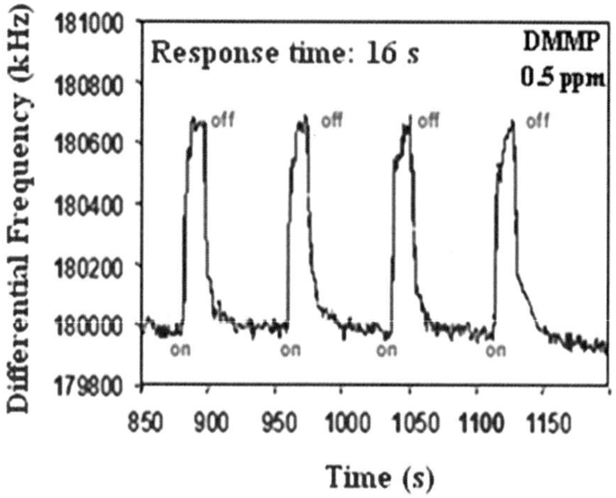

Fig. 14 Transient response curve of the SAW sensor for determining reproducibility

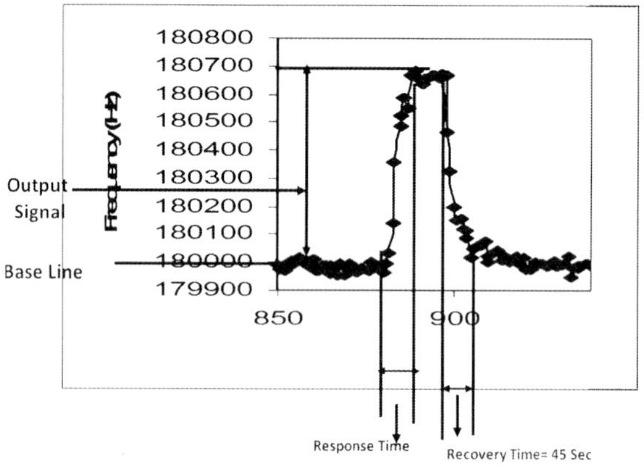

Fig. 15 Transient response curve for determining response and recovery time

curve with the variation of toxic gas in different concentration is shown in Fig. 16. It is observed that as the concentration level increases, the frequency shift of the detection electronics circuit increases. But the increase in frequency shift is nonlinear. This is normally the situation that most of the SAW devices show nonlinear response curve. This may be due to the fact that adsorption and desorption phenomenon do not occur in similar rate. For direct display, this nonlinearity can be compensated by analog or digital method of compensation.

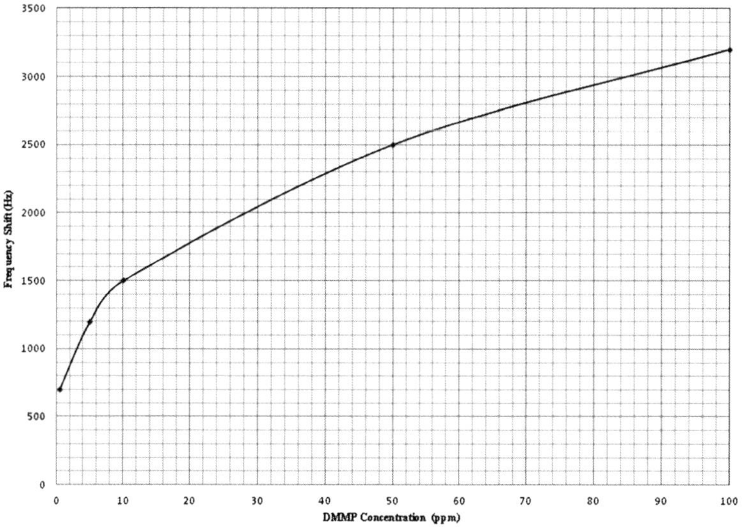

Fig. 16 Frequency response of the sensor at different concentration of DMMP

9 Conclusion

The present work deals with theory, design and application of SAW device for chemical war fare agent detection. A CW sensor with high frequency SAW device has been realized and tested. For this purpose SAW device has been fabricated on STX-quartz using standard photolithography. Fabricated device has been packaged and characterized using vector network analyzer. A high frequency amplifier has been designed and fabricated for SAW oscillator realization. SAW oscillator has been tested using the spectrum analyzer. For realizing the sensor a selective interface (polymer) film has been optimized on SAW device. A sensor head had been developed for the packaging of the device with the provision of gas inlet and out let port. Sensing experiments with DMMP a (stimulant of CW agents) had been carried. Sensor shows very fast response time which is the prime requirement for CW detection. Also sensor shows good sensitivity as it can detects in sub ppm range with good amount of frequency shift and it also shows good base line stability as it return to its base frequency after purging the sensor. Hence this work developed insight about integral approach for a product realization for CW agent detection.

References

1. B.A. Auld, *Acoustic Fields and Waves in Solids*, vol. 1 (Wiley, New York, 1973)
2. H. Matthews, *Surface Wave Filters: Design, Construction, and Use* (Wiley, New York, 1977)
3. A.A. Oliver (ed.), *Acoustic Surface Waves* (Springer-Verlag, Berlin, 1978)

4. C.K. Campbell, *Surface Acoustic Wave Devices for Mobile and Wireless Communications* (Academic, San Diego, 1989)
5. D.S. Ballantine, R.M. White, S.J. Martin, A.J. Ricco, E.T. Zellers, G.C. Frye, H. Wohltjen, *Acoustic Wave Sensors: Theory, Design, and Physico-Chemical Applications* (Academic Press, San Diego, 1997)
6. D.P. Morgan, *Surface-Wave Devices for Signal Processing* (Elsevier, Amsterdam, 1985)
7. S. Datta, *Surface Acoustic Wave Devices* (Prentice-Hall, Englewood Cliffs, 1986)
8. E.A. Ash, Fundamentals of signal processing devices, topics in applied physics, in *Acoustic Surface Waves*, vol. 24, ed. by A.A. Oliner (Springer, New York, 1987), pp. 97–185
9. H. Wohltjen, Mechanism of operation and design considerations for surface acoustic wave device vapour sensors. Sens. Actuators **5**, 307–325 (1984)
10. K. Hashimoto, *Surface Acoustic Wave Devices in Telecommunications: Modeling and Simulation* (Springer, New York, 2000)
11. H. Wohltjen, R. Dessy, Surface acoustic wave probe for chemical analysis. Anal. Chem. **51**, 1458–1475 (1979)
12. H. Wohltjen, Surface acoustic wave microsensors. in *Proceedings 4th International Conference Sensors and Actuators, Transducers*, vol. 87, pp. 471–477 (1987)
13. S.J. Martin, K.S. Schweizer, A.J. Ricco, T.E. Zipperian, Gas sensing with surface acoustic wave devices. in *Proceedings 3rd International Conference Solid State, Sensors and Actuators, Transducers* (1985), pp. 71–73
14. W.H. King, Piezoelectric sorption sensor. Anal. Chem. **36**(9), 1735–1739 (1964)
15. J.W. Grate, A. Snow, D.S. Ballantine, H. Wohltjen, M.H. Abraham, R.A. McGill, P. Sasson, Determination of partition coefficients from surface acoustic wave vapor sensor responses and correlation with gas-liquid chromatographic partition coefficients. Anal. Chem. **60**, 869–875 (1988)
16. E.T. Zellers, R.M. White, Selective surface-acoustic-wave styrene vapor sensor with regenerable reagent coating. in *Proceedings 4th International Conference Solid State Sensor and Actuators, Transducers* (1987), pp. 459–461
17. S.L. Rose-Pehrsson, J.W. Grate, D.S. Ballantine, P.C. Jurs, Detection of hazardous vapors including mixtures using pattern recognition analysis of responses from surface acoustic wave devices. Anal. Chem. **60**, 2801–2811 (1988)
18. D. Avramov, M. Rapp, A. Voigt, U. Stahl, M. Dirschka, Comparative studies on polymer coated SAW and STW resonators for chemical gas sensor applications. in *Proceedings of the IEEEFCS* (2000), pp. 58–65
19. J. Allan, B. Vrana, R. Greenwood, G.A. Mills, B. Roig, C. Gonzalez, A "toolbox" for biological and chemical monitoring requirements for the European Union's water framework directive. Talanta **69**, 302–322 (2006)
20. F. Bender, K. Länge, A. Voigt, M. Rapp, Improvement of surface acoustic wave gas and biosensor response characteristics using a capacitive coupling technique. Anal. Chem. **76**(13), 3837–3840 (2004)
21. K.T. Tang, DJ. Yao, C.M. Yang, H.C. Hao, J.S. Chao, C.H. Li, P.S.Gu, *A Portable Electronic Nose Based on Bio-Chemical Surface Acoustic Wave (SAW) Array with Multiplexed Oscillator and Readout Electronics*. (ISOEN, Brescia, 2009)
22. T. Chuang, R.M. White, Sensors utilizing thin-membrane SAW oscillators. in *Proceedings IEEE Ultrasonic, Symposium* (1981) , pp. 159–162
23. S.Y. Liao, *Microwave Circuit Analysis and Amplifier Design*. (Prentice-Hall Inc, Upper Saddle River, 1986)
24. Data sheet INA- 12063 1.5 GHz Low Noise Self-Biased Transistor Amplifier, Agilent Technologies, www.semiconductor.agilent.com
25. E.A. Gerber, A. Ballato, *Precision Frequency Control Volume 2 Oscillators and Standards*. (Academics Press Inc, New York, 1985)
26. M.W. Medley, *Microwave and RF Circuits: Analysis, Synthesis and Design*. (ArtHouse Inc, Norwood, 1993)
27. E. John, *Fundamentals of RF Design*. (Wiley, New York, 2001)

28. H.C. Hao, et al., Development of a portable electronic nose based on chemical surface acoustic wave array with multiplexed oscillator and readout electronics. Sensors and Actuators B: Chemical **146**(2) 545–553 (2010)
29. G. Sauerbrey, Verwendung von Schwingquarzen zur Wägung dünner Schechter und zur Mikrowägung. Zeitschrift für Physik **155**(2), 206–222 (1959)
30. D.L. Dermody et al., Interactions between organized, surface confined monolayers and vapor-phase probe molecules. II. Synthesis, characterization, and chemical sensitivity of self-assembled polydiacetylene/calix[n]arene bilayers. J. Am. Chern. Soc. **118**, 11912–11917 (1996)
31. H.C. Yang et al., Molecular interactions between organized, surface-confined monolayers and vapor-phase probe molecules. Langmuir **12**, 726–735 (1996)
32. C. Wang et al., A piezoelectric quartz crystal sensor array self-assembled calixarene bilayers for detection of volatile organic amine in gas. Talanta **57**, 1181–1188 (2002)
33. B. Drafts, Acoustic wave technology sensors. IEEE Trans. Microw. Theory Tech. **49**, 795–802 (2001)
34. T. Islam, U. Mittal, A. T. Nimal, M. U. Sharma, Surface acoustic wave (SAW) vapour sensor using 70 MHz SAW oscillator*In proceedings Sixth International Conference on Sensing Technology (ICST-2012)*, pp. 112–114 (2012)

Chapter 12
Electronic and Electromechanical Tester of Physiological Sensors

E. Sazonov, T. Haskew, A. Price, B. Grace and A. Dollins

Abstract Physiological sensors for respiration, cardiovascular activity and electrodermal activity have been historically used in polygraph devices and sleep laboratories. Periodic testing of these sensors is important to maintain predictable performance of the measurement equipment. This paper describes an Electronic and Electromechanical Tester (EET) for physiological sensors that allows for accurate and repeatable reproduction of the recorded or computer-generated physiological signals. The tester is interfaced to a personal computer via USB and contains four time-synchronous channels: two electromechanical simulators for testing abdominal and thoracic respiratory sensors, an electromechanical simulator for testing cardiovascular blood pressure cuff sensor, and an electronic simulator for testing electrodermal sensors. All of the simulated physiological channels apply direct physical actuation to the corresponding sensors. The validation of the EET demonstrated high accuracy and repeatability of the simulated physiological signals.

Keywords Physiological sensors · Electronic and electrochemical tester

1 Introduction

Polygraph equipment has been used by law enforcement and security forces since the 1920s [1]. At this time, polygraph testimonies are admissible in court in 19 states in the United States of America and polygraphs are actively used as an investigative tool in Canada [2]. Due to a long standing debate and controversy surrounding the scientific basis of polygraph testing [3, 4] polygraphs and polygraph evidence are

E. Sazonov (✉) · T. Haskew · A. Price · B. Grace
The University of Alabama, Tuscaloosa, AL 35487, USA
e-mail: sazonov@eng.ua.edu

A. Dollins
National Center for Credibility Assessment, Fort Jackson, SC 29207, USA

A. Mason et al. (eds.), *Sensing Technology: Current Status and Future Trends II*, Smart Sensors, Measurement and Instrumentation 8, DOI: 10.1007/978-3-319-02315-1_12,

less commonly used and typically not admissible in court in many other countries. Nevertheless, polygraphs are still commonly used as a screening tool for hiring new employees into security-sensitive positions.

The polygraph monitors several physiological indicators (such as skin conductivity, breathing, pulse and blood pressure) during an interview in which the subject is asked a series of questions. The expectation is that deceptive answers will register different physiological responses from those corresponding to truthful answers. Modern polygraphs are computerized and create digital recordings of all physiological signals captured during an interview.

Skin conductivity (also called electrodermal response or Galvanic Skin Response, GSR) is thought to be an indicator of psychological, emotional or physiological arousal. Measurement of skin conductivity is typically performed by an ohmmeter in which a constant 0.5 V is applied between the electrodes and the resulting current flow is measured by amplifying the voltage across a resistor in series with the skin [5].

Respiration in polygraph equipment is typically monitored using pneumatic pressure gauges. The process of breathing stretches abdominal and thoracic gauges and produces variations in air pressure within the gauge which are then captured by an air pressure sensor within the polygraph [6]. Piezoelectric or capacitive breathing sensors incorporated into elastic belts could also be used.

Cardiovascular activity such as heart rate and variation in blood pressure is typically examined by a sphygmomanometer (with the cuff inflated to 60 mmHg). Blood pressure is the force exerted on the outside walls of blood vessels throughout the body as the heart contracts and relaxes. The pressurized cuff of the sphygmomanometer captures cuff-pressure oscillation [7]. A pressure transducer within the polygraph instrument captures and records this change in pressure.

2 Motivation

Since polygraph testimony is still in wide usage, it is essential to understand how different polygraph instruments respond to identical excitations. This will not only aide in correctly identifying physiological responses between various subjects, but will also validate and calibrate physiological sensors and thus ensure reliability of obtained readings. The Electronic and Electromechanical Tester (EET) presented in this chapter simulates four physiological channels used in polygraph equipment (two channels of respiration, cardiovascular activity and electrodermal response) by providing a direct mechanical or electrical response equivalent to that of a human attached to the respective sensors. In this sense, the EET provides a simulation of the real physiological processes and provides an accurate and repeatable way of testing polygraph equipment by reproducing signals originating from the human body as well as computer-generated signals with high accuracy and repeatability. In addition, the EET can be used to test and validate other physiological measurement systems

Table 1 EET system requirements

Physiological source	Measurement sensor	Actuation transducer	Actuation resolution	Actuation value range	Actuation min. rate	Actuation smax. rate
Respiratory	Pneumatic pressure gauge	Screw-driven linear actuation	≥12 bits per full range of motion	0–5 cm	0 breaths per min	30 breaths per min
Cardiovascular	Sphygmomanometer	Screw-driven linear actuation	≥12 bits per full range of motion	20–200 mmHg	40 beats per min	200 beats per min
Galvanic skin response	0.5 V resistance measurement circuit	Switch-based resistive circuit	≥12 bits per full range of motion	1–4096 kOhm	DC	60 Hz

that capture respiration, cardiovascular activity and GSR, such as equipment used in sleep laboratories.

3 System Requirements

The requirements for the EET were formulated based on the fundamental properties of breathing, GSR and blood pressure, such as range and frequency of events. Table 1 lists measurement sensors (on polygraph), actuators (on EET) and desired ranges and resolutions of the actuation. The common sampling frequency of 60 Hz was set to provide an update rate sufficient for simulation of all physiological simulators in the system.

4 EET Architecture

The system architecture for the EET is shown in Fig. 1. A computer (1) with a graphical user interface reads the files containing recordings of physiological signals and sends them to the microcontroller (3) through a USB-to-serial converter (2). The graphical user interface supports loading, visualization and playback of the previously recorded physiological signals and generation of standard waveforms such as sinusoids of various frequency and amplitude. The interface to the physiological simulator is implemented as a virtual serial port connection though a USB port using an FT232R chip [8]. The chip supports bidirectional data exchange with rates up

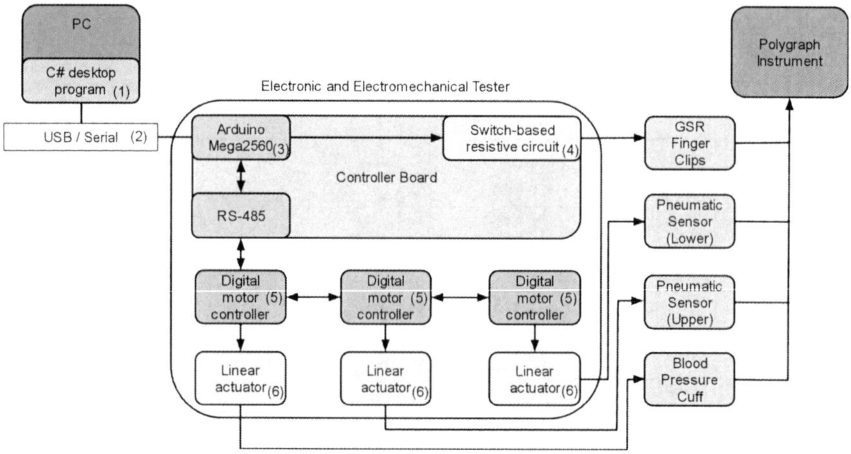

Fig. 1 Architecture of EET

to 3 Megabit per second (375 kbytes/s), which is sufficient for the application. With simultaneous playback of 4 channels at 60 Hz sampling rate and 16 bit resolution, the required data rate is below 4,000 bits/s. The microcontroller (3) uses a ring buffer with a fixed size of 30 samples to buffer the data stream from the C# desktop program to the microcontroller. The short periods of recordings are stored temporarily on internal memory. The microcontroller also maintains the sampling clock at which all actuators are updated. In this manner, all timing issues (such as jitter) that may arise by using a non-real-time OS (such as Microsoft Windows) are resolved, as exact timing is maintained using hardware timers on the microcontroller. The microcontroller also converts the raw waveforms into commands sent to digital motor controllers (5) that drive linear actuators (6) for the breathing and cardiovascular channels. The microcontroller also controls a switch-based resistive circuit (4) that simulates GSR. This channel is controlled using two digital ports (16 digital I/O pins) from the microcontroller. The microcontroller used in the design is an ATmega2560 [9] (on Arduino Mega2560 board) with 54 digital general purpose I/O pins, 16 analog inputs, 4 UARTs (hardware serial ports), and 16 MHz clock frequency.

5 Electrodermal Simulator

The electrodermal simulator was implemented using a binary switch-based resistive circuit. Two digital ports on the ATmega2560 were used to control the 16-bit binary resistive circuit. The entire resistive range, up to 4.096 MOhm, was implemented using a series of resistors controlled by digital switches. A switch-based resistive circuit proved to create a lower switching noise during resistance transitions than commercially available digital potentiometers. The 16-bit digitization of

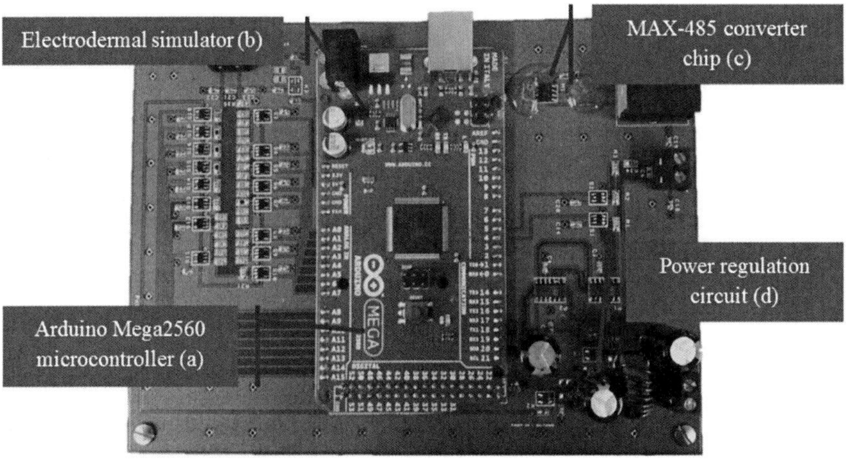

Fig. 2 EET controller board

the simulated skin resistance allows for 64 Ohm resolution, which is negligible compared to the typical values (KOhms or more) of skin resistance [10]. The output leads of the electrodermal simulator are connected to two BNC connectors. On the end of the BNC connectors are snap buttons, to which the clips of the polygraph are attached. Due to the high impedance of the GSR channel, it is particularly sensitive to switching noise that may leak from the power supply or digital communication lines. To reduce transient and switching noise, an LC circuit was used to reduce noise from the power supply and a LP3985 low-dropout regulator [11] with power supply rejection ratio of 50 dB was used to provide regulated power to the controller. Digital noise from USB communications was reduced by using an Acromag USB isolator [12]. A picture of the controller board is shown in Fig. 2. The controller board houses the microcontroller board (a), the binary switch-based resistive circuit of the electrodermal simulator (b), an RS-232–RS-485 converter (c), and a filter/power regulator circuit (d).

6 Respiratory Simulator

The respiratory simulator contained two identical channels providing mechanical excitation to the abdominal and thoracic pneumatic sensors. Each channel was implemented as a digitally controlled linear actuator driven by an Ezi-Servo EzM-42-XL-A stepper motor [13]. This is a high-precision and high-speed stepper motor providing 10,000 pulses per revolution of angular position feedback. The digital motor controller was a FASTECH Ezi-Servo Plus-R motor drive [13] that accepted commands via an RS485 interface from the microcontroller. The angular position commands

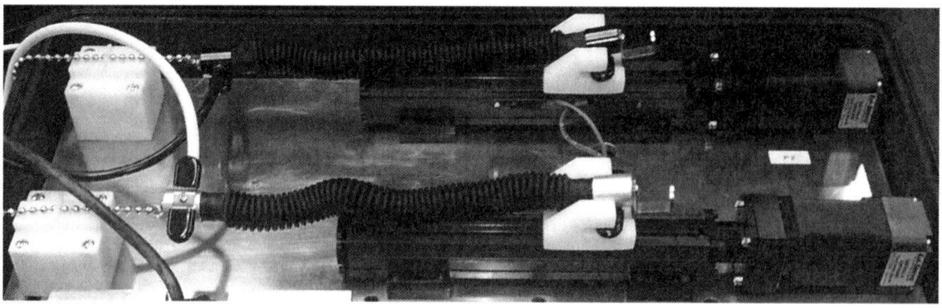

Fig. 3 Respiratory channels with pneumatic pressure sensors attached

were issued to the closed-loop position controller for the motor, which drove a lead screw linear actuator with a 10.16 mm screw lead and an available 121 mm stroke length [14]. One end of the pneumatic pressure sensor was attached to the carriage on the actuator while the other end of the pneumatic pressure sensor was attached to a stationary support. A sizing chain that is found on the majority of pneumatic pressure sensors attaches to the stationary support to accommodate for various sizes of pressure sensors. The respiratory subsystem implementation is shown in Fig. 3 below.

Each position command issued to the motor controller caused the pneumatic pressure sensor to expand and contract to simulate respiratory movement. Note that while the motor position was controlled in a close-loop manner, the loop was not closed on the linear displacement. In a reasonable assumption, all mechanical dynamics from the motor shaft to the pressure transducer were neglected.

For the purposes of sizing the motor and determining the screw lead during the design process, a sinusoidal desired position was assumed with the maximum peak-to-peak displacement of 5 cm and a frequency, f_r, of 0.5 Hz. In this case, the frequency of the sinusoidal desired position is denoted as:

$$\omega_r = 2\pi * f_r = 2\pi * 0.5 = \pi \; rad/s \tag{1}$$

The upper limit required from the system requirements must be met by the motor and linear actuator. With this extreme case in mind, the position, velocity and acceleration profiles were created.

$$x_r = 5 \sin(\omega_r t) = 5 \sin(\pi t) \text{ cm} \tag{2}$$

$$\nu_r = \frac{dx_r}{dt} = 5\pi \cos(\pi t) \text{ cm/s} \tag{3}$$

$$a_r = \frac{d^2 x_r}{dt^2} = \frac{d\nu_r}{dt} = -5\pi^2 \sin(\pi t) \text{ cm/s}^2 \tag{4}$$

Based on these profiles, $v_{r,max} = 0.157$ m/s and $a_{r,max} = 0.494$ m/s^2. The acceleration profile, along with actuator friction and carriage mass, m_c, dictated the maximum required linear force.

$$F = m_c a_r = (0.06)\,(0.494) = 0.03\,\mathrm{N} \tag{5}$$

The load acquired from the pneumatic pressure sensors was negligible. This force and velocity data was sufficient to find a screw lead and motor such that the requirements were within the torque speed curve of the motor. Of the screw leads available, a 10.16 mm screw lead was chosen.

$$F\nu_r = T\omega_r \tag{6}$$

$$T = F\frac{\nu_r}{\omega_r} = F\frac{l}{2\pi} = \frac{(0.03)\,(0.0106)}{2\pi} = 3.2 * 10^{-4}\ N - m = 0.32\ mN - m \tag{7}$$

The selected motor provides a maximum torque of 650 mN-m and is capable of delivering 20 mN-m of torque at a speed of 3,000 rpm, which corresponds to a linear velocity of 50 cm/s.

7 Cardiovascular Simulator

The cardiovascular simulator consisted of a single channel providing mechanical excitation to the cuff of the sphygmomanometer. The cardiovascular channel was implemented as a digitally controlled linear actuator similar but not identical to that of the respiratory channels. The controller accepted angular position commands, which drove a lead screw linear actuator with a 5 mm screw lead and an available 71 mm stroke length. As with the respiratory channels, the angular position of the motor was controlled in a closed-loop manner, the loop was not closed on linear displacement.

A custom pressure cuff support was used to simulate cardiovascular activity in the brachial artery (Figs. 4 and 5). A stationary support (Fig. 4, left) contains a half-cylinder made of PVC plastic that is attached across two beams. The mobile support consists of a horizontal half-cylinder mounted to the carriage of the linear actuator. Linear displacement of the mobile support creates a proportional change in the cuff's air pressure.

Each position command issued to the motor controller caused the cuff to expand and contract to simulate cardiovascular activity. The implemented configuration of the cardiovascular channel supports ensures that an evenly distributed force is applied to the cuff, minimizing bend and flex within the supports. The large carriage mount and side rails aide in reducing bend and flex within the supports while running the simulator at high frequencies and amplitudes.

To determine the size of the motor and screw lead for the cardiovascular channel, a sinusoidal desired position was assumed with the maximum peak-to-peak displacement of 5 mm and a frequency, f_c, of 3.33 Hz.

Fig. 4 Cardiovascular channel implementation

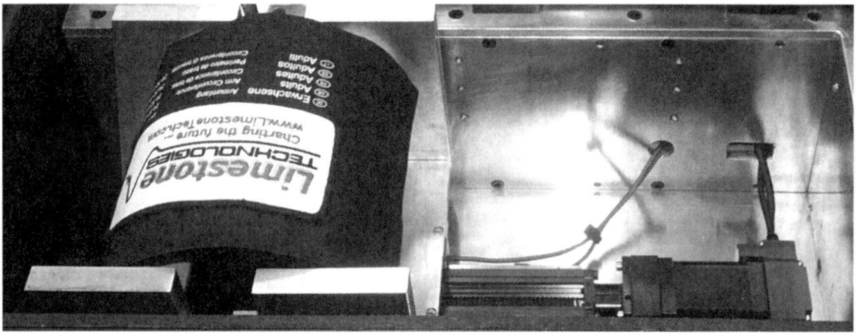

Fig. 5 Cardiovascular subsystem with blood pressure cuff attached

$$\omega_c = 2\pi * f_c = 2\pi * 3.33 = \frac{20\pi}{3} rad/s \tag{8}$$

From this information, position, velocity and acceleration profiles were created for the cardiovascular channel.

$$x_c = 5\ \sin\left(\omega_c t\right) = 5\ \sin\left(\frac{20\pi t}{3}\right)\ \text{mm} \tag{9}$$

$$\nu_c = \frac{dx_c}{dt} = \frac{100\pi}{3}\cos\left(\frac{20\pi}{3}t\right)\ \text{mm/s} \tag{10}$$

$$a_c = \frac{d^2x_c}{dt^2} = \frac{d\nu_c}{dt} = -\frac{2000\pi^2}{3}\sin\left(\frac{20\pi}{3}t\right)\ \text{mm/s}^2 \tag{11}$$

Based on these profiles, $\nu_{c,max}$= 0.105 m/s and $a_{c,max}$= 6.58 m/s^2. Because the EET must be able to accommodate pressures up to 200 mmHg, the load acquired from the pressure cuff on the cardiovascular channel cannot be neglected. It was experimentally determined that the maximum required linear force for the cardiovascular channel would be 275.8 N. This force and velocity data was sufficient to find a screw lead and motor such that the requirements were within the torque speed curve of the motor. A 5 mm screw lead was chosen for the cardiovascular channel.

$$F\nu_c = T\omega_c \tag{12}$$

$$T = F\frac{\nu_c}{\omega_c} = F\frac{l}{2\pi} = \frac{(275.8)\,(0.005)}{2\pi} = 0.219\,N - m = 219\,mN - m \tag{13}$$

The selected motor for the respiratory channels provided a maximum torque of 650 mN-m and is capable of delivering 20 mN-m of torque at a speed of 3,000 rpm, which corresponds to a linear velocity of 50 cm/s. This motor was also selected for the cardiovascular channel. Based on the screw lead, torque, velocity and acceleration requirements, it is satisfactory to use the same motor in all electromechanical channels.

8 Validation

The EET prototype was validated in a series of tests evaluating accuracy and repeatability of the playback of simulated physiological signals. Reference computer-generated signals and human recordings obtained on a polygraph were played back on the EET. The signals reproduced by the EET were recorded by an LX4000 Computerized Polygraph [15] made by Lafayette Instruments using Lafayette Polygraph System LX software version 11.1.4 and formatted to proprietary DACA file format using pREFORMAT Software Version 1.02. The rate at which the polygraph samples sensor data is 30 samples per second. When each vendor polygraph file is reformatted to the DACA file format, each file is converted to a relative scale ranging from 0 to 10,000 DACA units. Because these values are dimensionless, there is no universal conversion from DACA formatting to physical values. Rather, an individual linear transformation must be computed for each file. Default scaling for the playback of human recordings is set to each DACA unit corresponded to approximately 5 pulses of the stepper motor for the respiratory channel, 1 pulse of the stepper motor for the cardiovascular channel and 420 Ω for the GSR channel. Thus, the peak-to-peak amplitude of 10,000 DACA units translates to 50 mm physical displacement of the carriage on the linear actuator for the respiratory channel, 5 mm physical displacement of the carriage on the linear actuator for the cardiovascular channel and range of 4.2M Ω for the GSR channel.

The test suite used to validate the EET consisted of 9 computer generated signals and 1 human recording. Each of the 9 computer generated signals and the human recording were run 10 times each. The set of computer generated signals consisted of sinusoids with varying peak-to-peak amplitudes and frequencies for each channel.

Table 2 Amplitude and frequency range for computer generated signals

Simulator channel	Amplitude	Frequency (Hz)
Galvanic skin response	840 kΩ	0.25
	1.68 MΩ	0.5
	3.36 MΩ	1
Respiratory	40 mm	0.1
	20 mm	0.3
	10 mm	0.5
Cardiovascular	5 mm	0.667
	2.5 mm	2
	1.25 mm	3.333

For all amplitudes, each separate channel was run at 3 varying frequencies. The amplitudes of the sinusoids covered 80, 40 and 20 % of the full dynamic range of each simulator channel. Table 2 shows the peak-to-peak amplitude and frequency range for each channel.

The amplitudes and frequency ranges were carefully chosen to demonstrate the lower and upper limits of the simulator based on the system requirements. These frequencies are typical lower and upper limits of human capabilities. For the GSR channel, the response is usually a long lasting waveform of a simple shape [16] with small variations in resistance compared to the entire resistive range [10]. These smaller variations usually occur at a higher frequency than a variation that will cover a large resistive range. Since the set of test signals includes large variations in resistive range, it is adequate to test the EDA channel at lower frequencies. The normal breathing rate for an adult is approximately 14 breaths per min, or 0.233 Hz [17]. The frequencies chosen for the respiratory channel is representative of the typical breathing rate for an adult. A maximum displacement of 40 mm is adequate for the displacement seen by an adult wearing a pneumatic breathing sensor. The frequency range tested on the cardiovascular channel is also representative of typical physiological processes of an adult. Maximum linear displacement of 5 mm for the cardiovascular channel is satisfactory to accurately reproduce waveforms being played on the EET. Initially, the cuff is inflated to a pressure of 60 mmHg. With this initial pressure, a linear displacement of 5 mm produces an output of approximately 120 mmHg.

The human recording consisted of different waveforms in all channels of the recording. The human recording was obtained from a sample recording provided with the polygraph used in the experiments. Each physiological signal in the human recording falls within a typical range for amplitude as well as frequency for each respective channel.

Accuracy of the reproduction was evaluated as absolute average error E_{REF}. The reference signals (DACA files played on the EET) were aligned in time with the recordings captured on the polygraph (further referred as test signals) and trimmed to identical length (160 s for computer-generated test signal recordings, and 275 s for the human recording). The E_{REF} was computed as an average of the absolute difference between the reference and test signals averaged over 10 experiments:

$$E_{REF} = \frac{1}{N} \sum_{N} \left(\frac{\sum_{1}^{M} |x(t) - r(t)|}{M} \right) \tag{14}$$

where N = 10 is number of experiments, M is the length of a recording, x(t) is the test signal and r(x) is the reference signal.

Repeatability of the reproduction was evaluated in several ways. First, to disregard any differences between the reference and test signals caused by the transfer characteristics of the polygraph, absolute average error E_{MEAN} was computed as:

$$E_{MEAN} = \frac{1}{N} \sum_{N} \left(\frac{\sum_{1}^{M} |x(t) - m(t)|}{M} \right) \tag{15}$$

where m(t) is the mean signal obtained by point-to-point averaging of the test signals,

$$m(t) = \frac{1}{N} \sum_{N} x(t) \tag{16}$$

Second, average standard deviation of the test signals relative to their mean was computed as:

$$STD_{MEAN} = \frac{1}{N} \sum_{N} std\left(|x(t) - m(t)|\right) \tag{17}$$

9 Results

Figure 6 shows the assembled prototype of EET. The actuator and supports labeled (a) belong to the thoracic respiratory channel. The actuator and supports labeled (b) belong to the abdominal respiratory channel. The cardiovascular channel is labeled (c) and the GSR channel is labeled (d).

Figures 7–13 demonstrate the comparison of recorded signals being played back on EET (original signals) versus. signals captured from the EET by the polygraph (reproduced signals). A few charts from each channel were chosen to visually show the reproduction of signals on the EET. By visual comparison it is clear that the difference between the original and the reproduced signals is minimal. The summary of errors representing accuracy and repeatability of reproduction for computer- generated signals is shown in Table 3 for the galvanic skin response channel, Table 4 for the respiratory channel and Table 5 for the cardiovascular channel. Table 6 provides a summary of error representing accuracy and repeatability of reproduction for human recorded signals for all EET channels. The results are shown as error in DACA units (% of full range).

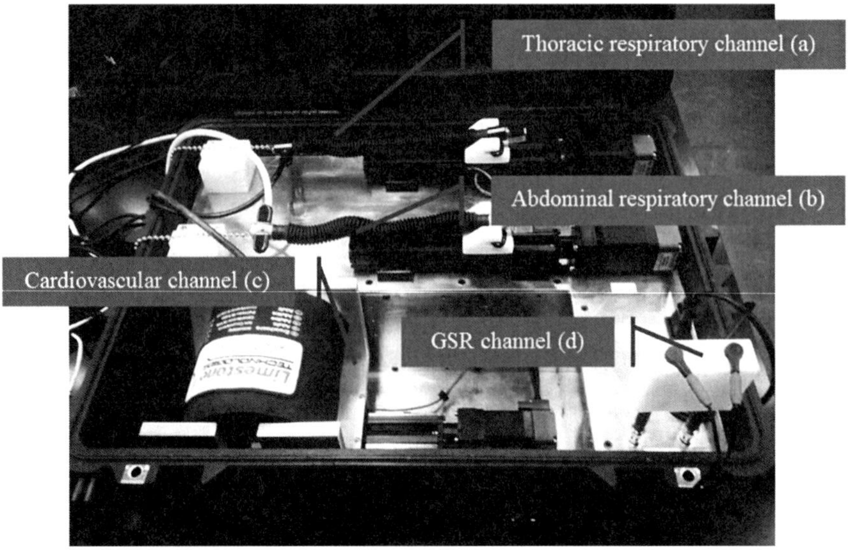

Fig. 6 Fully assembled EET

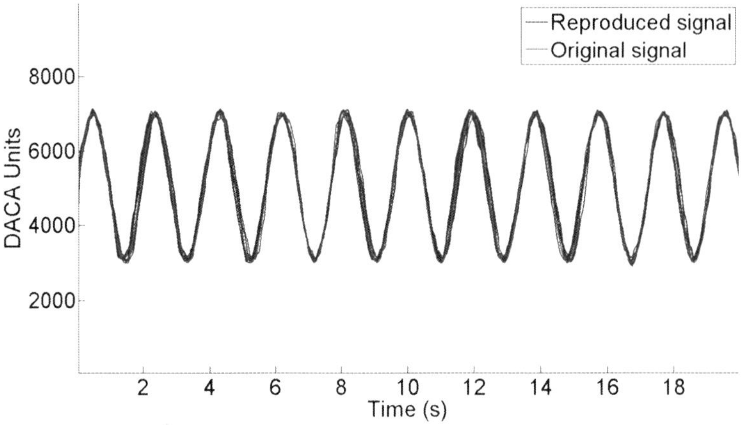

Fig. 7 Partial recording of sinusoid of 1.68 MΩ actuation range at 0.5 Hz for GSR channel

Figures 10–13 demonstrate the comparison of recorded signals being played back on EET (original signals) versus. signals captured from the EET by the polygraph (reproduced signals) for the human recording.

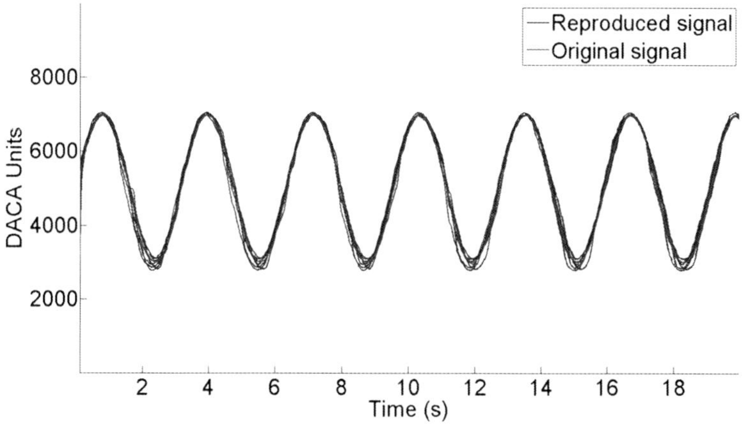

Fig. 8 Partial recording of sinusoid of 20 mm actuation range at 0.3 Hz for respiratory channel

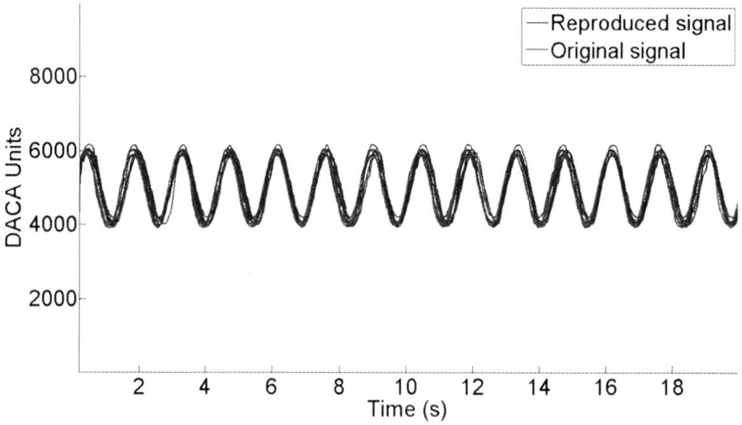

Fig. 9 Partial recording of sinusoid of 1.25 mm actuation range at 0.667 Hz for cardiovascular channel

10 Discussion

This chapter describes an Electronic and Electromechanical Tester of physiological sensors used in polygraph equipment, sleep laboratories and other applications. The EET was implemented to provide direct actuation to the sensors that are placed on humans during data acquisition and in this way simulate physical manifestations of physiological processes rather than their electrical equivalents. The benefit of such implementation is that transfer characteristics of each individual sensor contribute the signals captured by the polygraph and allow for detection of faulty or otherwise damaged sensors.

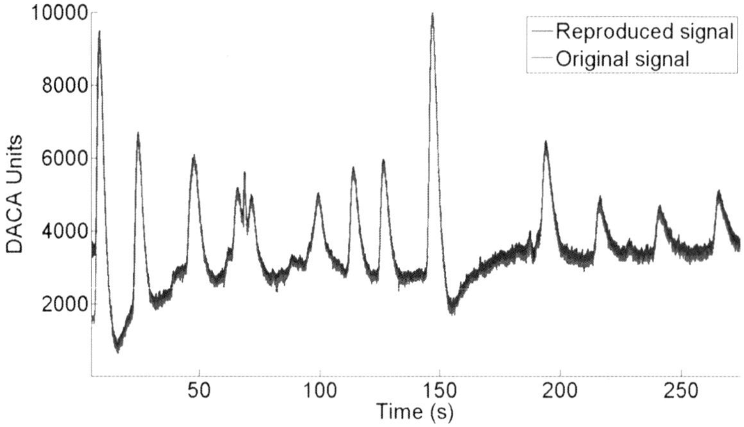

Fig. 10 Full-length human recording of the GSR channel

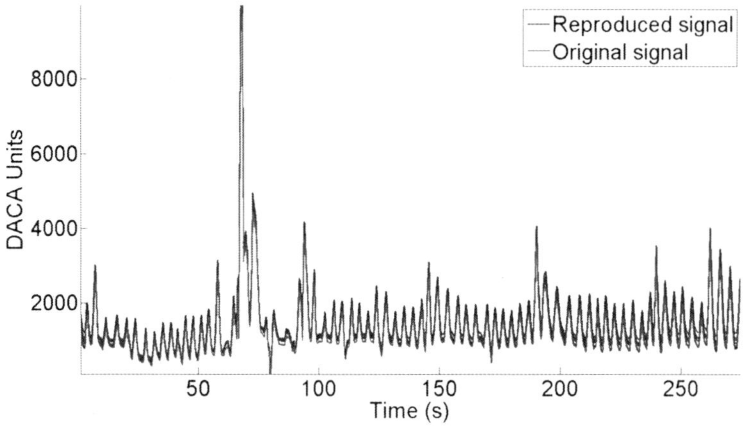

Fig. 11 Full-length human recording of respiratory channel

The prototype EET has proven to be sufficiently accurate and repeatable. As the data in the validation section shows, the average absolute error for the GSR simulator was less than 3.98 % for all computer generated signals and 1.02 % for the real signal. The average error for the respiratory channel was less than 3.74 % for all computer generated signals and approximately 1.60 % for the real signal. The cardiovascular channel had higher average absolute errors (less than 12.2 % for computer-generated signals and less than 2.1 % for real signal) than the other channels. This has to do with the high frequency of the sinusoids being played, representing the extreme ranges of human performance (such as heart rate of 200 bps), as well as the relatively low sample rate of the polygraph (30 Hz), that may not be sufficient for accurate representation of high-frequency inputs. Figure 14 illustrates the undersampling by the polygraph by demonstrating the signal captured on the cardiovascular channel

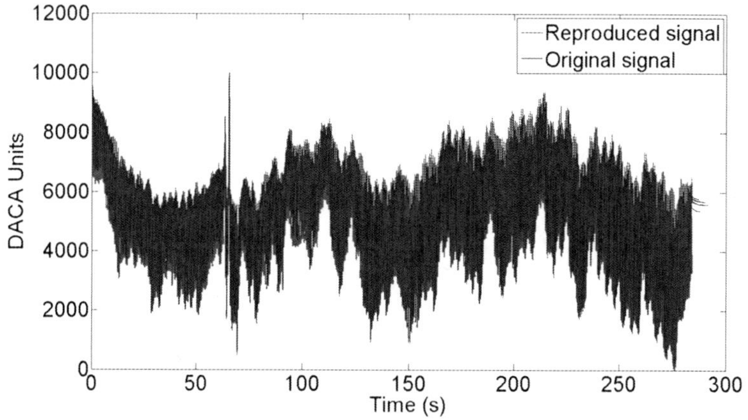

Fig. 12 Full-length human recording of the cardiovascular channel

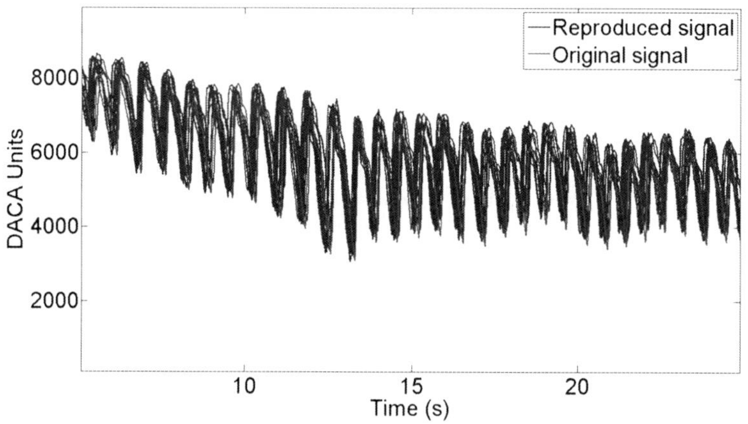

Fig. 13 Fragment of human recording of the cardiovascular channel showing greater detail of the waveform

with the maximum amplitude (5 mm) and frequency of 3.333 Hz. The original signal being played can be seen in Fig. 15. Although the physical linear displacement did not vary per sinusoidal cycle, it appears to do so on the graph.

Similarly, the average difference from the mean ranges was between 0.48 and 2.90 % with the standard deviation in the range of 0.43–2.91 % both for the GSR channel and respiratory channel. For the cardiovascular channel, the standard deviation ranged from 0.43 to 3.92 %. These values indicate that the reproduced physiological signal is a close match to the original signal and even if the polygraph-captured signal is slightly different from the reference, it is highly repeatable, which is desired. The range to all graphs was calibrated by an initial pulse at the beginning of each chart. The initial pulse consisted of a square wave lasting 1.5 s. The range of the pulse was

Table 3 Results for computer-generated simulation of galvanic skin response

Actuation range		Frequency		
		0.1 Hz (%)	0.5 Hz (%)	1 Hz (%)
840 kΩ	E_{REF}	116.4 (1.16)	175.3 (1.75)	194.8 (1.95)
	E_{MEAN}	61.8 (0.62)	83.9 (0.84)	94.5 (0.95)
	STD_{MEAN}	48.7 (0.49)	69.1 (0.70)	79.6 (0.80)
1.68 MΩ	E_{REF}	236.2 (2.36)	286.6 (2.87)	288.2 (2.88)
	E_{MEAN}	97.1 (0.97)	169.3 (1.69)	154.3 (1.54)
	STD_{MEAN}	85.3 (0.85)	142.0 (1.42)	173.4 (1.73)
3.36 MΩ	E_{REF}	346.6 (3.47)	321.8 (3.22)	398.4 (3.98)
	E_{MEAN}	137.9 (1.38)	150.9 (1.51)	290.5 (2.9)
	STD_{MEAN}	133.4 (1.33)	117.5 (1.18)	231.1 (2.31)

Table 4 Results for computer-generated simulation of respiration

Acuation range (mm)		Frequency		
		0.1 Hz (%)	0.3 Hz (%)	0.5 Hz (%)
10	E_{ERF}	71.4 (0.71)	92.3 (0.92)	112.9 (1.13)
	E_{MEAN}	61.4 (0.61)	84.5 (0.85)	92.3 (0.92)
	STD_{MEAN}	43.6 (0.43)	59.6 (0.60)	64.8 (0.65)
20	E_{ERF}	85.4 (0.85)	188.0 (1.88)	212.8 (2.13)
	E_{MEAN}	47.6 (0.48)	131.1 (1.31)	210.4 (2.10)
	STD_{MEAN}	49.2 (0.49)	100.7 (1.01)	166.1 (1.66)
40	E_{ERF}	195.2 (1.95)	217.6 (2.18)	374.1 (3.74)
	E_{MEAN}	94.1 (0.94)	131.3 (1.31)	367.9 (3.68)
	STD_{MEAN}	75.1 (0.75)	112.6 (1.13)	290.9 (2.91)

Table 5 Results for computer-generated simulation of the cardiovascular channel

Actuation range (mm)		Frequency		
		0.667 Hz (%)	2 Hz (%)	3.333 Hz (%)
1.25	E_{ERF}	109.7 (1.10)	187.6 (1.88)	504.3 (5.04)
	E_{MEAN}	62.3 (0.62)	93.6 (0.94)	317.4 (3.17)
	STD_{MEAN}	43.9 (0.43)	79.3 (0.79)	295.7 (2.96)
2.5	E_{ERF}	212.5 (2.12)	301.7 (3.02)	694.9 (6.95)
	E_{MEAN}	175.8 (1.76)	189.6 (1.90)	382.5 (3.83)
	STD_{MEAN}	112.6 (1.13)	162.2 (1.62)	312.2 (3.12)
5	E_{ERF}	258.9 (2.59)	341.8 (3.42)	1217.5 (12.18)
	E_{MEAN}	145.3 (1.45)	170.9 (1.71)	487.0 (4.87)
	STD_{MEAN}	116.6 (1.17)	143.5 (1.44)	392.1 (3.92)

the maximum actuation range for each channel. For example, upon initialization, the respiratory channel carriage would travel to 50 mm displacement for 1 s, to 0 mm displacement for 0.5 s, and would then start reproducing the respective waveform.

Table 6 Results for previously recorded human signal

Channel		Error Measure (%)
GSR channel	E_{ERF}	101.7 (1.02)
	E_{MEAN}	56.9 (0.57)
	STD_{MEAN}	27.8 (0.28)
Breathing channel	E_{ERF}	159.7 (1.60)
	E_{MEAN}	26.4 (0.26)
	STD_{MEAN}	159.7 (1.60)
Cardiovascular channel	E_{ERF}	209.2 (2.09)
	E_{MEAN}	91.8 (0.92)
	STD_{MEAN}	180.1 (1.80)

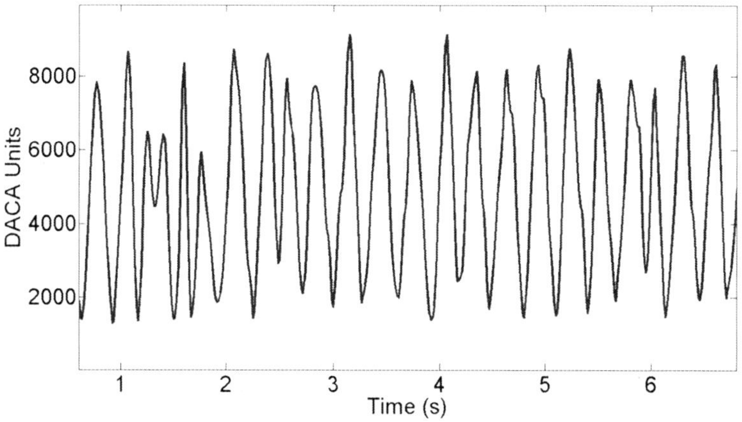

Fig. 14 Partial recording of sinusoid of 5 mm actuation range at 3.333 Hz for cardiovascular channel

This initial pulse served two purposes. First, it calibrated the range of all graphs. Therefore, when running sinusoids at 80 % of the maximum amplitude, the output graph would display sinusoids ranging from 1,000 to 9,000 DACA units. Also, the initial pulse signal was used to align each output signal in time. This was used to calculate error statistics.

Beside from being accurate and repeatable, EET demonstrated its suitability for the purpose of testing physiological sensors by detecting a faulty sensor in the one of the breathing channels. Figure 16 illustrates a decaying sinusoid captured by the faulty sensor. Ideally, this should replicate a sinusoid of constant amplitude. Figure 16 also illustrates a limitation of polygraphs that auto range their signals during acquisition and save these signals in dimensionless files without specifying conversion coefficients. For example, the peak-to-peak amplitude of the sinusoid provided by the EET in Fig. 16 never varied during the test. Nevertheless, the fault in the sensor resulted in about 20 % perceived reduction in the amplitude for the sinusoid. Therefore, some of the difference between the reference and test signals in this study

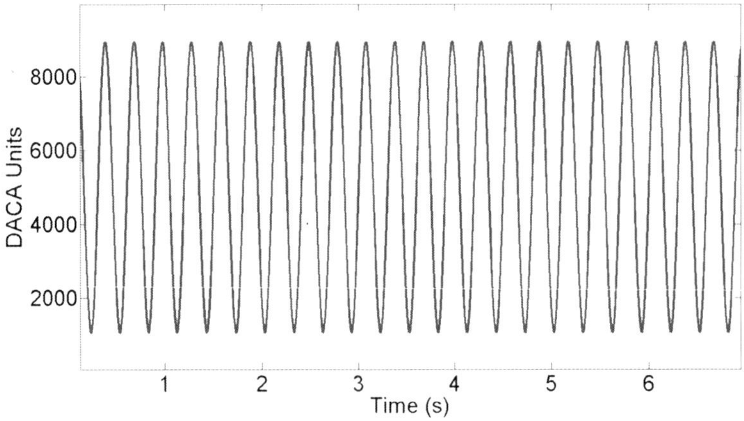

Fig. 15 Partial signal of sinusoid of 5 mm actuation range at 3.333 Hz for cardiovascular channel

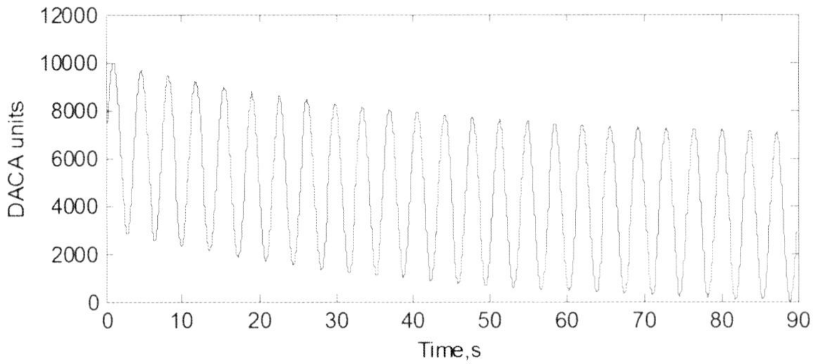

Fig. 16 Signal obtained from leaking breathing sensor

may have been caused by non-ideality of the polygraph sensors. This needs further investigation with more precise equipment allowing for capture of the reproduced signals in proper physical rather than dimensionless units.

The errors in reproduction may indicate some unaccounted factors in the transfer characteristic of EET-polygraph system. Future work on EET will include identification of these transfer characteristics. Overall, the proposed tester of physiological sensors proved to be an accurate and repeatable device capable of polygraph sensor testing.

11 Conclusions

An electronic and electromechanical tester of physiological sensors was implemented to provide direct actuation to the polygraph sensors that are placed on humans during data acquisition and in this way simulate physical manifestations of physiological processes rather their electrical equivalents. Testing of the EET demonstrated sufficiently high accuracy and repeatability of the reproduced signals, supporting its use for validation of polygraph equipment. The EET was able to successfully electromechanically replicate physiological signals originating from the human body.

References

1. K. Alder, *The Lie Detectors: The History of an American Obsession*. (Simon and Schuster, UK, 2007)
2. Wikipedia contributors, *Polygraph, Wikipedia, the free encyclopedia*. (Wikimedia Foundation, USA, 2012)
3. L. Saxe, D. Dougherty, T. Cross, The validity of polygraph testing: scientific analysis and public controversy. Am. Psychol. **40**(3), 355–366 (1985)
4. W.G. Iacono, Forensic lie detection?: procedures without scientific basis. J. forensic psychol. pract. **1**(1), 75–86 (2000)
5. D.C. Fowles, M.J. Christie, R. Edelberg, W.W. Grings, D.T. Lykken, P.H. Venables, Publication recommendations for electrodermal measurements. Psychophysiology **18**(3), 232–239 (1981)
6. L.A. Geddes, The truth shall set you free [development of the polygraph]. IEEE Eng. Med. Biol. Mag. **21**(3), 97–100 (Jun. 2002)
7. P. Zhu, L. Chen, Design of Automated Oscillometric Electronic Sphygmomanometer Based on MSP430F449, in *2010 International Conference on Future Information Technology and Management. Engineering*, vol. 1, pp. 463–466, Oct 2010
8. Future Technology Devices International Ltd. [Online]. Available: http://www.ftdichip.com/Products/ICs/FT232R.htm. Accessed 16 May 2013
9. ATmega1280- Atmel Corporation. [Online]. Available: http://www.atmel.com/devices/atmega1280.aspx. Accessed 12 Aug 2012
10. W. Boucsein, *Electrodermal Activity*, 2nd edn. (Springer Publishing, New York, 2012)
11. Texas Instruments. [Online]. Available: http://www.ti.com/product/lp3985. Accessed 16 May 2013
12. Acromag. [Online]. Available: http://www.acromag.com/catalog/746. Accessed 16 May 2013
13. Fastech. [Online]. Available: http://www.fastech.co.kr/bbs/data/product/942952012121709201901.pdf. Accessed 16 May 2013
14. Miniature Linear Actuators. [Online]. Available: http://www.pbclinear.com/ML-Miniature-Linear-Actuators. Accessed 13 Aug 2012
15. LX4000 Polygraph System | Polygraph from Lafayette Instrument Company. [Online]. Available: http://www.lafayettepolygraph.com/product_list.asp?subcatid=41. Accessed 12 Aug 2012
16. M. Tarvainen, A. Koistinen, M. Valkononen-Korhonen, J. Partanen, P. Karjalainen, Analysis of galvanic skin responses with principal components and clustering techniques. IEEE Trans. Biomed. Eng. **48**(10), 1071–1079 (Oct. 2001)
17. A.B. DuBois, A.W. Brody, D.H. Lewis, B.F. Burgess, Oscillation mechanics of lungs and chest in man. J. Appl. Physiol. **8**(6), 587–594 (May 1956)

About the Editors

Alex Mason graduated from the University of Liverpool, UK, with a first class honors degree in Computer and Multimedia Systems, after which he went on to complete a Ph.D. in Wireless Sensor Networks and their Industrial Applications at Liverpool John Moores University (LJMU), UK. Upon completing his Ph.D. in 2008, he concentrated for 2 years solely on research, working on aspects of non-invasive and non-destructive sensing for the healthcare, automotive and defense sectors.

Mason currently holds the position of Reader in Smart Technologies within the School of Built Environment at LJMU, after holding a Senior Lecture post prior to this since 2010. He has continued research in healthcare and defense, in addition to new areas such as water quality monitoring. Since becoming more involved in Built Environment issues, he has also developed an interest in Structural Health Monitoring and is currently working closely with the UK Defense Science and Technology Laboratories in this field.

Mason is responsible for supervising a number of Ph.D. students in the areas of sensing and renewable energy technologies, has coauthored over 100 publications (including 4 patents), has helped to organize national and international conferences, and gave a number of invited talks on his work. He is also an active member of the IET in the UK, and has achieved Chartered Engineer status with them.

A. Mason et al. (eds.), *Sensing Technology: Current Status and Future Trends II*,
Smart Sensors, Measurement and Instrumentation 8, DOI: 10.1007/978-3-319-02315-1,

Subhas Chandra Mukhopadhyay graduated from the Department of Electrical Engineering, Jadavpur University, Calcutta, India in 1987 with a Gold medal and received the Master of Electrical Engineering degree from Indian Institute of Science, Bangalore, India in 1989. He obtained the Ph.D. (Eng.) degree from Jadavpur University, India in 1994 and Doctor of Engineering degree from Kanazawa University, Japan in 2000.

He has authored/co-authored over 300 papers in different international journals and conferences, edited eleven conference proceedings. He has also edited ten special issues of international journals and fifteen books with Springer-Verlag as guest editor. He is currently the Series editor for the Smart Sensing, Measurements and Instrumentation of Springer-Verlag.

He is a Fellow of IEEE, a Fellow of IET (UK), a Topical editor of IEEE Sensors journal. He is also an Associate Editor for IEEE Transactions on Instrumentation and Measurements and a Technical Editor of IEEE Transactions on Mechatronics. He is a Distinguished Lecturer of IEEE Sensors council. He is Chair of the Technical Committee 18, Environmental Measurements of the IEEE Instrumentation and Measurements Society. He is in the editorial board of many international journals. He has organized many international conferences either as a General Chair or Technical programme chair.

Krishanthi Padmarani Jayasundera graduated from University of Peradeniya, Sri Lanka with honors degree in Chemistry. She obtained her both Master and Ph.D. in Organic Chemistry from Kanazawa University, Japan. After completing PhD she worked as researcher nearly 12 years in New Zealand involving various projects related to organic synthesis. Currently, she is working as a Postdoctoral researcher with the Institute of Fundamental Sciences, Massey University, New Zealand. She has published over 30 papers in different international journals and conference proceedings.

Nabarun Bhattacharyya is Associate Director in Centre for Development of Advanced Computing (C-DAC), Kolkata, India, which is a premier R&D Institute under Department of Information Technology, Government of India. He is a Ph.D. from Jadavpur University, Kolkata, India. He has authored more than 100 papers in peer-reviewed journals and conferences and two book chapters on the topics electronic nose, electronic vision and electronic tongue. His research areas focus on machine olfaction, soft computing, pattern recognition embedded systems for agricultural and environmental applications. He is a member of IEEE.